The Handbook of Environmental Chemistry

Founded by Otto Hutzinger

Editors-in-Chief: Damià Barceló • Andrey G. Kostianoy

Volume 19

The Handbook of Environmental Chemistry
Recently Published and Forthcoming Volumes

Emerging and Priority Pollutants in Rivers: Bringing Science into River Management Plans
Volume Editors: H. Guasch, A. Ginebreda, and A. Geiszinger
Vol. 19, 2012

Global Risk-Based Management of Chemical Additives I: Production, Usage and Environmental Occurrence
Volume Editors: B. Bilitewski, R.M. Darbra, and D. Barceló
Vol. 18, 2012

Polyfluorinated Chemicals and Transformation Products
Volume Editors: T.P. Knepper and F.T. Lange
Vol. 17, 2011

Brominated Flame Retardants
Volume Editors: E. Eljarrat and D. Barceló
Vol. 16, 2011

Effect-Directed Analysis of Complex Environmental Contamination
Volume Editor: W. Brack
Vol. 15, 2011

Waste Water Treatment and Reuse in the Mediterranean Region
Volume Editors: D. Barceló and M. Petrovic
Vol. 14, 2011

The Ebro River Basin
Volume Editors: D. Barceló and M. Petrovic
Vol. 13, 2011

Polymers – Opportunities and Risks II: Sustainability, Product Design and Processing
Volume Editors: P. Eyerer, M. Weller, and C. Hübner
Vol. 12, 2010

Polymers – Opportunities and Risks I: General and Environmental Aspects
Volume Editor: P. Eyerer
Vol. 11, 2010

Chlorinated Paraffins
Volume Editor: J. de Boer
Vol. 10, 2010

Biodegradation of Azo Dyes
Volume Editor: H. Atacag Erkurt
Vol. 9, 2010

Water Scarcity in the Mediterranean: Perspectives Under Global Change
Volume Editors: S. Sabater and D. Barceló
Vol. 8, 2010

The Aral Sea Environment
Volume Editors: A.G. Kostianoy and A.N. Kosarev
Vol. 7, 2010

Alpine Waters
Volume Editor: U. Bundi
Vol. 6, 2010

Transformation Products of Synthetic Chemicals in the Environment
Volume Editor: A.B.A. Boxall
Vol. 2/P, 2009

Contaminated Sediments
Volume Editors: T.A. Kassim and D. Barceló
Vol. 5/T, 2009

Biosensors for the Environmental Monitoring of Aquatic Systems
Bioanalytical and Chemical Methods for Endocrine Disruptors
Volume Editors: D. Barceló and P.-D. Hansen
Vol. 5/J, 2009

Environmental Consequences of War and Aftermath
Volume Editors: T.A. Kassim and D. Barceló
Vol. 3/U, 2009

Emerging and Priority Pollutants in Rivers

Bringing Science into River Management Plans

Volume Editors: Helena Guasch · Antoni Ginebreda · Anita Geiszinger

With contributions by

S. Agbo · J. Akkanen · A. Arini · C. Barata · D. Barceló · E. Becares · S. Blanco · B. Bonet · C. Bonnineau · F. Cassió · W. Clements · N. Corcoll · A. Cordonier · M. Coste · T.T. Duong · L. Faggiano · A. Feurtet-Mazel · C. Fortin · S. Franz · C. Gallampois · A. Ginebreda · M. Gros · H. Guasch · A. Jelić · J.V.K. Kukkonen · M. Laviale · I. Lavoie · M.T. Leppänen · J.C. López-Doval · K. Mäenpää · A. Moeller · S. Morin · I. Muñoz · A. Munné · L. Olivella · C. Pascoal · M. Petrović · F. Pérès · N. Prat · L. Proia · M. Ricart · A.M. Romaní · S. Sabater · F. Sans-Piché · M. Schmitt-Jansen · A. Serra · H. Segner · T. Slootweg · C. Solà · L. Tirapu · A. Tlili · E. Tornés · M. Vilanova

Editors
Helena Guasch
Institute of Aquatic Ecology
University of Girona
Girona
Spain

Antoni Ginebreda
IDAEA-CSIC
Department of Environmental Chemistry
Barcelona
Spain

Anita Geiszinger
Institute of Aquatic Ecology
University of Girona
Girona
Spain

The Handbook of Environmental Chemistry
ISSN 1867-979X e-ISSN 1616-864X
ISBN 978-3-642-25721-6 e-ISBN 978-3-642-25722-3
DOI 10.1007/978-3-642-25722-3
Springer Heidelberg Dordrecht London New York

Library of Congress Control Number: 2012932738

Printed on acid-free paper

Springer is part of Springer Science+Business Media (www.springer.com)

Editors-in-Chief

Advisory Board

The Handbook of Environmental Chemistry Also Available Electronically

The Handbook of Environmental Chemistry is included in Springer's eBook package *Earth and Environmental Science*. If a library does not opt for the whole package, the book series may be bought on a subscription basis.

For all customers who have a standing order to the print version of *The Handbook of Environmental Chemistry,* we offer free access to the electronic volumes of the Series published in the current year via SpringerLink. If you do not have access, you can still view the table of contents of each volume and the abstract of each article on SpringerLink (www.springerlink.com/content/110354/).

You will find information about the

- Editorial Board
- Aims and Scope
- Instructions for Authors
- Sample Contribution

at springer.com (www.springer.com/series/698).

All figures submitted in color are published in full color in the electronic version on SpringerLink.

Aims and Scope

Since 1980, *The Handbook of Environmental Chemistry* has provided sound and solid knowledge about environmental topics from a chemical perspective. Presenting a wide spectrum of viewpoints and approaches, the series now covers topics such as local and global changes of natural environment and climate; anthropogenic impact on the environment; water, air and soil pollution; remediation and waste characterization; environmental contaminants; biogeochemistry; geo-ecology; chemical reactions and processes; chemical and biological transformations as well as physical transport of chemicals in the environment; or environmental modeling. A particular focus of the series lies on methodological advances in environmental analytical chemistry.

Series Preface

With remarkable vision, Prof. Otto Hutzinger initiated *The Handbook of Environmental Chemistry* in 1980 and became the founding Editor-in-Chief. At that time, environmental chemistry was an emerging field, aiming at a complete description of the Earth's environment, encompassing the physical, chemical, biological, and geological transformations of chemical substances occurring on a local as well as a global scale. Environmental chemistry was intended to provide an account of the impact of man's activities on the natural environment by describing observed changes.

While a considerable amount of knowledge has been accumulated over the last three decades, as reflected in the more than 70 volumes of *The Handbook of Environmental Chemistry*, there are still many scientific and policy challenges ahead due to the complexity and interdisciplinary nature of the field. The series will therefore continue to provide compilations of current knowledge. Contributions are written by leading experts with practical experience in their fields. *The Handbook of Environmental Chemistry* grows with the increases in our scientific understanding, and provides a valuable source not only for scientists but also for environmental managers and decision-makers. Today, the series covers a broad range of environmental topics from a chemical perspective, including methodological advances in environmental analytical chemistry.

In recent years, there has been a growing tendency to include subject matter of societal relevance in the broad view of environmental chemistry. Topics include life cycle analysis, environmental management, sustainable development, and socio-economic, legal and even political problems, among others. While these topics are of great importance for the development and acceptance of *The Handbook of Environmental Chemistry*, the publisher and Editors-in-Chief have decided to keep the handbook essentially a source of information on "hard sciences" with a particular emphasis on chemistry, but also covering biology, geology, hydrology and engineering as applied to environmental sciences.

The volumes of the series are written at an advanced level, addressing the needs of both researchers and graduate students, as well as of people outside the field of "pure" chemistry, including those in industry, business, government, research establishments, and public interest groups. It would be very satisfying to see these volumes used as a basis for graduate courses in environmental chemistry. With its high standards of scientific quality and clarity, *The Handbook of*

Environmental Chemistry provides a solid basis from which scientists can share their knowledge on the different aspects of environmental problems, presenting a wide spectrum of viewpoints and approaches.

The Handbook of Environmental Chemistry is available both in print and online via www.springerlink.com/content/110354/. Articles are published online as soon as they have been approved for publication. Authors, Volume Editors and Editors-in-Chief are rewarded by the broad acceptance of *The Handbook of Environmental Chemistry* by the scientific community, from whom suggestions for new topics to the Editors-in-Chief are always very welcome.

Damià Barceló
Andrey G. Kostianoy
Editors-in-Chief

Volume Preface

The enduring changes in the aquatic environment and the increasing input of contaminants require research on novel conceptual and methodological approaches in relating chemical pollution and ecological alterations in ecosystems. Improving environmental risk assessment based on the analysis of priority pollutants or other preselected contaminants and extending the risk evaluation to new pollutants are essential for a better understanding of the causes of ecological quality loss and the cause–effect relationships of pollution.

At the same time, a great effort has been undertaken by European Member States to implement the Water Framework Directive. The ultimate goal of this Directive is the achievement of the "good quality status" of water bodies in EU river basins by 2015, it being understood as the combination of both "good ecological and chemical status." Whereas the connection between these two dimensions of water quality is accepted as one of the underlying premises of the WFD, there is still a lot to know on how it is produced. But in any case, there is little doubt that it has practical consequences for a proper river basin management. Therefore, it is of great interest to bring the increasing pool of scientific knowledge to water managers, providing a link between the scientific research and management practices aiming to evaluate the effects of emerging and priority pollutants in river ecosystems. With this aim, the Marie Curie Research Training Network KEYBIOEFFECTS organized the workshop "Emerging and Priority Pollutants: Bringing science into River Basin Management Plans" (Girona, Spain, 2010).

This book provides an overview of the main outcomes of the KEYBIOEFFECTS project as they were reflected in the aforementioned workshop. It includes scientific advances concerning the sampling, analyses, occurrence, bioavailability, and effects caused by emerging and priority pollutants in European rivers, the current status of the River Management Plans in Europe, and the applicability of the newly developed techniques for water monitoring purposes. These scientific advances are presented in the context of the Water Framework Directive evaluating their missing gaps and providing the basics for filling them.

A special attention is dedicated to report the occurrence and elimination of emerging pollutants such as pharmaceuticals during conventional wastewater treatment. Assessing the bioavailability of organic contaminants is also presented, highlighting the difficulties for regulation, more specially in the case of emerging contaminants. The book presents an extensive set of newly developed methods to assess ecological integrity in multistressed rivers. Different ecological perspectives: heterotrophic, phototrophic, and macroinvertebrate community indicators, laboratory and field investigations, as well as multibiomarker approaches are reviewed providing, in each case, the pros of cons for their application. Finally, a specific case study of river quality status assessment performed by a river basin water authority following the principles of the Water Framework Directive is presented.

It is not always evident how science returns its value to society. We hope that the results presented in this book will serve as a good example of how scientific research is able to provide support to issues of public concern, as it is the management of the water cycle and hence contributing to the preservation of ecosystems health and human welfare.

Girona, Spain Helena Guasch
Barcelona, Spain Antoni Ginebreda
Girona, Spain Anita Geiszinger

Contents

Occurrence and Elimination of Pharmaceuticals During Conventional Wastewater Treatment

Aleksandra Jelić, Meritxell Gros, Mira Petrović, Antoni Ginebreda, and Damià Barceló

Abstract Pharmaceuticals have an important role in the treatment and prevention of disease in both humans and animals. Since they are designed either to be highly active or interact with receptors in humans and animals or to be toxic for many infectious organisms, they may also have unintended effects on animals and microorganisms in the environment. Therefore, the occurrence of pharmaceutical compounds in the environment and their potential effects on human and environmental health has become an active subject matter of actual research.

There are several possible sources and routes for pharmaceuticals to reach the environment, but wastewater treatment plants have been identified as the main point of their collection and subsequent release into the environment, via both effluent wastewater and sludge. Conventional systems that use an activated sludge process are still widely employed for wastewater treatment, mostly because they produce effluents that meet required quality standards (suitable for disposal or recycling purposes), at reasonable operating and maintenance costs. However,

A. Jelić (✉) • A. Ginebreda
Institute of Environmental Assessment and Water Research (IDAEA-CSIC), C/Jordi Girona, 18-26, 08034 Barcelona, Spain
e-mail: aljqam@cid.csic.es

M. Gros
Catalan Institute for Water Research (ICRA), c/Emili Grahit 101, 17003 Girona, Spain

M. Petrović
Catalan Institute for Water Research (ICRA), c/Emili Grahit 101, 17003 Girona, Spain

Catalan Institution for Research and Advanced Studies (ICREA), Passeig Lluis Companys 23, 80010 Barcelona, Spain

D. Barceló
Institute of Environmental Assessment and Water Research (IDAEA-CSIC), C/Jordi Girona, 18-26, 08034 Barcelona, Spain

Catalan Institute for Water Research (ICRA), c/Emili Grahit 101, 17003 Girona, Spain

King Saud University (KSU), P.O. Box 2455, 11451 Riyadh, Saudi Arabia

H. Guasch et al. (eds.), *Emerging and Priority Pollutants in Rivers*,
Hdb Env Chem (2012) 19: 1–24, DOI 10.1007/978-3-642-25722-3_1,

this type of treatment has been shown to have limited capability of removing pharmaceuticals from wastewater. The following chapter reviews the literature data on the occurrence of these microcontaminants in wastewater influent, effluent, and sludge, and on their removal during conventional wastewater treatment.

Keywords Pharmaceuticals • Removal • Sludge • Wastewater

Contents

Abbreviations

HRT Hydraulic retention time
NSAIDs Nonsteroidal anti-inflammatory drugs
SRT Solid retention time
WWTP Wastewater treatment plant

1 Introduction

Pharmaceuticals are a large and diverse group of compounds designed to prevent, cure, and treat disease, and improve health. Hundreds of tons of pharmaceuticals are dispensed and consumed annually worldwide. The usage and consumption are increasing consistently due to the discoveries of new drugs, the expanding population, and the inverting age structure in the general population, as well as due to expiration of patents with resulting availability of less expensive generics [1]. After intake, these pharmaceutically active compounds undergo metabolic processes in organisms. Significant fractions of the parent compound are excreted in unmetabolized form or as metabolites (active or inactive) into raw sewage and wastewater treatment systems. Municipal sewage treatment plant effluents are discharged to water bodies or reused for irrigation, and biosolids produced are reused in agriculture as soil amendment or disposed to landfill. Thus body metabolization and excretion followed by wastewater treatment are considered to be the primary pathway of pharmaceuticals to the environment. Disposal of drug leftovers to sewage and trash is another source of entry, but its relative significance is unknown with respect to the overall levels of pharmaceuticals in the environment [2].

Continual improvements in analytical equipment and methodologies enable the determination of pharmaceuticals at lower concentration levels in different environmental matrices. Pharmaceuticals and their metabolites in surface water and aquatic sediment were subject of numerous studies concerning pharmaceuticals in the environment [3–5]. Several studies investigated the occurrence and distribution of pharmaceuticals in soil irrigated with reclaimed water [6–8] and soil that received biosolids from urban sewage treatment plants [9, 10]. These studies indicated that the applied wastewater treatments are not efficient enough to remove these micropollutants from wastewater and sludge, and as a result they find their way into the environment (Fig. 1). Once they enter the environment, pharmaceutically active compounds can produce subtle effects on aquatic and terrestrial organisms, especially on the former since they are exposed to long-term continuous influx of wastewater effluents. Several studies investigated and reported on it [11–13]. No evidence exists linking the presence of pharmaceuticals in the environment to human health risks; still complex mixtures may have long-term unseen effects, especially on tissues other than those on which the pharmaceuticals were designed to act.

Therefore, the occurrence of pharmaceutical compounds in the environment and their potential effects on human and environmental health, as well as the extent to which they can be eliminated during wastewater treatment, have become active subject matter of actual research. Since the concern about the discharge of pharmaceuticals (and other emerging contaminants, as well) into wastewater is relatively recent, it is not strange that they are not yet covered by the currently existing regulation.

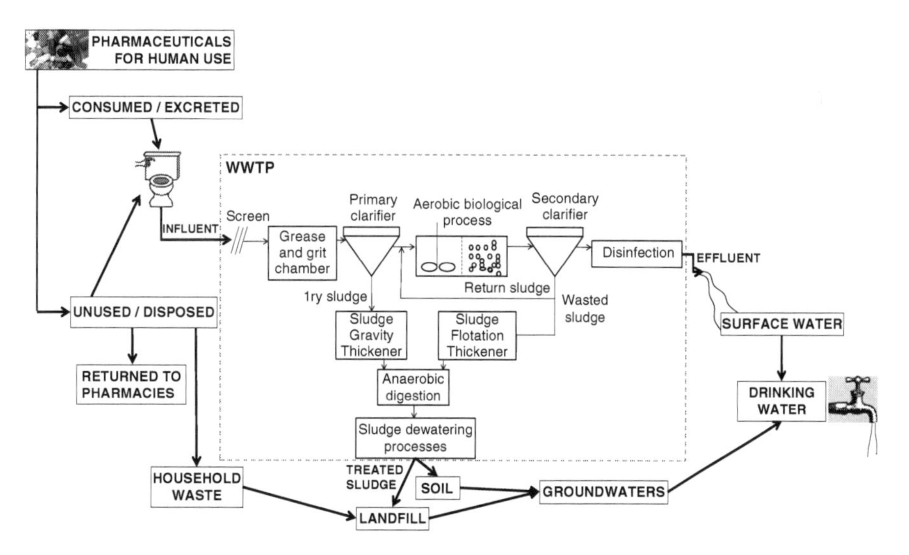

Fig. 1 Routes of release of pharmaceuticals for human use to the environment with a schematic diagram of a conventional WWTP

2 Activated Sludge Process for Treatment of Wastewater

Wastewater treatment systems that use activated sludge processes have been employed extensively throughout the world, mostly because they produce effluents that meet required quality standards (suitable for disposal or recycling purposes), at reasonable operating and maintenance costs. Figure 1 shows a schematic diagram of a conventional wastewater treatment. All design processes include preliminary treatment consisting of bar screen, grit chamber, and oil and grease removal unit [14], typically followed by primary gravity settling tank, in all but some of the smaller treatment facilities. The primary-treated wastewater enters into a biological treatment process—usually an aerobic suspended growth process—where mixed liquor (i.e., microorganisms responsible for the treatment, along with biodegradable and nonbiodegradable suspended, colloidal, and soluble organic and inorganic matter) is maintained in liquid suspension by appropriate mixing methods. During the aeration period, adsorption, flocculation, and oxidation of organic matter occur. After enough time for appropriate biochemical reactions, mixed liquor is transferred to a settling reactor (clarifier) to allow gravity separation of the suspended solids (in form of floc particles) from the treated wastewater. Settled solids are then returned to the biological reactor (i.e., return activated sludge) to maintain a concentrated biomass for wastewater treatment. Microorganisms are continuously synthesized in the process; thus some of suspended solids must be wasted from the system in order to maintain a selected biomass concentration in the system. Wasting is performed by diverting a portion of the solids from the biological reactor to solid-handling processes. The most common practice is to waste sludge from the return sludge line because return activated sludge is more concentrated and requires smaller waste sludge pumps. The waste sludge can be discharged to the primary sedimentation tanks for co-thickening, to thickening tanks, or to other sludge-thickening facilities, in order to increase the solid content of sludge by removing a portion of the liquid fraction. Through the subsequent processes such as digestion, dewatering, drying, and combustion, the water and organic content is considerably reduced, and the processed solids are suitable for reuse or final disposal. To achieve better effluent water quality, further treatment steps - tertiary treatment - can be added to the above outlined generla process, e.g. activated carbon adsorption, additioanl nutrient removal etc.

3 Occurrence of Pharmaceuticals During Conventional Wastewater Treatment

3.1 Occurrence of Pharmaceuticals in Wastewater Influent and Effluent

More than 10,000 prescription and over-the-counter pharmaceuticals are registered and approved for usage today, with around 1,300 unique active ingredients (Orange book, FDA). This is a versatile group of compounds that differ in the mode of action, chemical structure, physicochemical properties, and metabolism. They are typically classified using the Anatomical Therapeutic Chemical Classification System (ATC system) according to their therapeutic application and chemical structure. Because of the volume of prescription, the toxicity, and the evidence for presence in the environment, nonsteroidal anti-inflammatory drugs (NSAIDs), antibiotics, beta-blockers, antiepileptics, blood lipid-lowering agents, antidepressants, hormones, and antihistamines were the most studied pharmaceutical groups [15].

Even though a number of research publications have been focused on the occurrence, fate, and effects of pharmaceuticals in the environment, we have data on the occurrence of only 10% of the registered active compounds, and very little information on their effects in the environment. There is even less information regarding the occurrence and fate of the transformation/degradation products (active or not) of pharmaceuticals. Both the qualitative and the quantitative analysis of pharmaceuticals in the environmental matrices are definitely a starting point for the establishment of new regulations for the environmental risk assessment of pharmaceutical products.

The pharmaceuticals find their way to the environment primarily via the discharge of raw and treated sewage from residential users or medical facilities. Through the excretion via urine and feces, extensively metabolized drugs are released into the environment. But the topically applied pharmaceuticals (when washed off) along with the expired and unused ones (when disposed directly to trash or sewage) pose a direct risk to the environment because they enter sewage in their unmetabolized and powerful form [2]. Even though the production of drugs is governed by rigorous regulations, pharmaceutically active substances are frequently released with the waste from drug manufacturing plants [16–18].

The occurrence of the pharmaceutical compounds in wastewater treatment plants has been investigated in several countries around the world (Austria, Canada, England, Germany, Greece, Spain, Switzerland, USA, etc). More than 150 pharmaceuticals belonging to different therapeutic groups have been detected in concentration ranging up to the μg/L level in sewage water. Their environmental occurrence naturally depends on the rate of production, the dosage and frequency of administration and usage, the metabolism and environmental persistence, as well as the removal efficiency of wastewater treatment plants (WWTPs). Figures 2 and 3 show the occurrence of the selected, most investigated pharmaceuticals in wastewater influent and effluent, as found in the literature.

NSAIDs are the most used class of drugs for the treatment of acute pain and inflammation. They are administered both orally and topically and available as prescription and over-the-counter (nonprescription) drugs. High consumption and way of administration of NSAIDs result in elevated concentration reported in the effluent from WWTPs. Among the most studied NSAIDs during wastewater treatments are ibuprofen, diclofenac, naproxen, ketoprofen, and mefenamic acid [19]. The compounds usually detected in the highest concentrations in the influent of WWTPs are ibuprofen, naproxen, and ketoprofen (in range of some μg/L)

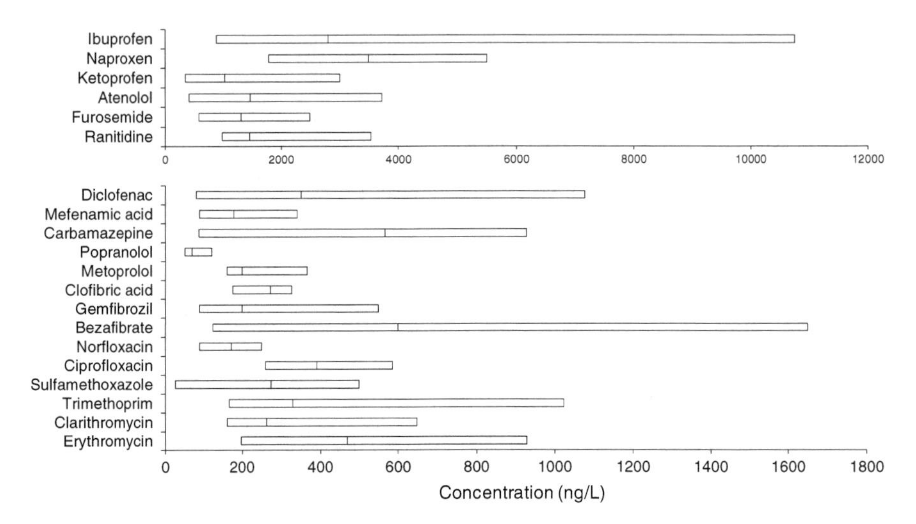

Fig. 2 Concentrations of the pharmaceuticals in wastewater influent (25th, median, and 75th percentile, for the values found in the literature)

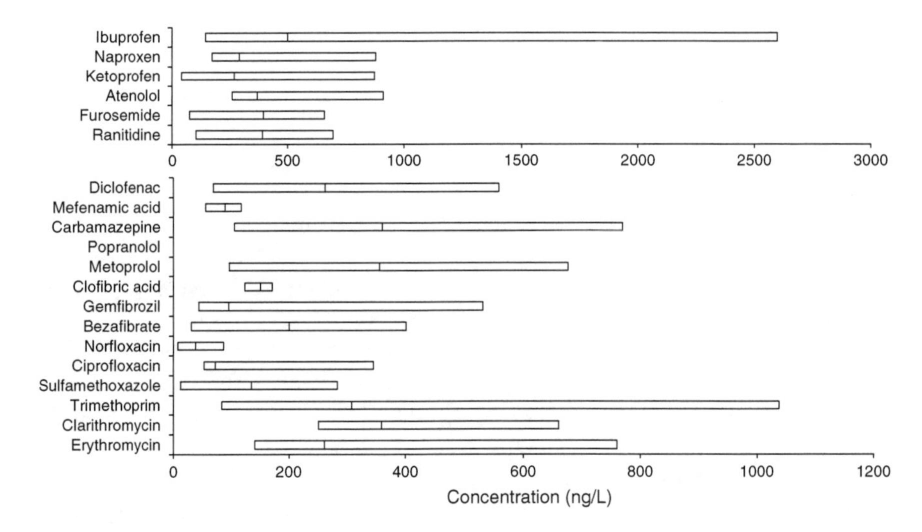

Fig. 3 Concentrations of the pharmaceuticals in wastewater effluent (25th, median, and 75th percentile, for the values found in the literature)

[20–22]. Even though the concentrations of these compounds are markedly lowered at the effluent, they are far from negligible. Heberer et al. [23] identified diclofenac as one of the most important pharmaceuticals in the water cycle, with low μg/L concentrations in both row and treated wastewater (3.0 and 2.5 μg/L at the influent and effluent, respectively).

Beta-blockers are another very important class of prescription drugs. They are very effective in treating cardiovascular diseases. As NSAIDs, beta-blockers are not highly persistent, but they are present in the environmental due to their high volume of use. Due to same mode of action of beta-blockers, it has been found that the mixture of beta-blockers showed concentration addition indicating a mutual specific nontarget effect on algae [24]. These compounds are generally found in aqueous phase because of their low sorption affinity and elevated biodegradability [25]. Atenolol, metoprolol, and propranolol have been frequently identified in wastewaters, where atenolol was detected in the highest concentrations, in some cases ranging up to 1 μg/L [26–28]. As a result of the incomplete removal during conventional wastewater treatment, these compounds were also found in surface waters in the ng/L to low mg/L range ([29, 30]; Ternes et al. 1998; [27]).

Lipid-lowering drugs, with statins and fibrates particularly, are used in the treatment and prevention of cardiovascular disease. In the last decade, statins became the drug of choice to lower cholesterol levels and their usage is increasing. According to the National Center for Health Statistics of USA [32], from 1988–1994 to 2003–2006, the use of statin drugs by adults aged 45 years and over increased almost tenfold, from 2 to 22%. Among lipid-lowering drugs and pharmaceuticals, in general, clofibric acid is one of the most frequently detected in the environment and one of the most persistent drugs with an estimated persistence in the environment of 21 years [15]. It has been detected in the ng/L range concentrations in influent, without big difference in the concentrations at the effluent. Many analogues of clofibrate, such as gemfibrozil, bezafibrate, and fenofibrate, were detected in the samples of sewage plants in concentrations up to low μg/L at the influent [20, 27, 33]. Among statins, atorvastatin, mevastatine, and prevastatine were detected in various environmental matrices including raw and treated wastewater as well as surface water near the points of discharge [20, 30, 34–36].

Antibiotics are destined to treat diseases and infection caused by bacteria. They are among the most frequently prescribed drugs for humans and animals in modern medicine. Beta-lactams, macrolides, sulfonamides, fluoroquinolones, and tetracyclines are the most important antibiotic groups used in both human and veterinary medicine. High global consumption of up to 200,000 tons per year [37] and high percentage of antibiotics that may be excreted without undergoing metabolism (up to 90%) result in their widespread presence in the environment [38]. Unmetabolized pharmaceutically active forms of antibiotics concentrated in raw sludge may promote the development of bacterial resistance. Bacteria in raw sludge are more resistant than bacteria elsewhere [39]. Many active antibiotic substances were found in raw sewage matrices, including both aqueous and solid phase. Beta-lactams are among the most prescribed antibiotics. Despite their high usage, they readily undergo hydrolysis, and thus have been detected in very low concentrations in treated wastewater, or not at all detected [40]. Sulfonamides, fluoroquinolone, and macrolide antibiotics show the highest persistence and are frequently detected in wastewater and surface waters [38]. Sulfamethoxazole is one of the most detected sulfonamides [41–45] that was reported with various concentrations and up to ca. 8μg/L (in raw influent in China) [46]. Sulfamethoxazole is often administrated in combination with trimethoprim, and commonly analyzed together [47]. Trimethoprim exhibits high persistence with little removal being effected by WWTPs, and thus is ubiquitously detected in ranges from very low ng/L to 1 μg/L in wastewater influent and effluent [41, 48]. Fluoroquinolone antibiotics ciprofloxacin and norfloxacin have been frequently detected in various streams in low μg/L [22, 41, 49, 50]. Even though it is greatly reduced during treatment, ciprofloxacin was found to be present in effluent wastewater at average concentration from 0.1 to 0.6 μg/L [48, 51, 52]. The class of tetracyclines, widely used broad-spectrum antibiotics, with chlortetracycline, oxytetracycline, and tetracycline as mostly used, was detected in raw and treated sewage in many studies in the ng/L [53] to μg/L concentrations [54]. Tetracyclines and fluoroquinolones form stable complexes with particulates and metal cations, showing the capacity to be more abundant in the sewage sludge [55, 56]. Some of the most prescribed antibiotics—macrolides clarithromycin, azithromycin, roxithromycin, and dehydro-erythromycin—were found in various environmental matrices in a variety of concentrations from very low ng/L to few μg/L [57–59]. High effluent concentrations were reported for dehydro-erythromycin, i.e., up to 2.5 μg/L [51, 54, 60]. In the final US EPA report CCL-3 from September 2009, erythromycin was one of three pharmaceuticals included as priority drinking water contaminant, based on health effects and occurrence in environmental waters [61].

According to the National Center for Health Statistics of USA [32], the usage of *antidiabetic drugs* by adults aged 45 years and over increased about 50%, from 7% in 1988–1994 to 11% in 2003–2006, as a consequence of an increase in the detection of antidiabetics in the environment over time. Still, only few data are reported on the occurrence and fate of antidiabetics. Glyburide (also glibenclamide) was found to be ubiquitous in both aqueous and solid phase of sewage treatment [20, 30, 62].

Histamine H2-receptor antagonists are used in the treatment of peptic ulcer and gastro-esophageal reflux disease. Certain preparations of these drugs are available OTC in various countries. Among H2-receptor antagonists, cimetidine and ranitidine have been frequently detected in wastewater and sludge. As for all the other pharmaceuticals, the reported concentration varies from very low ng/L to a few μg/L [20, 21, 63].

While antiepileptic carbamazepine is one of the most studied and detected pharmaceuticals in the environment, there is not much information on the occurrence and fate of other of psychoactive drugs in WWTPs. Carbamazepine is one of the most widely prescribed and very important drug for the treatment of epilepsy, trigeminal neuralgia, and some psychiatric diseases (e.g., bipolar affective disorders [64, 65]. In humans, following oral administration, it is metabolized to pharmacologically active carbamazepine-10,11-epoxide, which is further hydrolyzed to inactive carbamazepine-10, 11-trans-dihydrodiol, and conjugated products which are finally excreted in urine. Carbamazepine is almost completely transformed by metabolism with less than 5% of a dose excreted unchanged [66]. Still, glucuronide conjugates of carbamazepine can presumably be cleaved during wastewater treatment, so its environmental concentrations increase [27]. In fact, carbamazepine and its metabolites have been detected in both wastewaters and biosolids [67].

Carbamazepine is heavily or not degraded during wastewater treatment and many studies have found it ubiquitous in various environment matrices (groundwater, river, soil) [34, 62, 68–71]. The concentrations of carbamazepine vary from one plant to another, and they are usually around hundreds ng/L, and in some cases also few μg/L [27, 72].

3.2 Occurrence of Pharmaceuticals in Sewage Sludge

Sludge, originating from the wastewater treatment processes, is the semisolid residue generated during the primary (physical and/or chemical), the secondary (biological), and the tertiary (often nutrient removal) treatment. In the last years the quantities of sludge have been increasing in EU because of the implementation of the Directive 91/271/EEC on urban wastewater treatment. There was nearly nine million tons of dry matter produced in 2005 in EU Member states. Similar regulation in USA was introduced by US EPA regulations 40 Code of Federal Regulations Part 503 (40 CFR 503) that established the minimum national standards for the use and disposal of domestic sludge. By this regulation, sludge was classified into two different microbiological types according to the extent of pathogen removal achieved by the sludge treatment process, i.e., Class A (usage without end-use restrictions) and Class B (controlled/limited disposal). The amount of sludge generated in 2006 was estimated to be more than eight million tons of which 50% were land applied [73].

Due to the physical–chemical processes involved in the treatment, the sludge tends to concentrate heavy metals and poorly biodegradable trace organic compounds as well as potentially pathogenic organisms (viruses, bacteria, etc.) present in waste waters. Sludge is, however, rich in nutrients such as nitrogen and phosphorous and contains valuable organic matter that is useful when soils are depleted or subject to erosion. It has been used in agriculture over a long time. In EU, since 1986, the utilization of sewage sludge has been ruled by the EU Directive (86/278/EEC), which encouraged the use of sludge regulating its use with respect to the quality of sludge, the soil on which it is to be used, the loading rate, and the crops that may be grown on treated land. Nonc of the regulations cover the question of pharmaceuticals and other emerging pollutants that may be transported to soil after land application of biosolids, having the potential to enter surface water, leach into groundwater, or be accumulated by vegetation or other living organisms.

Most of the studies on the fate of pharmaceuticals in WWTPs focused only on the aqueous phase, and concentrations of the compounds in sludge were rarely determined mainly due to the demanding efforts required in the analysis in this difficult matrix. Out of 117 publications studied by Miege et al. [19], only 15 reported the concentrations of pharmaceuticals in sludge and 1 in suspended solid, and none of these papers reported the removal obtained taking into account both aqueous and solid phases of WWTPs. Still, the screening of sewage sludge showed that these micropollutants are very present in this medium [74–77]. High aqueous-phase removal rates for some compounds would suggest very good removal of these compounds during the wastewater treatment. But only a certain percent of the total mass of input is really lost (i.e., biodegradaded) during the treatment. The rest accumulates in sludge or ends up discharged with the effluent. As shown in Fig. 4, sorption of some pharmaceuticals analyzed in the study of Jelic et al. [20]

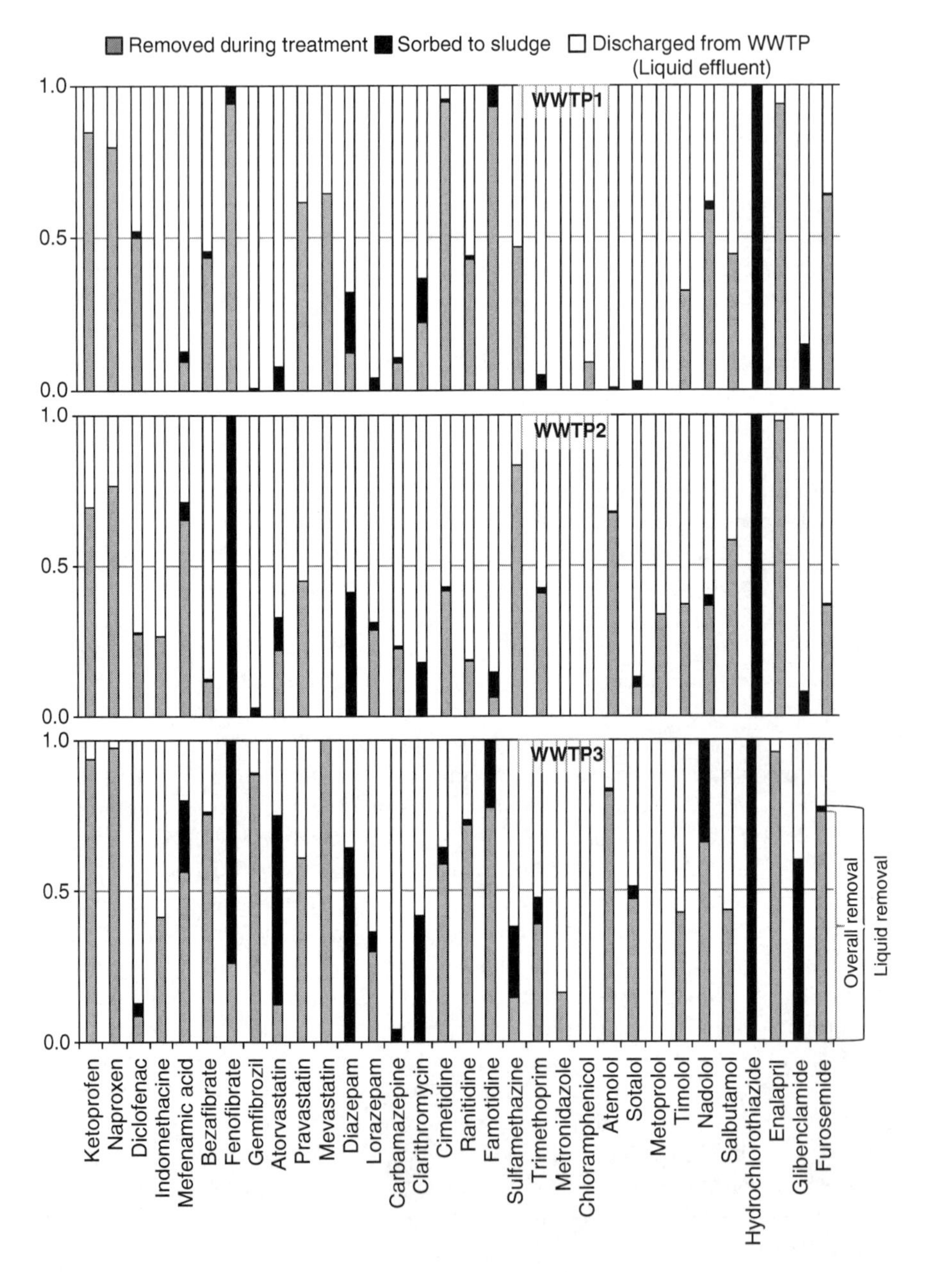

Fig. 4 Normalized mass loads of the selected pharmaceuticals entering three studied WWTPs, i.e., fraction discharged with effluent, sorbed to sludge, and removed during treatment

(e.g., atorvastatin, clarithromycin) contributed to the elimination from the aqueous phase with more than 20% related to the amount of these compounds at the influent. In the figure, term *liquid removal* indicates a difference in the loads of pharmaceuticals between influent and effluent wastewater, and *overall removal*, a difference in the loads that enter (i.e., influent) and exit (including both effluent and sludge loads) the plants. The difference between overall and liquid removal is the fraction that sorbed to sludge matter.This example clearly indicates the importance of the analysis of sludge when studying wastewater treatment performances.

The sorption behavior of pharmaceuticals can be very complex and difficult to assess. These compounds can absorb onto bacterial lipid structure and fat fraction of the sewage sludge through hydrophobic interactions (e.g., aliphatic and aromatic groups), adsorb onto often negatively charged polysaccharide structures on the outside of bacterial cells through electrostatic interactions (e.g., amino groups), and/or they can bind chemically to bacterial proteins and nucleic acids [78]. Also the mechanisms other than hydrophobic partitioning, such as hydrogen bonding, ionic interactions, and surface complexation, play a significant role in the sorption of pharmaceuticals in sludge. Therefore, also the compounds with low LogK_{ow} and LogK_d values may easily sorb onto sludge. Despite their negative K_{ow}, fluoroquinolones have a high tendency for sorption because of their zwitterionic character (pK_{aCOOH} = 5.9 – 6.4, pK_{aNH2} = 7.7 – 10.2) [49]. These antibiotics were reported to be present in the highest concentration in sewage sludge samples from various WWTPs [50, 74, 79]. Also tetracycline and sulfonamides exhibit strong sorption onto sludge particles, higher than expected based on their hydrophobicity.

In general, data on the occurrence of pharmaceuticals in sludge are sparse, where the group of antibiotics was mostly analyzed and found to be the most abundant. Kinney et al. [7] measured erythromycin-H_2O, sulfamethoxazole, and trimethoprim in microgram per kilogram concentrations in nine different biosolids. Out of the 72 pharmaceuticals and personal care products targeted in the study of EPA in its 2001

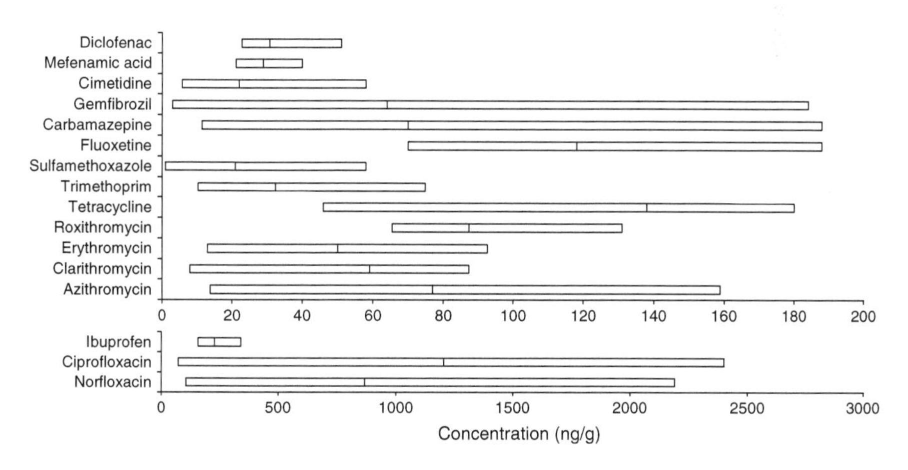

Fig. 5 Concentrations of the pharmaceuticals in sludge (25th, median, and 75th percentile, for the values found in the literature)

National Sewage Sludge Survey, 38 (54%) were detected at concentrations ranging from the low ng/g to the μg/g range. All the analyzed antibiotics constituted about 29% of the total mass of pharmaceuticals per sample [76]. Only few studies reported data on the occurrence of anti-inflammatories and some other therapeutic classes (psychoactive drugs, lipid-lowering drugs, etc.) [20, 79–82]. Figure 5 summarizes some literature data on the occurrence of several pharmaceuticals in sewage sludge in various WWTPs.

4 Removal of Pharmaceuticals During Conventional Wastewater Treatment

Municipal wastewater treatment plants were basically designed to remove pathogens and organic and inorganic suspended and flocculated matter. Even though the new treatment technologies have been developed to deal with health and environmental concerns associated with findings of nowadays research, the progress was not as enhanced as the one of the analytical detection capabilities and the pharmaceutical residues remain in the output of WWTPs. When speaking about pharmaceuticals, the term *removal* refers to the conversion of a pharmaceutical to a compound different than the analyzed one (i.e., the parent compound). Thus, it accounts for all the losses of a parent compound produced by different mechanisms of chemical and physical transformation, biodegradation, and sorption to solid matter. All these processes (mainly biodegradation, sorption, and photodegradation) are limited in some way for the following reasons: (a) pharmaceuticals are designed to be biologically stable; (b) the sorption depends on the type and properties of the suspended solids (sludge); (c) and even though they are photoactive, because many of them have aromatic rings, heteroatoms, and other functional groups that could be susceptible to photodegradation; they may also give the products of environmental concern.

Removal, as difference in the loads between influent and effluent, has negative values in some cases. The explanation for this could be found in sampling protocols [83], not only because they could be inadequate, but because of the nature of disposal of pharmaceuticals. The fact is that the substances arrive in a small number of wastewater packets to the influent of WWTP, in unpredictable amounts and time intervals; thus the influent loads, especially, are easily systematically underestimated.

Even though the analysis of effluent and sludge yields more certain results, because they come from stabilization processes, the sampling in general may result in underestimated and even negative removals. Furthermore, the negative removal can be explained by the formation of unmeasured products of human metabolism and/or transformation products (e.g., glucuronide conjugate, methylates, and glycinates) that passing through the plant convert back to the parent compounds. This can be considered as a reasonable assumption since the metabolites and some derivates of pharmaceuticals are well known (e.g., hydroxy and epoxy-derivatives of carbamazepine, 4-trans-hydroxy and 3-cis-hydroxy derivatives of glibenclamide, ortho- and parahydroxylated derivatives of atorvastatin, etc.) [67, 84, 85]. During complex metabolic processes in human body and biochemical in wastewater treatment, various scenarios of transformation from parent compound to metabolite and derivatives and vice versa can occur. Generally speaking, metabolites tend to increase the water solubility of the parent compound. Two different strategies are followed to achieve that objective: chemical modification of the parent molecule through addition of polar groups like OH, COOH, etc.; alternatively, the parent structures can be linked reversibly to polar highly soluble biological molecules, such as glucuronic acid, forming the so-called conjugates. Metabolites can be just as active as their parent compounds. Therefore, the occurrence of metabolites and transformation products and pathways should be included in the future studies in order to obtain accurate information on the removal of pharmaceuticals during treatment and to determine treatment plant capabilities.

The extent to which one compound can be removed during wastewater treatment is influenced by chemical and biological properties of the compound, but also of wastewater characteristics, operational conditions, and treatment technology used. Therefore, as shown in the following examples, high variations in elimination may be expected, and no clear and definitive conclusion can be made on the removal of any particular compound, and even less on the fate of a therapeutic group. Some operating parameters such as hydraulic retention time (HRT), solid retention time (SRT), redox conditions, and temperature may affect removal of pharmaceuticals during conventional treatment [127]. Of all the operating parameters, SRT is the most critical parameter for activated sludge design as SRT affects the treatment process performance, aeration tank volume, sludge production, and oxygen requirements. It has been proved that longer SRT, especially, influences and improves the elimination of most of the pharmaceuticals during sewage treatment [43, 68, 86]. WWTPs with high SRTs allow the enrichment of slowly growing bacteria and consequently the establishment of a more diverse biocoenosis with broader physiological capabilities (e.g., nitrification or the capability for certain elimination pathways) than WWTPs with low SRTs [68].

NSAIDs have been the most studied group in terms of both occurrence and removal during wastewater treatment. As noted, the pharmaceuticals are grouped according to the therapeutic applications; thus high variations in removal rates were observed within the group due to the differences in chemical properties. While ibuprofen and naproxen are generally removed with very high efficiency, diclofenac is only barely removable during conventional treatment. The removal rates of ibuprofen and naproxen are commonly higher than 75% and 50%, respectively [87–92]. Diclofenac shows rather low and very inconsistent removal rates, between 0 and 90% [88, 89, 91, 92]. Its persistence is attributed to the presence of chlorine group in the molecule. Some studies on removal during wastewater treatment showed no influence of SRT on the removal of diclofenac [68, 93, 94]. The removal of ibuprofen, ketoprofen, indomethacin, acetaminophen, and mefenamic acid is reported to be very high (>80%) or even complete for SRT typical for nutrient removal ($10 < SRT < 20$ days) [88, 90, 94].

Beta-blockers were reported to be only partially eliminated by conventional biological treatment [128, 26, 27, 31]). The data on their removal are very inconsistent and the removal rates vary from less than 10% up to 95% depending on the treatment. Maurer et al. proved that the elimination of beta-blockers in WWTPs depends on the HRT, which could be a good explanation for the variable removal reported in the literature. The highest average elimination rates can be observed for atenolol and sotalol (around 60%). But for the same compounds low removals were reported as well. Maurer et al. [95] and Wick et al. [96] reported removal rates lower than 30% for sotalol, and Castiglioni et al. [87] reported a removal of 10% for atenolol during the winter months. In the same study, atenolol achieved better elimination in summer (i.e., 55%) due to a higher microbial activity [87]. In most cases, metoprolol showed very low removal rates, i.e., 10–30% [26, 31, 97]. The occurrence and microbial cleavage of conjugates are well known to influence the mass balance in WWTPs [98–100]. The microbial cleavage of conjugates of metoprolol could be responsible for an underestimation of its removal efficiency. For propranolol as well, various removal rates, mostly moderately low [96, 101], were observed. This compound is by far the most lipophilic beta-blocker and the only one with a bioaccumulation potential. Sludge–water partition coefficients were found to be less than 100 L/kg_{ss} for sotalol and atenolol, and 343 L/kg_{ss} for propanolol [95]. For $K_{d,s}$ values less than 500 L/kg_{ss} the removal by sorption in a WWTP with a typical sludge production of 0.2 g_{ss}/L is less than 10% [102] and hence does not significantly contribute to the removal of beta-blockers in activated sludge units [96]. Therefore, the partial removal of beta-blockers can be assumed to be due to biotransformation.

No significant to medium removal was reported for all the *lipid regulators*. Zorita et al. [22] reported a medium removal of 61% during primary and biological treatment, where during the latter one no removal was observed. Lower removal rates were reported by some other authors, <35% [87, 103, 104]. No or low removal of clofibric acid was observed in different WWTPS of Berlin [105]. Winkler et al. [106] found no evidence for biotic degradation of clofibric acid. Radjenovic et al. [36] showed that clofibric acid may be removed more efficiently with MBR (72–86%) compared to the activated sludge process (of 26–51%). The removal rates of fibrates bezafibrate, fenofibrate, and gemfibrozil vary between 30 and 90% [88, 94, 107]. Various removal rates during biological treatment in ten WWTPs in Spain were reported in the studies of Jelic et al. [20] and Gros et al. [30]. It was found that sorption to sludge particulate contributes to the removal values with ca. 20% [20].

Results from various studies showed that *anticonvulsant carbamazepine* is recalcitrant to biological treatment and it is not removed during either conventional wastewater treatment or membrane bioreactor treatments [34, 62, 69, 71, 91, 108]. Physicochemical processes such as coagulation-flocculation and flotation did not give better results concerning its degradation [109–111]. It was constantly found at higher concentrations at the effluent of WWTPs. Knowing that the activated sludge has glucuronidase activity, which allows the cleavage of the glucorinic acid moiety in WWTP [112], rational explanation for the increase in concentration is conversion of CBZ glucorinides and other conjugated metabolites to the parent compound by enzymatic processes in a WWTP. No influence of SRT on the removal of CBZ during conventional wastewater treatment was noticed [68, 93]. Except for carbamazepine, information on the occurrence and fate of other psychoactive drugs in WWTPs is very scarce. This is probably because of the low therapeutic dose resulting generally in low concentrations in the environment [63, 113, 114]. Conventional treatment achieves a removal lower than 10% in case of diazepam [72]. Zorita et al. [22] reported very high removal efficiency in case of fluoxetine and its active metabolite norfluoxetine. Still, the lowest observed effect concentration of fluoxetine for zooplankton and benthic organisms is close to the maximal measured WWTP effluent concentrations [115]. Low effluent concentrations can be due to the fact that fluoxetine rapidly passes to solid phase where it appears to be very persistent [116, 117]. Additionally, it was found that it has a high bioaccumulation potential when detected in wild fish [118].

The removal of several other drugs such as the *histamine H2-receptor antagonists* cimetidine, famotidine, and ranitidine varied from low to very high. Radjenovic et al. [36] reported rather poor and unstable removal of histamines during conventional treatment (15–60%). Castiglioni et al. [87] found that the removal of ranitidine depends on season, and showed 39% removal in winter and 84% in summer. High removal of ranitidine during activate sludge treatment (89%) was observed in a study of Kasprzyk-Hordern et al. [21].

Antibiotics cover a broad range of chemical classes, and it is very difficult to characterize their behavior during activated sludge process due to varying removal efficiencies reported from studies undertaken worldwide. Due to their limited biodegradability and sorption properties, sulfonamides and trimethoprim appear to be only partially removed by conventional wastewater treatment. The removal of these antibiotics has been reported to vary significantly [41–44, 97, 119]. The explanation could be found in different operational parameters such as HRT, SRT, and temperature and also in the fact that sulfonamides are easily transformed from their parent compounds to their metabolites and vice versa; thus the removal efficiencies may be easily or underestimated or overestimated. In case of trimethoprim, only minor removal was noticed during primary and biological treatment, but the advanced treatment [43] and nitrification organisms appear to be capable of degrading trimethorpim [120, 121]. This suggests an important role for aerobic conditions for the biotransformation of trimethoprim. Consistent with this, removal efficiency of trimethoprim appears to be enhanced by long SRT during biological treatment, which is conducive to nitrification [122]. Also macrolide antibiotics are often incompletely removed during biological wastewater treatment. Studies from different conventional WWTPs have revealed that the removal of macrolides varied from high but negative values, to around 50% [43, 88]. Karthikeyan and Meyer [59] found that 43 to 99% of erythromycin was removed by activated sludge process and aerated lagoons. In the study of Kobayashi et al. [123], 50% of clarithromycin and azithromycin were removed from three conventional WWTPs. Gobel et al. [43] proposed gradual release of the macrolides (e.g., clarithromycin) from feces particles during biological treatment as an explanation for the possible negative removal rates for these antibiotics. Sorption of macrolides to wastewater biomass is attributed to hydrophobic interactions (high partitioning coefficient). But knowing that the surface of activated sludge is predominantly negatively charged , and under typical wastewater conditions, the basic dimethylamino group (p*K*a > 8.9) is protonated, sorption could occur due to cation exchange interaction as well [124]. Greater adsorption of azithromycin to biomass compared to clarithromycin has been reported [123]. Varying removal was reported for fluoroquinolone antibiotics, as well. [125] found that norfloxacin and ciprofloxacin were removed with 78% and 80%, respectively, where around 40% was removed during the biological treatment. Similar removal was reported by Zorita et al. [22] during secondary and tertiary treatment. The predominant removal mechanism of fluoroquinolones has been suggested to be adsorption to sludge and/or flocs rather than biodegradation [22, 49, 122, 125]. Ciprofloxacin and norfloxacin sorbed to sludge independently of changes in pH during wastewater treatment, and more than 70% of the total amount of these compounds passing through the plant was ultimately found in the digested sludge [125]. These findings indicated sludge as the main reservoir of fluoroquinolones that may potentially release the antibiotics into the environment when applied to agricultural land. Karthikeyan and Meyer [59] reported removal efficiency of 68% for tetracycline, and Yang et al. [45] have reported removals of 78 and 67% for chlortetracycline and doxycycline, respectively, during activated sludge processes. Also some tetracyclines have significant potential for absorption onto solids due to a combination of non-hydrophobic mechanisms, such as ionic interactions, metal complexation, hydrogen bond formation, or polarization [126]. Their removal is not so affected by HRT, but SRT appears to significantly influence the removal during biological treatment [53].

5 Conclusion

Various studies showed that even though the conventional WWTPs meet the regulatory requirements for wastewater treatment (Directive 91/271/EEC), they are only moderately effective in removing pharmaceuticals. The removal of pharmaceuticals will naturally depend on their chemical and biological properties, but also on wastewater characteristics, operational conditions, and treatment technology used. Thus, high variations in removal may be expected, and no clear and definitive conclusion can be made on the removal of any particular compound. Of all the operating parameters, the solid retention time (SRT) is the most critical parameter for activated sludge process and it has been proved that longer SRT improves the elimination of most of the pharmaceuticals during sewage treatment. Estimation of removal rates may be underestimated due to at least three factors: removal efficiency was calculated from the mean concentration values; the metabolites and transformation products of pharmaceuticals and their amounts were not defined; and the time-proportional sampling may not be perfectly suitable for pharmaceutical analysis, especially on influent. Still, the fact is that a number of pharmaceuticals were detected in effluent wastewater and sludge samples. These pharmaceuticals cover a wide range of physical-chemical properties and biological activities. Although the chronic toxicity effects of such a mixture are unknown and thus the risk that it could pose to the environment could not be fully assessed, their presence must not be ignored. More information on quality, quantity, and toxicity of pharmaceuticals and their metabolites is definitely needed especially when attempting to reuse wastewater and dispose sludge to agricultural areas and landfills.

References

1. Daughton CG (2003) Cradle-to-cradle stewardship of drugs for minimizing their environmental disposition while promoting human health. I. Rational for and avenues toward a green pharmacy. Environ Health Perspect 111:757–774
2. Ruhoya ISR, Daughton CG (2008) Beyond the medicine cabinet: an analysis of where and why medications accumulate. Environ Int 34:1157–1169
3. Bartelt-Hunt SL, Snow DD, Damon T, Shockley J, Hoagland K (2009) The occurrence of illicit and therapeutic pharmaceuticals in wastewater effluent and surface waters in Nebraska. Environ Pollut 157:786–791
4. Nilsen EB, Rosenbauer RR, Furlong ET, Burkhardt MR, Werner SL, Greaser L, Noriega M (2007) Pharmaceuticals, personal care products and anthropogenic waste indicators detected in streambed sediments of the lower Columbia River and selected tributaries, National Ground Water Association, Paper 4483, p 15
5. Vazquez-Roig P, Segarra R, Blasco C, Andreu V, Picó Y (2010) Determination of pharmaceuticals in soils and sediments by pressurized liquid extraction and liquid chromatography tandem mass spectrometry. J Chromatogr A 1217:2471–2483
6. Gielen GJHP, Heuvel MRvd, Clinton PW, Greenfield LG (2009) Factors impacting on pharmaceutical leaching following sewage application to land. Chemosphere 74:537–542
7. Kinney CA, Furlong ET, Werner SL, Cahill JD (2006) Presence and distribution of wastewater-derived pharmaceuticals in soil irrigated with reclaimed water. Environ Toxicol Chem 25:317–326
8. Ternes TA, Bonerz M, Herrmann N, Teiser B, Andersen HR (2007) Irrigation of treated wastewater in Braunschweig, Germany: an option to remove pharmaceuticals and musk fragrances. Chemosphere 66:894–904
9. Carbonell G, Pro J, Gómcz N, Babín MM, Fernández C, Alonso E, Tarazona JV (2009) Sewage sludge applied to agricultural soil: ecotoxicological effects on representative soil organisms. Ecotoxicol Environ Saf 72:1309–1319
10. Lapen DR, Topp E, Metcalfe CD, Li H, Edwards M, Gottschall N, Bolton P, Curnoe W, Payne M, Beck A (2008) Pharmaceutical and personal care products in tile drainage following land application of municipal biosolids. Sci Total Environ 399:50–65
11. Cleuvers M (2004) Mixture toxicity of the anti-inflammatory drugs diclofenac, ibuprofen, naproxen, and acetylsalicylic acid. Ecotoxicol Environ Saf 59:309–315
12. Nentwig G, Oetken M, Oehlmann J (2004) Effects of pharmaceuticals on aquatic invertebrates—the example of carbamazepine and clofibric acid. In: Kümmerer K (ed) Pharmaceuticals in the environment. Sources, fate, effects and risks, 2nd edn. Springer, Berlin, pp 195–207
13. Schnell S, Bols NC, Barata C, Porte C (2009) Single and combined toxicity of pharmaceuticals and personal care products (PPCPs) on the rainbow trout liver cell line RTL-W1. Aquat Toxicol 93:244–252

14. Metcalf and Eddy Inc (2003) Wastewater engineering – treatment and reuse, 4th edn. Tata McGraw Hill, New Delhi
15. Khetan SK, Collins TJ (2007) Human pharmaceuticals in the aquatic environment: a challenge to green chemistry. Chem Rev 107:2319–2364
16. Larsson DGJ, de Pedro C, Paxeus N (2007) Effluent from drug manufactures contains extremely high levels of pharmaceuticals. J Hazard Mater 148:751–755
17. Li D, Yang M, Hu J, Ren L, Zhang Y, Li K (2008) Determination and fate of oxytetracycline and related compounds in oxytetracycline production wastewater and the receiving river. Environ Toxicol Chem 27:80–86
18. Li D, Yang M, Hu J, Zhang Y, Chang H, Jin F (2008) Determination of penicillin G and its degradation products in a penicillin production wastewater treatment plant and the receiving river. Water Res 42:307–317
19. Miège C, Choubert JM, Ribeiro L, Eusèbe M, Coquery M (2009) Fate of pharmaceuticals and personal care products in wastewater treatment plants - Conception of a database and first results. Environ Pollut 157:1721–1726
20. Jelic A, Gros M, Ginebreda A, Cespedes-Sánchez R, Ventura F, Petrovic M, Barcelo D (2011) Occurrence, partition and removal of pharmaceuticals in sewage water and sludge during wastewater treatment. Water Res 45:1165–1176
21. Kasprzyk-Hordern B, Dinsdale RM, Guwy AJ (2009) The removal of pharmaceuticals, personal care products, endocrine disruptors and illicit drugs during wastewater treatment and its impact on the quality of receiving waters. Water Res 43:363–380
22. Zorita S, Mårtensson L, Mathiasson L (2009) Occurrence and removal of pharmaceuticals in a municipal sewage treatment system in the south of Sweden. Sci Total Environ 407:2760–2770
23. Heberer T, Reddersen K, Mechlinski A (2002) From municipal sewage to drinking water: fate and removal of pharmaceutical residues in the aquatic environment in urban areas. Water Sci Technol 46:81
24. Escher BI, Bramaz N, Richter M, Lienert J (2006) Comparative ecotoxicological hazard assessment of beta-blockers and their human metabolites using a mode-of-action-based test battery and a QSAR approach†. Environ Sci Technol 40:7402–7408
25. Ramil M, El Aref T, Fink G, Scheurer M, Ternes TA (2009) Fate of beta blockers in aquatic-sediment systems: sorption and biotransformation. Environ Sci Technol 44:962–970
26. Lee H-B, Sarafin K, Peart TE (2007) Determination of [beta]-blockers and [beta]2-agonists in sewage by solid-phase extraction and liquid chromatography-tandem mass spectrometry. J Chromatogr A 1148:158–167
27. Ternes TA (1998) Occurrence of drugs in German sewage treatment plants and rivers. Water Res 32:3245
28. Vieno N, Tuhkanen T, Kronberg L (2007) Elimination of pharmaceuticals in sewage treatment plants in Finland. Water Res 41:1001–1012
29. Bendz D, Paxéus NA, Ginn TR, Loge FJ (2005) Occurrence and fate of pharmaceutically active compounds in the environment, a case study: Höje River in Sweden. J Hazard Mater 122:195–204
30. Gros M, Petrovic M, Ginebreda A, Barceló D (2010) Removal of pharmaceuticals during wastewater treatment and environmental risk assessment using hazard indexes. Environ Int 36:15–26
31. Vieno NM, Tuhkanen T, Kronberg L (2006) Analysis of neutral and basic pharmaceuticals in sewage treatment plants and in recipient rivers using solid phase extraction and liquid chromatography-tandem mass spectrometry detection. J Chromatogr A 1134:101–111
32. Health, U.S. (2010) Health, United States, 2009: with special feature on medical technology. National Center for Health Statistics, Hyattsville, MD
33. Gracia-Lor E, Sancho JV, Hernández F (2010) Simultaneous determination of acidic, neutral and basic pharmaceuticals in urban wastewater by ultra high-pressure liquid chromatography-tandem mass spectrometry. J Chromatogr A 1217:622–632

34. Metcalfe CD, Koenig BG, Bennie DT, Servos M, Ternes TA, Hirsch R (2003) Occurrence of neutral and acidic drugs in the effluents of Canadian sewage treatment plants. Environ Toxicol Chem 22:2872–2880
35. Miao X-S, Metcalfe CD (2003) Determination of cholesterol-lowering statin drugs in aqueous samples using liquid chromatography-electrospray ionization tandem mass spectrometry. J Chromatogr A 998:133–141
36. Radjenovic J, Petrovic M, Barceló D (2009) Fate and distribution of pharmaceuticals in wastewater and sewage sludge of the conventional activated sludge (CAS) and advanced membrane bioreactor (MBR) treatment. Water Res 43:831–841
37. Kümmerer K (2003) Significance of antibiotics in the environment. J Antimicrob Chemother 52:5–7
38. Huang CH, Renew JE, Smeby KL, Pinkerston K, Sedlak DL (2001) Assessment of potential antibiotic contaminants in water and preliminary occurrence analysis. Water Resour Update 120:30–40
39. Jones OAH, Voulvoulis N, Lester JN (2004) Potential ecological and human health risks associated with the presence of pharmaceutically active compounds in the aquatic environment. Crit Rev Toxicol 34:335–350
40. Le-Minh N, Khan SJ, Drewes JE, Stuetz RM (2010) Fate of antibiotics during municipal water recycling treatment processes. Water Res 44:4295–4323
41. Brown KD, Kulis J, Thomson B, Chapman TH, Mawhinney DB (2006) Occurrence of antibiotics in hospital, residential, and dairy effluent, municipal wastewater, and the Rio Grande in New Mexico. Sci Total Environ 366:772–783
42. Choi K-J, Kim S-G, Kim C-W, Kim S-H (2007) Determination of antibiotic compounds in water by on-line SPE-LC/MSD. Chemosphere 66:977–984
43. Göbel A, McArdell CS, Joss A, Siegrist H, Giger W (2007) Fate of sulfonamides, macrolides, and trimethoprim in different wastewater treatment technologies. Sci Total Environ 372: 361–371
44. Levine AD, Meyer MT, Kish G (2006) Evaluation of the persistence of micropollutants through pure-oxygen activated sludge nitrification and denitrification. Water Environ Res 78:2276–2285
45. Yang S, Cha J, Carlson K (2005) Simultaneous extraction and analysis of 11 tetracycline and sulfonamide antibiotics in influent and effluent domestic wastewater by solid-phase extraction and liquid chromatography-electrospray ionization tandem mass spectrometry. J Chromatogr A 1097:40–53
46. Peng X, Wang Z, Kuang W, Tan J, Li K (2006) A preliminary study on the occurrence and behavior of sulfonamides, ofloxacin and chloramphenicol antimicrobials in wastewaters of two sewage treatment plants in Guangzhou, China. Sci Total Environ 371:314–322
47. Gobel A, Thomsen A, McArdell CS, Joss A, Giger W (2005) Occurrence and Sorption Behavior of Sulfonamides, Macrolides, and Trimethoprim in Activated Sludge Treatment. Environ Sci Technol 39:3981–3989
48. Watkinson AJ, Murby EJ, Costanzo SD (2007) Removal of antibiotics in conventional and advanced wastewater treatment: Implications for environmental discharge and wastewater recycling. Water Res 41:4164–4176
49. Golet EM, Xifra I, Siegrist H, Alder AC, Giger W (2003) Environmental exposure assessment of fluoroquinolone antibacterial agents from sewage to soil. Environ Sci Technol 37:3243
50. Lindberg RH, Wennberg P, Johansson MI, Tysklind M, Andersson BAV (2005) Screening of human antibiotic substances and determination of weekly mass flows in five sewage treatment plants in Sweden. Environ Sci Technol 39:3421–3429
51. Alder AC, McArdell CS, Golet EM, Ibric S, Molnar E, Nipales NS, Giger W (2001) Occurrence and fate of fluoroquinolone, macrolide, and sulfonamide antibiotics during wastewater treatment and in ambient waters in Switzerland. In: Daughton CG, Jones-Lepp TM (eds) Pharmaceuticals and personal care products in the environment, scientific and regulatory

issues, American Chemical Society Symposium Series 791. American Chemical Society, Washington, DC, pp 56–69
52. Golet EM, Alder AC, Hartmann A, Ternes TA, Giger W (2001) Trace determination of fluoroquinolone antibacterial agents in urban wastewater by solid-phase extraction and liquid chro matography with fluorescence detection. Anal Chem 73:3632
53. Kim S, Eichhorn P, Jensen JN, Weber AS, Aga DS (2005) Removal of antibiotics in wastewater: effect of hydraulic and solid retention times on the fate of tetracycline in the activated sludge process. Environ Sci Technol 39:5816–5823
54. Yang S, Carlson K (2003) Evolution of antibiotic occurrence in a river through pristine, urban and agricultural landscapes. Water Res 37:4645–4656
55. Alexy R, Kümpel T, Kümmerer K (2004) Assessment of degradation of 18 antibiotics in the Closed Bottle Test. Chemosphere 57:505–512
56. Daughton CG, Ternes TA (1999) Pharmaceuticals and personal care products in the environment: Agents of subtle change? Environ Health Perspect 107:907–938
57. Göbel A, Thomsen A, McArdell CS, Alder AC, Giger W, Theiβ N, Loeffler D, Ternes TA (2005) Extraction and determination of sulfonamides, macrolides, and trimethoprim in sewage sludge. J Chromatogr A 1085:179–189
58. Gulkowska A, Leung HW, So MK, Taniyasu S, Yamashita N, Yeung LWY, Richardson BJ, Lei AP, Giesy JP, Lam PKS (2008) Removal of antibiotics from wastewater by sewage treatment facilities in Hong Kong and Shenzhen, China. Water Res 42:395–403
59. Karthikeyan KG, Meyer MT (2006) Occurrence of antibiotics in wastewater treatment facilities in Wisconsin, USA. Sci Total Environ 361:196–207
60. Hirsch R, Ternes T, Haberer K, Kratz KL (1999) Occurrence of antibiotics in the aquatic environment. Sci Total Environ 225:109
61. Richardson SD (2010) Environmental mass spectrometry: emerging contaminants and current issues. Anal Chem 82:4742–4774
62. Radjenovic J, Petrovic M, Barceló D (2007) Analysis of pharmaceuticals in wastewater and removal using a membrane bioreactor. Anal Bioanal Chem 387:1365–1377
63. Kolpin DW, Furlong ET, Meyer MT, Thurman EM, Zaugg SD, Barber LB, Buxton HT (2002) Pharmaceuticals and other organic wastewater contaminants in US streams, 1999-2000: a national reconnaissance. Environ Sci Technol 36:1202
64. Fertig EJ, Mattson RH (2008) Carbamazepine in the book: Epilepsy: A comprehensive textbook, vol 2. Lippincott Williams and Wilkins, Philadephia, PA, pp 1543–1556
65. Yoshimura R, Yanagihara N, Terao T, Minami K, Toyohira Y, Ueno S, Uezono Y, Abe K, Izumi F (1998) An active metabolite of carbamazepine, carbamazepine-10,11-epoxide, inhibits ion channel-mediated catecholamine secretion in cultured bovine adrenal medullary cells. Psychopharmacology 135:368–373
66. Shorvon S, Perucca E, FIsh D, Dodson E (2004) Carbamazepine in: the treatment of epilepsy. Blackwell Science, Maiden, MA, pp 131–132
67. Miao X-S, Yang J-J, Metcalfe CD (2005) Carbamazepine and its metabolites in wastewater and in biosolids in a municipal wastewater treatment plant. Environ Sci Technol 39: 7469–7475
68. Clara M, Kreuzinger N, Strenn B, Gans O, Kroiss H (2005) The solids retention time – a suitable design parameter to evaluate the capacity of wastewater treatment plants to remove micropollutants. Water Res 39:97–106
69. Clara M, Strenn B, Kreuzinger N (2004) Carbamazepine as a possible anthropogenic marker in the aquatic environment: investigations on the behaviour of Carbamazepine in wastewater treatment and during groundwater infiltration. Water Res 38:947–954
70. Joss A, Keller E, Alder AC, Gbel A, McArdell CS, Ternes T, Siegrist H (2005) Removal of pharmaceuticals and fragrances in biological wastewater treatment. Water Res 39:3139
71. Zhang Y, Geißen S-U, Gal C (2008) Carbamazepine and diclofenac: removal in wastewater treatment plants and occurrence in water bodies. Chemosphere 73:1151–1161

72. Ternes TA, Janex-Habibi M-L, Knacker T, Kreuzinger N, Siegrist H (2005) Assessment of technologies for the removal of pharmaceuticals and personal care products in sewage and drinking water facilities – POSEIDON Project
73. Kinney CA, Furlong ET, Kolpin DW, Burkhardt MR, Zaugg SD, Werner SL, Bossio JP, Benotti MJ (2008) Bioaccumulation of pharmaceuticals and other anthropogenic waste indicators in earthworms from agricultural soil amended with biosolid or swine manure. Environ Sci Technol 42:1863–1870
74. Lillenberg M, Yurchenko S, Kipper K, Herodes K, Pihl V, Sepp K, Lõhmus R, Nei L (2009) Simultaneous determination of fluoroquinolones, sulfonamides and tetracyclines in sewage sludge by pressurized liquid extraction and liquid chromatography electrospray ionization-mass spectrometry. J Chromatogr A 1216:5949–5954
75. Lindberg RH, Fick J, Tysklind M (2010) Screening of antimycotics in Swedish sewage treatment plants – Waters and sludge. Water Res 44:649–657
76. McClellan K, Halden RU (2010) Pharmaceuticals and personal care products in archived U.S. biosolids from the, 2001 EPA national sewage sludge survey. Water Res 44:658–668
77. Radjenovic J, Jelic A, Petrovic M, Barcelo D (2009) Determination of pharmaceuticals in sewage sludge by pressurized liquid extraction (PLE) coupled to liquid chromatography-tandem mass spectrometry (LC-MS/MS). Anal Bioanal Chem 393:1685–1695
78. Meakins NC, Bubb JM, Lester JN (1994) Fate and behaviour of organic micropollutants during wastewater treatment processes: a review. Int J Environ Pollut 4:27–58
79. Chenxi W, Spongberg AL, Witter JD (2008) Determination of the persistence of pharmaceuticals in biosolids using liquid-chromatography tandem mass spectrometry. Chemosphere 73:511–518
80. Nieto A, Borrull F, Marcé RM, Pocurull E (2007) Selective extraction of sulfonamides, macrolides and other pharmaceuticals from sewage sludge by pressurized liquid extraction. J Chromatogr A 1174:125–131
81. Okuda T, Yamashita N, Tanaka H, Matsukawa H, Tanabe K (2009) Development of extraction method of pharmaceuticals and their occurrences found in Japanese wastewater treatment plants. Environ Int 35:815–820
82. Spongberg AL, Witter JD (2008) Pharmaceutical compounds in the wastewater process stream in Northwest Ohio. Sci Total Environ 397:148–157
83. Ort C, Lawrence MG, Rieckermann Jr, Joss A (2010) Sampling for pharmaceuticals and personal care products (PPCPs) and illicit drugs in wastewater systems: are your conclusions valid? a critical review. Environ Sci Technol 44:6024–6035
84. Aviram M, Rosenblat M, Bisgaier CL, Newton RS (1998) Atorvastatin and gemfibrozil metabolites, but not the parent drugs, are potent antioxidants against lipoprotein oxidation. Atherosclerosis 138:271–280
85. Shipkova M, Wieland E (2005) Glucuronidation in therapeutic drug monitoring. Clin Chim Acta 358:2–23
86. Suárez S, Ramil M, Omil F, Lema JM (2005) Removal of pharmaceutically active compounds in nitrifying-denitrifying plants. Water Sci Technol 52:9–14
87. Castiglioni S, Bagnati R, Fanelli R, Pomati F, Calamari D, Zuccato E (2006) Removal of pharmaceuticals in sewage treatment plants in Italy. Environ Sci Technol 40:357–363
88. Clara M, Strenn B, Gans O, Martinez E, Kreuzinger N, Kroiss H (2005) Removal of selected pharmaceuticals, fragrances and endocrine disrupting compounds in a membrane bioreactor and conventional wastewater treatment plants. Water Res 39:4797
89. Gómez MJ, Martínez Bueno MJ, Lacorte S, Fernández-Alba AR, Agüera A (2007) Pilot survey monitoring pharmaceuticals and related compounds in a sewage treatment plant located on the Mediterranean coast. Chemosphere 66:993–1002
90. Jones OAH, Voulvoulis N, Lester JN (2005) Human pharmaceuticals in wastewater treatment processes. Crit Rev Environ Sci Technol 35:401–427
91. Joss A, Keller E, Alder AC, Göbel A, McArdell CS, Ternes T, Siegrist H (2005) Removal of pharmaceuticals and fragrances in biological wastewater treatment. Water Res 39:3139–3152

92. Lindqvist N, Tuhkanen T, Kronberg L (2005) Occurrence of acidic pharmaceuticals in raw and treated sewages and in receiving waters. Water Res 39:2219–2228
93. Kreuzinger N, Clara M, Strenn B, Kroiss H (2004) Relevance of the sludge retention time (SRT) as design criteria for wastewater treatment plants for the removal of endocrine disruptors and pharmaceuticals from wastewater. Water Sci Technol 50:149–156
94. Lishman L, Smyth SA, Sarafin K, Kleywegt S, Toito J, Peart T, Lee B, Servos M, Beland M, Seto P (2006) Occurrence and reductions of pharmaceuticals and personal care products and estrogens by municipal wastewater treatment plants in Ontario, Canada. Sci Total Environ 367:544–558
95. Maurer M, Escher BI, Richle P, Schaffner C, Alder AC (2007) Elimination of [beta]-blockers in sewage treatment plants. Water Res 41:1614–1622
96. Wick A, Fink G, Joss A, Siegrist H (2009) Fate of beta blockers and psycho-active drugs in conventional wastewater treatment. Water Res 43:1060–1074
97. Paxéus N (2004) Removal of selected non-steroidal anti-inflammatory drugs (NSAIDs), gemfibrozil, carbamazepine, β-blockers, trimethoprim and triclosan in conventional wastewater treatment plants in five EU countries and their discharge to the aquatic environment. Water Sci Technol 5:253–260
98. Andersen H, Siegrist H, Halling-Sørensen B, Ternes TA (2003) Fate of estrogens in a municipal sewage treatment plant. Environ Sci Technol 37:4021–4026
99. Belfroid AC, Van der Horst A, Vethaak AD, Schäfer AJ, Rijs GBJ, Wegener J, Cofino WP (1999) Analysis and occurrence of estrogenic hormones and their glucuronides in surface water and waste water in The Netherlands. Sci Total Environ 225:101–108
100. Ternes TA, Kreckel P, Mueller J (1999) Erratum: Behaviour and occurrence of estrogens in municipal sewage treatment plants – II. Aerobic batch experiments with activated sludge. Sci Total Environ 228:89–99
101. Alder AC, Bruchet A, Carballa M, Clara M, Joss A, Löffler D, McArdell CS, Miksch K, Omil F, Tuhkanen T, Ternes TA (2006) Consumption and occurrence. In: Ternes TA, Joss A (eds) Human pharmaceuticals, hormones and fragrances. IWA Publishing, London, pp 15–54
102. Ternes TA, Herrmann N, Bonerz M, Knacker T, Siegrist H, Joss A (2004) A rapid method to measure the solid-water distribution coefficient (Kd) for pharmaceuticals and musk fragrances in sewage sludge. Water Res 38:4075
103. Kosjek T, Heath E, Petrovic M, Barceló D (2007) Mass spectrometry for identifying pharmaceutical biotransformation products in the environment. TrAC Trends Anal Chem 26:1076–1085
104. Matamoros V, Caselles-Osorio A, García J, Bayona JM (2008) Behaviour of pharmaceutical products and biodegradation intermediates in horizontal subsurface flow constructed wetland. A microcosm experiment. Sci Total Environ 394:171–176
105. Heberer T (2002) Tracking persistent pharmaceutical residues from municipal sewage to drinking water. J Hydrol 266:175
106. Winkler M, Lawrence JR, Neu TR (2001) Selective degradation of ibuprofen and clofibric acid in two model river biofilm systems. Water Res 35:3197–3205
107. Ternes TA, Meisenheimer M, McDowell D, Sacher F, Brauch H Jr, Haist-Gulde B, Preuss G, Wilme U, Zulei-Seibert N (2002) Removal of pharmaceuticals during drinking water treatment. Environ Sci Technol 36:3855–3863
108. Miao X-S, Metcalfe CD (2003) Determination of carbamazepine and its metabolites in aqueous samples using liquid chromatography – electrospray tandem mass spectrometry. Anal Chem 75:3731–3738
109. Carballa M, Omil F, Lema JM, Llompart M, Garcia-Jares C, Rodriguez I, Gomez M, Ternes T (2004) Behaviour of pharmaceuticals, cosmetics and hormones in a sewage treatment plant. Water Res 38:2918
110. Reif R, Suárez S, Omil F, Lema JM (2008) Fate of pharmaceuticals and cosmetic ingredients during the operation of a MBR treating sewage. Desalination 221:511–517

111. Suárez S, Carballa M, Omil F, Lema JM (2008) How are pharmaceutical and personal care products (PPCPs) removed from urban wastewaters? Rev Environ Sci Biotechnol 7:125–138
112. Ternes TA, Kreckel P, Mueller J (1999) Behaviour and occurrence of estrogens in municipal sewage treatment plants – II. Aerobic batch experiments with activated sludge. Sci Total Environ 225:91–99
113. Kümmerer K (2004) Pharmaceuticals in the environment: sources, fate effects and risks. Springer, Berlin
114. Zorita S, Mårtensson L, Mathiasson L (2007) Hollow-fibre supported liquid membrane extraction for determination of fluoxetine and norfluoxetine concentration at ultra trace level in sewage samples. J Sep Sci 30:2513–2521
115. Fent K, Weston AA, Caminada D (2006) Ecotoxicology of human pharmaceuticals. Aquat Toxicol 76:122–159
116. Kwon J-W, Armbrust KL (2006) Laboratory persistence and fate of fluoxetine in aquatic environments. Environ Toxicol Chem 25:2561–2568
117. Redshaw C, Cooke M, Talbot H, McGrath S, Rowland S (2008) Low biodegradability of fluoxetine HCl, diazepam and their human metabolites in sewage sludge-amended soil. J Soils Sedim 8:217–230
118. Chu S, Metcalfe CD (2007) Analysis of paroxetine, fluoxetine and norfluoxetine in fish tissues using pressurized liquid extraction, mixed mode solid phase extraction cleanup and liquid chromatography-tandem mass spectrometry. J Chromatogr A 1163:112–118
119. Gros M, Petrovic M, Barceló D (2006) Development of a multi-residue analytical methodology based on liquid chromatography-tandem mass spectrometry (LC-MS/MS) for screening and trace level determination of pharmaceuticals in surface and wastewaters. Talanta 70:678–690
120. Batt AL, Bruce IB, Aga DS (2006) Evaluating the vulnerability of surface waters to antibiotic contamination from varying wastewater treatment plant discharges. Environ Pollut 142: 295–302
121. Pérez S, Eichhorn P, Aga DS (2005) Evaluating the biodegradability of sulfamethazine, sulfamethoxazole, sulfathiazole, and trimethoprim at different stages of sewage treatment. Environ Toxicol Chem 24:1361–1367
122. Batt AL, Kim S, Aga DS (2007) Comparison of the occurrence of antibiotics in four full-scale wastewater treatment plants with varying designs and operations. Chemosphere 68:428–435
123. Kobayashi Y, Yasojima M, Komori K, Suzuki Y, Tanaka H (2006) Removal characteristics of human antibiotics during wastewater treatment in Japan. Water Pract Technol 1:1–9
124. Carberry J, Englande A (1983) Sludge characteristics and behavior. Martinus Nijhoff, Boston, MA
125. Lindberg RH, Olofsson U, Rendahl P, Johansson MI, Tysklind M, Andersson BAV (2006) Behavior of fluoroquinolones and trimethoprim during mechanical, chemical, and active sludge treatment of sewage water and digestion of sludge. Environ Sci Technol 40: 1042–1048
126. Tolls J (2001) Sorption of veterinary pharmaceuticals in soils: a review. Environ Sci Technol 35:3397–3406
127. Suarez S, Carballa M, Omil F, Lema JM (2008) How are pharmaceutical and personal care products (PPCPs) removed from urban wastewaters? Reviews in Environmental Science and Biotechnology 7:125–138
128. Huggett DB, Khan IA, Foran CM, Schlenk D (2003) Determination of beta-adrenergic receptor blocking pharmaceuticals in united states wastewater effluent. Environmental Pollution 121:199–205

Bioavailability of Organic Contaminants in Freshwater Environments

Jarkko Akkanen, Tineke Slootweg, Kimmo Mäenpää, Matti T. Leppänen, Stanley Agbo, Christine Gallampois, and Jussi V.K. Kukkonen

Abstract It has been well established that in many cases total concentration of a chemical in the environment is not the best indicator for the potential threats that the chemical may cause. The term bioavailability describes the fact that only a fraction of a chemical in the environment is available for uptake by organisms, which then determines possible biological effects. Uptake of contaminants is a complex interplay among biological, chemical, and physical factors and processes. Properties of chemicals, environmental conditions, and characteristics of the organisms and the interaction among these ultimately dictate the exposure. This chapter represents the important factors determining bioavailability of organic contaminants in both the water phase and sediment in freshwater systems. In addition, we introduce techniques to measure bioavailability by passive sampling and present modeling tools for estimations. In the end, we offer a short insight about bioavailability in risk management.

Keywords Bioavailability • DOM • Modeling • Organic chemicals • Passive sampling • Sediment

J. Akkanen (✉) • K. Mäenpää • M.T. Leppänen • S. Agbo • J.V.K. Kukkonen
Department of Biology, University of Eastern Finland, Joensuu campus, P.O. Box 111, 80101 Joensuu, Finland
e-mail: jarkko.akkanen@uef.fi

T. Slootweg
ECT Oekotoxikologie GmbH, Flörsheim am Main, Germany

The Water Laboratory, P.O. Box 734, 2003 RS Haarlem, The Netherlands

C. Gallampois
Effect-Directed Analysis, Helmholtz Centre for Environmental Research – UFZ, Leipzig, Germany

H. Guasch et al. (eds.), *Emerging and Priority Pollutants in Rivers*,
Hdb Env Chem (2012) 19: 25–54, DOI 10.1007/978-3-642-25722-3_2,

Contents

1 Introduction

Bioavailability is an important concept when evaluating the fate and harmful effects of contaminants in the environment. It has been established that the total concentration of a given contaminant in a given environment does not translate well into uptake or toxicity in organisms living in that environment. Ecotoxicological effects due to organic chemicals are usually the result of uptake and bioaccumulation of the chemical from the ambient environment or food, followed by toxicodynamic processes which actually result in eliciting the final effect. Persistent compounds that are present in low environmental concentrations bioaccumulate in the organism toward equilibrium, and thus a certain body burden will be reached, which determines the severity of the toxic effects. A lower bioavailability results in a lower uptake by organisms and reduced adverse effects.

At first sight, the concept of bioavailability may appear quite straightforward; however, when giving more thought to processes behind the concept it becomes more complex. That is probably why the concept of bioavailability is not yet regularly applied in regulation or ecological risk management of chemicals. In this chapter, we follow the definition offered by Semple et al. [1] and supported by Ehlers and Loibner [2] in which the bioavailability is defined as the fraction of a contaminant that is free to be taken up by organisms (i.e., free to pass through biological membranes). In the aquatic environment, this is basically comparable to the freely dissolved concentration. In addition, we touch bioaccessibility, which is defined by the same authors as the fraction of a contaminant that is free to be taken up after desorption. Several characteristics of the environment and a contaminant affect the freely dissolved concentration and the bioaccessible fraction of the chemical. Bioavailability can be evaluated by means of traditional bioaccumulation tests, different chemical methods, or modeling.

The purpose of this chapter is to briefly introduce the factors controlling bioavailability of organic contaminants in the aquatic environment, manners to estimate bioavailability, and to take a stand on how this knowledge could be taken into account in water protection and risk management. Given the nature of this publication (only a chapter within a book), this is not a blanket review, but more like an introductory insight into the current knowledge on bioavailability of organic contaminants in the aquatic environment. The chapter also introduces techniques to measure and estimate bioavailability.

2 Bioavailability of Hydrophobic Organic Contaminants in Sediments

Sediments are an important sink for hydrophobic organic contaminants (HOCs) entering aquatic ecosystems [3]. Since HOCs have a potential to persist in the environment and could possibly bioaccumulate in organisms, it is important to know to which extent compounds are still bioavailable for uptake in organisms, once they are bound to sediment. Sorption of organic chemicals is an important process because it controls the fate and ecotoxicological risks of sediment-bound compounds. Several parameters have an influence on the sorption capacity and thus the bioavailability of compounds, for example sediment composition, compound properties, and ecological conditions [2]. This chapter gives an overview of these parameters and their role in bioavailability.

2.1 Sediment Structure and Sorbing Phases

Sediment organic matter (SOM) has been found to be the dominating sorbent for HOCs in sediment [4, 5]. SOM is not a homogeneous substrate, but contains different organic materials which all have different sorption capacities [6]. SOM can be grouped into an amorphous soft/rubbery carbon phase consisting of partly degraded and/or reconstituted biopolymers (e.g., polysaccharides, lignins), lipoproteins, amino acids, lipids, and humic substances, and a condensed or hard/glassy phase consisting of carbonaceous geosorbents (CG) like kerogen, black carbon (BC), and coal [7–11]. The properties of the different SOM compartments have been reviewed earlier, e.g., [2, 12–14]. In this part of the chapter, the most relevant characteristics for sorption are highlighted and the main differences between the sediment compartments are discussed. In addition to SOM, sorption of HOC by natural inorganic particles can also play a role in the overall sorption when the organic carbon content in sediment is relatively low ($<0.1\%$) [6].

SOM characteristics that have been demonstrated to influence sorption are the polarity (e.g., expressed as O/C atomic ratio), aromatic content, aliphatic content, and molecular weight [15–18]. Less polar (lower O/C atomic ratio) SOM is more hydrophobic, which appears to be the driving force for sorption [19]. In addition, aromatic and aliphatic carbon content and the molecular weight of SOM have positive relationships with sorption capacity [15, 20, 21].

Humic materials are dominant components in SOM [22]. They are primarily made up of aromatic rings which are joined by long-chain alkyl structures to form a flexible network [23]. Other organic components—carbohydrates, proteins, lipids, biocides, as well as inorganic compounds such as clay minerals and hydrous oxides—are bound to this alkyl benzene structure [24]. Humic substances are heterogeneous mixtures of macromolecules with molecular weights ranging from a few hundred to several hundred thousand Daltons [12] and relatively high O/C ratios. According to solubility in acid solutions, humic substances can be separated into three fractions, namely fulvic acid, humic acid, and humin [9]. In the order from fulvic acid to humic acid to humin, the material becomes more condensed, has higher molecular weights, and has a higher aliphacity resulting in stronger nonlinear sorption [25, 26].

Recent studies show that glassy CG are also important organic constituents in soils and sediments and their contents vary from a few percent to as high as 80% of the SOM content [173, 174]. Typical examples of CG are kerogen and BC. The molecular structures of kerogen and BC are very different from those of humic acids. In general, CG is characterized by condensed, rigid, and aromatic structures, with high carbon contents and relatively few polar functional groups [13]. The condensed domains are more thermostable and less polar than amorphous domains [6]. Due to the widely varied source materials and combustion conditions, physical and chemical properties of CG may vary greatly. For example, BC freshly formed from liquid fuel combustion may be highly aromatic and reduced (soot), whereas BC formed from incomplete combustion of biomass and coals often has textures and morphologies of parental materials (char) [12]. The structure of soot particles is globular structured and coarse, whereas char is less structured and contains deep narrow nanopores. Compared to soot, char has a higher specific surface area which increases the sorption capacity of this material [27].

2.2 Sorption Mechanisms

Since sediments are highly heterogeneous, HOC sorption varies also among sediment compartments. Each compartment has a characteristic sorption energy and sorption property. Sorption into humic substances is a relatively linear and fast and noncompetitive process. Humic substances are capable of protonation-deprotonation upon change of pH [12]. The π–π electron donor–acceptor interaction plays a major role for π-donor compound sorption in HS [28]. Because of its polarity, humic substances can be swollen by water molecules and become more flexible to allow chemicals to diffuse rapidly in and out according to the chemical gradient [12].

Sorption isotherms for CG are highly nonlinear and the rate of uptake can be slow [12]. The affinity of CG for HOC is very high: it can be up to three orders of magnitude stronger compared to humic substances [29]. At low aqueous HOC concentrations sorption to CG can completely dominate total sorption [13]. Both the rigid, planar, aromatic surface of CG and the abundant nano- and micropores in CG act like major sorption sites [13, 27]. On the surface of the exterior and pore surfaces, free high-energy sites are available to which compounds can bind by means of van der Waals interactions. Besides the binding to the surface, another mode of fixation for compound is the physical occlusion inside restricted pores or between the aromatic macrostructures during BC formation [27].

These differences in sorption between amorphous and hard geosorbents have led to a concept of dual-mode sorption for HOC [30] and depending upon the relative contents of the two SOM phases, sorption by sediments could range from linear partitioning to highly nonlinear adsorption [19, 31].

2.3 *Sorption–Desorption hysteresis*

In sediments it has been observed that sorption to SOM is often faster than the release of the compounds from SOM, a phenomenon that is called sorption–desorption hysteresis [32]. A rapid desorption phase is often followed by a slow and a very slow desorption phase which differ in their desorption rate constants [33, 34]. It is hypothesized that two pathways are important for the slower desorption of compounds. Firstly, during increasing contact time, bound molecules could move from reversible sorption sites to irreversible sites, and low-energy binding sites (reversible sites) may be converted to high-energy sites (irreversible sites) [35]. These sites are not readily accessed by tissues, cells, or enzymes and the compounds become therefore resistant to biological, physical, or chemical treatment. This means that the bioavailability of these sequestered compounds becomes governed by the very slow rate of release to more accessible sites [30, 36]. Secondly, compounds could diffuse through and along meso-, micro-, and nanopores that are abundantly available in CG, where they become irreversibly entrapped [37]. Nanopores with voids in the diameter range of 0.3–1.0 nm are in the size range of toxicologically significant organics [36].

Recently, it was also envisioned that the sorption of molecules may cause deformation of SOM, which results in the "lock-in" of the sorbate molecules [38–41]. Chemical sorption at sites with high energy or specific interaction is liable to cause sorbent structure rearrangement. Another possibility is that the small pores become blocked by other SOM molecules or minerals [36].

These processes resulting in sorption and slow desorption of HOC lead to decrease in the freely dissolved concentration in pore water through which organisms are in many cases exposed to HOCs [42, 43].

2.4 Properties of HOCs

In addition to sediment characteristics, HOC properties such as molecular weight, planarity, and polarity influence the sorption strength [11]. Polycyclic aromatic hydrocarbons (PAHs) with a higher molecular volume have a stronger sorption to SOM. This could be attributed to (1) steric hindrance: larger molecules are easily blocked within pores, and (2) a higher contact area between the compounds and sediment surface for Van der Waals interaction [29, 44, 45]. This high contact area is also a reason why planar compounds bind so strongly to SOM [11, 27]. The HOC concentration in an aquatic ecosystem also influences the sorption strength. At high loadings, sorption constants become lower. This can be explained by saturation of surface adsorption sites [27, 45].

For PAHs, which are ubiquitous contaminants, it was demonstrated that also the origin of compounds can influence the bioavailability in the environment [46–48]. PAHs can be divided into three categories: biogenic PAHs are formed by transformation of natural organic precursors in the environment, pyrogenic PAHs are formed by the burning of organic matter, and petrogenic PAHs are derived from fossil fuels [48]. Pyrogenic PAHs often contain three or more aromatic rings. They form complex assemblages and the unalkylated parent compound is mostly the dominant one. Petrogenic PAHs contain one or more methyl, ethyl, butyl, or occasionally higher alkyl substituents on one or more of the aromatic carbons. Mostly, these alkyl PAHs are more abundant than the parent compounds in petroleum [48]. Pyrogenic PAHs were found to bioaccumulate to a lesser extent in mussel from sediment than petrogenic PAHs, independent of their log K_{ow} [46]. This is probably due to the fact that pyrogenic PAHs are often released in association with combustion soot, which acts as a strong sorption medium [46, 48]. This reemphasizes the influence of the sorption phase on the bioavailability of compounds.

The sorption capacity is also dependent on physicochemical conditions in the sediment, like pH and temperature. Depending on pH, the protonation and deprotonation of functional groups result in a different net charge on the SOM surfaces. In alkaline environments, acidic groups of BC can for example react with hydroxyl, resulting in a decreased sorption capacity by functional groups on the BC surface [49].

The complexity of sediment aggregations makes it challenging to predict the overall sorption properties for sediments based on quantitative analyses of black carbon, kerogen, and humic acid and their respective sorption equilibriums. Predictions of overall sorption using sorption parameters for each of the pure phases overestimates the actual, experimentally measured sorption. This could be caused by attenuation of HOC sorption to BC by DOM molecules that block the pores, or due to surface competition effects by coadsorbing HOC or DOM molecules [27].

2.5 Ecological Conditions

A last important factor that influences the bioavailability of compounds is ecology, i.e., the life habits of organisms. The rapidly desorbing fraction of sediment-associated contaminants may be in equilibrium with those dissolved in interstitial and overlying water. From there compounds can be taken up via respiratory surfaces and skin [50]. Another uptake pathway is via ingestion of sediment particles. Feeding behavior (selectivity), burrowing, ingestion, digestion, and assimilation influence the extent of uptake by organisms [50]. Although several studies demonstrated that mainly the rapidly desorbing fraction is bioavailable for uptake by organisms, it was also shown that the slow desorbing fractions should not be neglected [51]. It is possible that the slow desorbing fraction attributes to a slow replenishment of depleted pore water [52] or that desorption becomes enhanced by digestive fluid during passage in the gut [53]. This may even lead to biomagnification between organic matter and primary consumer [54].

3 Bioavailability in the Water Phase

Water chemistry reflects the local geology and land use. Rock distribution, soil types, flow-path, precipitation, and evaporation are largely controlling the surface water characteristics. Several water quality characteristics affect the availability of contaminants in natural waters. In addition, the characteristics of the contaminant largely determine how strong the effects are or in which direction. Water hardness, pH, dissolved organic matter (DOM), and particulate organic matter (POM) are in key role controlling the availability of contaminants. At the same time we have to remember that properties of the organisms in turn determine how much is taken up or on which level the tissue concentrations will be.

3.1 Dissolved Organic Matter

DOM is present in all natural aquatic systems and is globally an important carbon pool as well as the carbon link between the terrestrial and aquatic environment. If we consider specifically natural DOM in aquatic freshwater systems the main sources are terrestrial ecosystems and wetlands [55]. The release of DOM to aquatic systems is controlled largely by the presence of mineral and organic soil types, vegetation, and climatic factors. On the other hand, on areas dominated by mineral soils and/or high population, wastewater effluents may be significant source of DOM. A part of DOM has a readily identifiable structure, but especially humic substances, which constitute up to 90% of DOM in boreal waters [56, 57], do not have an unambiguous and readily definable structure. However, recent advances in structural research have indicated that the large DOM macromolecules consist, in fact, of small individual building blocks, e.g., [58] and that the main components include carboxylic-rich alicyclics, heteropolysaccharides, and aromatic compounds, e.g., [59]. Major parts of the before mentioned components are derived from linear and cyclic terpenoids. Whereas hydrophobic fractions (i.e., humic substances) dominate in natural DOM, in wastewaters hydrophilic (aliphatic) structures are more common [60, 61]. The word matter in DOM refers to the fact that it contains also other elements besides carbon and thus complicates quantification. Therefore, dissolved organic carbon (DOC) concentration has been used for quantification.

How do we then separate dissolved and particulate phases? Usually membrane filters with nominal pore sizes between 0.2 and 0.7 μm, most commonly 0.45 μm, are being used to remove particulate material. In marine studies, mass of 1,000 Da is set as the limit for dissolved and colloidal material between 1,000 Da and 0.45 μm is considered colloidal [62]. Thus, in freshwater studies the DOM fraction contains both truly dissolved and colloidal material. In this chapter, DOM is defined as the material passing through the 0.45 μm filter.

The sorption and thus DOM controlled bioavailability of contaminants is affected by four major factors: properties of the contaminant, DOM concentration, DOM quality, and water chemistry. In freshwaters, the quality and quantity of DOM as well as contaminant properties are the key factors. However, most studies on DOM–contaminant interaction are focused on sorption phenomena using isolated fractions of DOM, mostly with humic and fulvic acids, which can be operationally separated by difference in aqueous solubility. Humic acids aggregate and precipitate at pH 2 and below, whereas fulvic acids are soluble at the whole pH range [56]. Isolated fractions have been used mainly because of the possibility to use similar material from one set of experiments to another. In addition, humic and fulvic acids have been considered to be main DOM fractions responsible for sorption of contaminants. However, this can be misleading in some cases as presented below. Less effort has been put to study the implications of sorption for bioavaibility, but current assumption is that only freely dissolved contaminants are bioavailable in water-only exposures [63].

Partition coefficients (K_{DOC}) have been traditionally used to express the sorption of HOCs to DOM. A variety of methods have been used to determine K_{DOC} values and most of them have certain limitations and comparison of values obtained with different methods is difficult. The methods can be divided into two groups: the ones that disturb and the ones that do not disturb the equilibrium in the distribution between water and DOM (e.g., [63, 64]). Lipophilicity (expressed as log K_{ow}) has been anticipated to be the main property of organic contaminants influencing the affinity to DOM. Therefore, Burkhard [65] has suggested a predictive relationship between contaminant lipophilicity and sorption to DOM to be $K_{DOC} = 0.08K_{ow}$. However, as Burkhard [65] also states and as the short review presented here indicates, the situation is more complex than the simple relationship shows.

3.1.1 Effects of Contaminant Properties on Sorption and Bioavailability in the Presence of DOM

When considering DOM–contaminant interactions, PAHs are the most studied group of organic contaminants. The magnitude of the sorption correlates with the lipophilicity of individual PAH congeners, higher lipophilicity resulting in higher sorption, e.g., [66–70]. However, this cannot be generalized to concern other contaminant groups; the sorption of PAHs to DOM is higher than that of other groups of contaminants when comparing congeners with equal lipophilicity. Whether or not this applies to alkylated PAH homologues is also unknown. For example despite higher lipophilicity the sorption of PCB77 has been shown to be up to order of a magnitude lower than that of benzo[*a*]pyrene (BaP) as tested in natural lake water DOM [63, 69]. The same study also showed that the difference in the sorption was more pronounced when using depletive methods, i.e., methods that affect the equilibrium reached in contaminant partitioning between DOM and water. This is probably due to the fact that the main difference is in the strongly sorbed fraction, which is much greater for BaP [71]. Nevertheless, the bioavailability followed the measured sorption, i.e., bioavailability of BaP was more affected by DOM. Similar findings have also been seen when comparing the sorption of DDT and PAHs to dissolved humic acid. Despite the clearly lower lipophilicity of pyrene the logarithm of partition coefficient (log K_{DOC}) was 5.5, whereas for DDT it was 5.2 [68].

Sorption of PCBs to DOM does not directly follow the hydrophobicity of the congeners [66, 72]. Ortho-substituted congeners appear to have lower sorption to DOM than the non-ortho-substituted ones. The difference in the K_{DOC} values between ortho- and non-ortho-substituted with equal number of chlorines is usually less than order of a magnitude [72, 73]. All polybrominated diphenyl ether (BDEs) congeners are nonplanar and their association with DOM is probably hydrophobicity driven, although the differences for example between the sorption of BDE47 and BDE99 to DOM are not that great despite the order of a magnitude difference in K_{ow} values [74]. However, there are no extensive studies on this group of compounds. Sorption of highly hydrophobic pyrethroid insecticides depends on pyrethroid in question, but the log K_{DOC} in natural water samples varies from 4.2 to 5.1, indicating significance of DOM in the environmental fate of these compounds [75, 76].

Usually, polar and/or ionizable organic contaminants with low lipophilicity do not have high affinity to DOM. For example the sorption of chlorophenols to DOM is quite low with chlorine substituent number dependent log K_{DOC} values up to 3 [64, 77]. It appears that only the unionized form can be sorbed by DOM [78]. Thus DOM does not have any major influence on the bioavailability of chlorophenols in the water phase at near or above neutral pH [77, 79]. Similarly bioavailability of atrazine (log $K_{ow} < 3$) in water is not significantly affected by DOM or other water quality characteristics [80]. However, there are also exceptions. Despite high water solubility of paraquat, DOM has a significant effect on its bioavailability [81], probably because of the positive charges through which the association with the negative charges in DOM (e.g., carboxylic and phenolic groups) is possible. Naturally the association of paraquat is driven by the ionic interactions with DOM.

Pharmaceuticals are one group of organic contaminants whose environmental concern is increasing rapidly. Yet their environmental behavior is quite modestly studied at the moment. However, so far the findings show that association of pharmaceuticals, such as fenbendazole, albendazole, thiabendazole, flubendazole carbamazepine, and naproxen, with DOM is quite low (usually $\log K_{DOC} < 3.5$) [82, 83]. Nevertheless, under high DOC concentrations the association can be environmentally significant.

3.1.2 Effects of Quality and Quantity of DOM on Sorption and Bioavailability

An increase in the concentration of DOM leads to changes in DOM conformation resulting in lower K_{DOC}. For example Akkanen and Kukkonen [63] showed a twofold decrease in the K_{DOC} with increasing DOM concentration from 1 to 20 mg C/L. Despite the decrease in the K_{DOC}, increasing DOM concentration leads to decrease in freely dissolved concentration of contaminants and thus reduction in bioavailability. Bioavailability of a contaminant in the water phase is mostly expressed as bioconcentration; the bioconcentration factor (BCF) is calculated as the concentration in organisms divided by the total concentration in water. The relationship between BCF and concentration of DOM is nonlinear [84–86]. The shape of the curve is such that up to a DOC concentration of 10–15 mg/L the bioavailability decreases steeply and thereafter the freely dissolved contaminant concentration changes less and thus bioavailability changes less as well. Similar nonlinear relationships have been shown to apply both for a dilution series of DOM from the same source and for a concentration series of DOM from various sources [63, 67, 80, 87]. The higher the K_{DOC} is, the steeper is the change in bioavailability at low levels of DOM. Therefore, it has been concluded that DOM concentration is an important factor controlling the bioavailability in the water phase [67, 80].

In addition to quantity, quality of DOM has also been shown to play a role in the bioavailability of HOCs. Especially, the sorption of PAHs has been shown to closely follow molecular weight and/or aromaticity (in many cases these two parameters intercorrelate) of DOM in natural freshwater samples. Higher aromaticity (measured as specific UV absorbance) leads to higher sorption and lower bioavailability [67, 69, 80]. However, this does not necessarily apply to other groups of HOCs. For example the sorption of PCBs in a natural water series did not follow the aromaticity of DOM in one study [67], whereas in another study the sorption was aromaticity depended, but not as sharply as PAHs [69]. Actually it has been shown that PCBs and PAHs tend to associate with different fractions of DOM [66]. The bioavailability of pyrethroid insecticides can be also affected by DOM. Yang et al. [76] found that the sorption correlated with the abundance of carboxylic groups in DOM. The sorption of BDEs appears to be less affected by the DOM properties [69, 74].

Wastewater-derived DOM and other biodegradable DOM (of algal and bacterial origin) can also reduce bioavailability of organic contaminants. The more biodegradable the material, the smaller the effect on bioavailability of PAHs [88]. In addition, with proceeding biodegradation of DOM, its aromaticity and the K_{DOC} values for PAHs tend to increase [89]. This means that the effect on bioavailability is greater for DOM that is more refractory toward degradation than for example humic acids are (high aromaticity and low degradability). As pointed out earlier, different groups of HOCs behave differently. For example the sorption of benzimidazole pharmaceuticals is approximately at the same level for hydrophobic humic acids and more hydrophilic sewage sludge-derived DOM [82]. On the other hand, pulp mill effluents can contain DOM that is a stronger sorbent DOM to different types of HOCs than natural [90].

In some of the previous studies low concentrations of DOM have increased uptake of contaminants compared to DOM-free conditions (e.g., [67, 91, 92]). The studied DOM samples included both humic acids and natural water samples. However, a systematic study by Haitzer et al. [79] using DOM from several sources could not reproduce these kinds of results. Related to this it has also been shown that DOM can promote biodegradation of PAHs, i.e., it enhances the bioavailability of PAHs to microbes [93, 94]. Despite the bioavailability reduction, the association with DOM increases solubility and transport of contaminants that are otherwise insufficiently soluble in water. Recent findings have indicated that dissolved humic acids have shown not only to facilitate transport of HOCs from solid phase to water phase or from solid phase to another through the unstirred boundary layer [73, 95, 96], but can also enhance uptake of solid phase associated HOCs in aquatic organisms [97]. Therefore, the overall role of DOM on availability in the aquatic environment is still rather unclear.

3.1.3 Effects of Water Chemistry on Sorption and Bioavailability in the Presence of DOM

Water chemistry affects the functionality of DOM in several ways. Ionic strength, presence of multivalent cations, and changes in pH all affect the three-dimensional conformation of DOM, which in turn affects the capacity to associate with HOCs [14, 98–103]. This is mainly due to neutralization and dissociation reactions in the functional groups that control the intra- and intermolecular repulsion by negative charges. Under alkaline conditions, the major functional groups (−COOH, −OH) are dissociated; and repulsion between negative charges hinders the formation of hydrophobic domains [104]. When the solution turns acidic, negative charges are neutralized, hydration occurs, and ultimately hydrophobic domains are formed. Consequently, the sorption of organic chemicals to DOM is greater at low pH, e.g., [77, 105].

If we consider neutral compounds, increasing ionic strength or salinity decreases the solubility of HOCs in water on the one hand and changes the structure of DOM to become less favorable for sorption on the other. High ionic strength compresses the humic molecules into tight spheres hindering the access to hydrophobic domains or even prevents the formation of the domains. This is comparable to the effect seen with DOM concentration where the sorption capacity decreases with increasing concentration. However, multivalent metal ions, such as Ca, Mg, and Al, may neutralize negative charges and reduce repulsion, thus creating and expanding the sorption domains resulting in increase of sorption [14, 104, 105]. The laboratory studies using NaCl to control ionic strength have shown an increase in sorption to dissolved humic substances, probably because the decrease in solubility (i.e., the salting-out effect) drives the HOC molecules to the hydrophobic domains of dissolved humic material exceeding the effect on DOM structure [106, 107]. However, most studies have shown a decrease in sorption of HOCs to dissolved humics with increasing ionic strength, e.g., [105, 108, 109]. Increasing water hardness, within the range found in European freshwaters [80], has been shown to affect the sorption capacity of DOM for HOCs [110]. This is probably due to conformational alterations in the DOM structure under changed ionic conditions. In estuary systems, the increase in salinity and ionic strength is far higher, also in brackish water environment such as the Baltic Sea [74]. In this study, the salinity gradient ranging from 0 to 5.5‰ caused a decrease in sorption of PAH and BDEs to humic and fulvic acids. In natural water samples, the increase was due to changes in salinity, DOM quality, and quantity.

The availability of many of ionizable compounds is more sensitive to changes in water chemistry than to the presence of DOM. For example, bioavailability of weak organic acids such as pentachlorophenol or dehydroabietic acid is more affected by pH than DOM, whereas bioavailability of neutral HOCs is irrespective of pH [77]. Fate and distribution of surfactants depend largely on their ionization behavior, which in turn is influenced by water chemistry. For example, increasing the aqueous concentrations of linear alkylbenzene sulphonate (LAS) under a corresponding increase in water hardness is known to reduce toxicity. This phenomenon is believed to be due to a reduction in the overall solubility of LAS under variable water chemistry [111].

3.1.4 Effects of Suspended Particles on Bioavailability

So far this part of the chapter considered mainly DOM and other water chemistry characteristics. However, depending on the nutrient status and for example seasonal events, particles can also be important in HOC partitioning and thus bioavailability in the water phase, e.g., [74]. Ecological differences among organisms result in differential effects of suspended particles. Suspended particles may for example increase uptake of HOCs by filter feeders compared to nonfilter feeders [112]. In addition, uptake of DOM-associated HOCs is low (or negligible) in filter feeders, but the uptake via particles can be more significant [63, 113]. The type of particles also plays a major role in this. Algal particles reduce bioavailability more than bacterial ones and activated carbon particles reduce bioavailability more efficiently than diatom or wood particles [172, 114].

4 Estimation of Bioavailability

4.1 *Passive Sampling*

Passive sampling is based on partitioning of HOCs toward an equilibrium between the freely dissolved fraction and the different sorbing phases such as SOM or biota lipids and a passive sampler. As stated previously, the freely dissolved concentration (C_{free}) is more informative than the total concentration (C_{total}), for the estimation of bioavailability of chemicals in aquatic environments [115, 116]. C_{free} is the fraction of chemical that can be readily bioaccumulated by aquatic organisms in contrast to C_{total}, which is a sum of both C_{free} and chemical bound to geosorbents (C_{bound}). Therefore, C_{free} mainly determines the exposure. This part of the chapter gives a brief overview of materials used in passive sampling of HOCs (Table 1).

Most of the passive samplers represent a fast equilibrating sorbent phase and therefore they are found to be effective for the measurement of different bioavailability parameters [116]. In regard to material and geometry, a wide variety of passive samplers are increasingly applied to different compartments, such as sediment, water, and biota, in the aquatic environment. The basis of most techniques is to equilibrate a thin layer of polymer, or lipid-like sorbent phase, in a matrix to be measured [150]. The analyte concentration in the polymer ($C_{polymer}$) is then measured and converted into useful information: (1) The freely dissolved concentration of the analyte can be determined as the ratio of $C_{polymer}$ and the analyte-specific polymer to water partition ratio ($K_{water,\ polymer}$) [121]. (2) Chemical activities can be determined using analyte-specific activity coefficients ($\gamma_{analyte}$) [132]. (3) $C_{polymer}$ can also be converted into the equilibrium partitioning concentration in lipids ($C_{lipid,partitioning}$) when applying partition ratios between lipid and polymer ($K_{lipid,polymer}$) [124, 131, 151]. Furthermore, the direct measurement of chemical concentrations in a given matrix can be attained by applying external calibration standards. For example, parallel equilibration of a sampler with sediment and external lipid standards results in direct estimates of $C_{lipid,partitioning}$ in organisms [124]. Since the approach is based on the assumption of equilibrium among different compartments in the system, proving that the equilibrium between the sampled matrix and the sampler is a key factor.

Semipermeable membrane devices (SPMD) were among the first passive samplers and were developed to measure the uptake of dissolved organic contaminants in water [43, 117, 118]. SPMD consists of a small amount of lipid, mostly triolein, which is sealed inside a polyethylene tube. When immerged into water phase, lipophilic compounds partition into the lipid phase through PE

Table 1 Examples of passive samplers and their applicability

Passive sampler	Main characteristics/applicability	Selected references
Semipermeable membrane device (SPMD)	Thin film of lipid (triolein) sealed inside PE tube. Suitable for a number of hydrophobic compounds (PAH, PCB, PCDD/F). Used in kinetic and equilibrium sampling from water and sediment. Nondepletive	Huckins et al. [117], Booij et al. [118], Leppänen and Kukkonen [43], Prest et al. [175], McCarthy et al. [119], Booij et al. [118], Lyytikäinen et al. [120]
Polydimethyl siloxane (PDMS)	Various forms: SPME fiber coating, tubing, disc and coated glass. Suitable for various organic contaminants (PCBs, PAHs, pharmaceuticals, pesticides etc.). Used in equilibrium sampling from water, sediment, tissue, and headspace. Nondepletive	Mayer et al. [121], Hawthorne et al. [122], Legind et al. [123], Mäenpää et al. [124], Zhou et al. [125], Zhang et al. [126], Jahnke and Mayer [127], Pehkonen et al. [11], Rusina et al. [128, 129], Mayer et al. [130], Jahnke et al. [131], Reichenberg et al. [132]
Ethylene vinyl acetate (EVA)	Various forms: coated glass and disc. Suitable for pesticides, chlorobenzenes, and PCBs. Used in equilibrium mode to sample from water, sediment, tissue, and headspace. Nondepletive	Meloche et al. [133], St George et al. [134], Wilcockson and Gobas [135]
Blue Rayon (BR)	Coated fiber. Suitable for PAHs and their derivatives nitro-PAHs, keto-PAHs, hydroxyl-PAHs, heterocyclic aromatic amines, etc., mainly planar compounds. Used in water sampling. Nondepletive	Hayatsu [136], Sakamoto and Hayatsu [137], Kummrow et al. [138], Watanabe et al. [139]
Polyacrylate (PA)	SPME fiber coating. Suitable for LAS sampling from water. Nondepletive	Rico-Rico et al. [140]
PDMS-divinylbenzene (PDMS-DVB)	SPME fiber coating. Suitable to sample synthetic musks from sewage sludge. Nondepletive	Wu and Ding [141]
Low-density polyethylene (LD-PE)	Plastic sheet. Suitable for chlorobenzenes, organochlorine pesticides, PCBs, and PAHs. Nondepletive	Hale et al. [142], Booij et al. [118], Friedman et al. [143]
Polyoxymethylene (POM)	Plastic plates. Suitable for variety of organic contaminants (PAH, pesticides, PCBs, PBDEs) from water and sediment. Nondepletive	Sormunen et al. [144], Jonker and Koelmans [145], van der Heijden and Jonker [146]
Tenax® resin beads	Plastic beads. Suitable for variety of organic contaminats (PAHs, pesticides, PCBs, PBDEs) from water and sediment sludge. Depletive	Cornelissen et al. [147], Trimble et al. [148], Sormunen et al. [144]
Cyclodextrin	Sugar derivative. Suitable for PAH measurements from sediment. Depletive	Dean and Scott [149]

membrane according to their chemical properties [175]. SPMD have been used widely ever since to measure C_{free} of various hydrophobic and polar organic contaminants [118, 119]. SPMD has also shown varying success when used to predict bioaccumulation from sediment for polychlorinated dibenzo-p-dioxins and furans (PCDD/F) [120] and PAHs and PCBs [43]. In contrast to the newer passive sampling techniques relying on equilibrium, SPMD is normally applied by measuring the kinetics of the chemical uptake [152, 175].

Solid phase microextraction (SPME) was introduced as a simple, fast, and cost-effective option to be used instead of liquid–liquid extraction for analyzing various compounds [153]. Since then polymer-coated fibers have been used for the determination of freely dissolved concentrations of chemicals in pore water and hence to estimate chemical bioavailability in sediments [11, 146]. Various materials can be used as a sorbent coating in the fibers. Polydimethyl siloxane (PDMS) is one of the most used coatings, and is typically used for hydrophobic nonpolar compounds, such as polychlorinated biphenyls (PCB) and polyaromatic hydrocarbons (PAHs) [121, 122]. Polyacrylate (PA) has been used to measure freely dissolved concentrations of linear alkylbenzenesulfonates (LAS) in seawater [140], and PDMS-divinylbenzene (PDMS-DVB) coated fibers have been found optimal to detect synthetic musks in sewage sludge [141]. Fibers can be equilibrated with the sample by directly inserting the SPME fiber into the sample [11] or in the headspace above the sample [123, 124]. A sophisticated method in which PDMS fibers are inserted into fish to measure bioconcentration of pharmaceuticals in vivo in free swimming fish [125, 126]. Recently, it has been shown that the sorptive properties of PDMS do not change upon contact with complex matrices, such as fish tissue, sediment, or foodstuff [127].

Blue Rayon (BR) consists of a blue pigment, copper phthalocyanine trisulfonate, which is covalently bounded to a fiber, rayon [136]. The usefulness of the BR lies in its ability to adsorb aromatic compounds with three or more fused rings including PAHs and their derivatives (nitro-PAHs, keto-PAHs, hydroxyl-PAHs), heterocyclic aromatic amines, etc. (mainly planar compounds). BR is commonly used in the field to monitor mutagens in surface waters [137], such as phenylbenzotriazole [139] and PAHs [138].

PDMS silicone rubber has a low transport resistance for HOCs [128, 129] and is therefore advantageous among the polymers used as passive samplers. PDMS has been used in different forms: besides used as a fiber coating, it has been used as tubing that can be equilibrated with lipid-rich samples, such as oils or tissue samples [130]. Analogous to the method where a fiber is equilibrated within tissue of fish, the method to equilibrate thin PDMS discs in fresh fish tissue has been successful with lipid-rich fish [131]. However, lean fish was found problematic to be equilibrated within the satisfactory time period probably due to slow diffusion of chemicals in the tissue. In addition to these methods, PDMS can be used to coat the interior of glass vials with thin layers of multiple thicknesses. The vials can be then equilibrated with sediment or soil to measure free chemical concentrations [124, 132]. Ethylene vinyl acetate (EVA) has been used in similar applications as PDMS. For example, EVA-coated glass has been applied to sample pesticides and

PCBs at equilibrium mode [133, 134]. The approach of equilibrating varying thicknesses of polymer with the sample results in proportionality between coating and analyte mass and takes into account several validity criteria, such as confirmation of equilibrium, and polymer abrasion or contamination [132]. Similar to headspace sampling with SPME fibers, submicrometer thin film discs of EVA have been equilibrated with poorly volatile hydrophobic compounds in headspace [135].

Low-density polyethylene (LD-PE) plastic has been successively used as a passive sampler to follow changes in PAH bioavailability after the addition of sorbent amendments into contaminated sediments [142], and to study the effect of temperature on partitioning processes between LD-PE and sorbed phase with various compounds [118]. Furthermore, LD-PE has been used to study the bioavailability of PCBs in sediments and was shown to bioaccumulate PCBs at the same extent as the lipid phase of Polychaete worms [143].

Polyoxymethylene (POM) is easy to handle and more durable toward mechanical abrasion than for example PDMS. POM in the solid phase extraction (POM-SPE) has been used to measure concentrations of variety of chemicals in sediment pore water in order to evaluate bioaccumulation in laboratory [144] and field conditions [145, 146]. A study comparing several passive sampling methods and materials to measure C_{free} found SPME fiber and POM-SPE yielding the best correlations with field-exposed Oligochaeta worms [146].

As a contrast to the nondepleting methods described above, another set of the sampling techniques consist of materials that keep the freely dissolved concentration in minimum by fast adsorption resulting in desorption from different sorption domains in the studied matrix. These approaches utilize for example Tenax® resin beads [144, 148] or cyclodextrin [149] that are used to define the bioaccessible chemical fraction, i.e., rapidly desorbing fraction [1, 116].

Thus the techniques, such as passive sampling, have shown to be useful in sampling HOCs in the aquatic environment where they exist in trace concentrations without a need to for example process large volumes of water. In addition, these methods have proven to be feasible for bioavailability estimations instead for more laborious methods.

4.2 Partitioning Models

As already stated many factors modify bioavailability of contaminants in aquatic environment and it is difficult to take all of those into account in mathematical models. Therefore, models represent a simplified description of the fate of contaminants in a system under study but, on the other hand, offer a useful method for mechanistic understanding of bioaccumulation. Two main approaches have been applied to predict and model bioaccumulation of contaminants [154]: equilibrium partitioning (EqP) and kinetic models. EqP describes contaminant distribution between environmental phases at steady state. Steady state is a state at which rates of uptake and elimination of a contaminant within an organism or tissue are equal.

We use the wording steady state when speaking of bioaccumulation rather than equilibrium due to the finite nature of steady states (steady concentrations) in biological systems. Equilibrium will be used with passive samplers that sense partitioning between the phases and offer information on theoretical equilibrium concentrations.

The EqP models have been successful in describing partitioning in the environment and identifying components acting as sorbing phases in environmental fate. Those are also effective screening tools based on simple information on physicochemical characteristics of contaminants and partitioning phases. Kinetic models are more appropriate when concentrations vary over time; multiple contaminant sources are responsible for accumulation and when progress of accumulation is of concern [154]. In some applications, like in food web models, both EqP and kinetic sections can be incorporated into the model [155]. In toxicology, kinetic models are usually physiological-based pharmacokinetic models (PBPK) where substance distribution and fate in a body are extended from single compartment to organ levels. The PBPK models have also been applied in aquatic toxicology [156, 157].

Kinetic bioaccumulation models usually employ compartments in mathematical formulations and may incorporate several sources and rate constants to describe relevant processes [158]:

$$\mathrm{d}C_a/\mathrm{d}t = k_1 \times C_{\text{free}} + k_f \times C_f - (k_2 + k_e + k_m + k_g) \times C_a,$$

where k_1 is the uptake clearance constant from water (ml/g/h), C_{free} is the contaminant concentration freely dissolved in a water (ng/ml), k_{f} is the uptake clearance constant from diet or solid material (g/g/h), C_{f} is the contaminant concentration in the ingested diet or solid material (ng/g), C_{a} is the contaminant concentration in an organism (ng/g), k_2 is the elimination rate constant (1/h) across the diffusional membranes, k_{e} is the egestion (fecal, urinal, and biliary elimination) rate constant (ng/g), k_{m} is the rate constant (1/h) for metabolism, and k_{g} is the rate constant (1/h) for the growth. Application of kinetic models for aquatic organisms usually contains only two compartments because of need of simplicity in parameterization for invertebrates. The uptake clearance constants sum all possible factors influencing uptake, like diffusional limitations of desorption in solid matrices (e.g., sediment particles). Therefore, tweaking of source concentration to account assumed bioavailable fraction is possible, e.g., [47, 159]. On the other hand, additional parameters like assimilation efficiency can be incorporated into the model, e.g., [160]. The ratio between uptake clearance constant and elimination rate constant(s) results in a partitioning ratio between organism and sources at steady state. For example, in kinetic sediment and soil bioaccumulation tests, concentrations are usually normalized to organisms' lipids and organic carbon content in solid matrix. Thus, the model parameters can be used to calculate a ratio, that is, actually the most typical equilibrium partitioning model (EqP) applied for solid matrices (biota sediment accumulation factor; BSAF).

Thermodynamic partitioning between sorbing phases has been acknowledged already a long time ago, e.g., [161] and the ratio between organisms' lipids and sediment organic carbon has been widely applied in bioaccumulation tests and environmental risk assessment [115]:

$$\text{BSAF} = (C_a/f_l)/(C_s/f_{OC}),$$

where f_l is the fraction of lipid in an organism, C_s is the concentration of a contaminant in sediment (ng/g), and f_{OC} is the fraction of organic carbon in solid matrix. Theoretically, if the affinities of sorbing phases (or fugacity capacities) for a given contaminant are equal, BSAF at steady state should be a uniform value across experiments and field observations. (McFarland and Clarke [162]) suggested that the ratio should be 1.7 based on fish and sediment data. However, large data evaluations have revealed that considerable variation exists in BSAF estimates [163, 164]. There are several reasons for these discrepancies which most likely relate to the variations in the properties of lipids and organic carbon as sorptive phases, the feeding behavior of organisms affecting through assimilation, and the estimation of the actual bioavailable fraction (C_s/f_{OC}) in the model. Variations in the quality of sorption domains in organic carbon fraction are at least partly related to black carbon content of the sediment, which has stronger sorbing capacity for hydrophobic contaminants than amorphous, noncondensed organic carbon [44].

Partitioning in a water column environment has been condensed to BCF that describes partitioning between an organism and water at steady state, and, similarly as with sediments, influencing factors has gained interest in the literature especially by Donald MacKay and coworkers [165]. The conceptual model for describing bioaccumulation uses three main phases (organisms' lipids, organic carbon, and water phase) between which contaminants should attain steady state [115]. The freely dissolved concentration in water has a main role in determining bioavailability and the fraction free (f) of the total contaminant can be estimated [85]:

$$f_{\text{free}} = \frac{1}{1 + K_{\text{DOC}} C_{\text{DOC}}},$$

where K_{DOC} is the partitioning coefficient between DOC and water and C_{DOC} is the concentration of the DOC (kg/L). If one assumes that bioconcentration is the main driving force in bioaccumulation in water column as well as in sediment systems, the estimation of freely dissolved concentration in water or pore water (C_{free}) could be used to calculate steady-state concentration in organisms' lipids (C_a), e.g., [115]:

$$C_a = \text{BCF} \times C_{\text{free}},$$

where BCF is a lipid–water partitioning coefficient (K_{lipid}). The BCF can be approximated from K_{ow} [44] or calculated from an empirical data, e.g., [79, 166]. The freely dissolved concentration is not an easily accessible value for hydrophobic contaminants. In sediments for example, it can be estimated iteratively from a model accounting noncondensed and condensed carbon as sorbing phases [44, 167]:

$$C_s = (f_{OC} - f_{BC})K_{AOC}C_{free} + f_{BC}K_{BC}C_{free}^{n},$$

where f_{BC} is the fraction of black carbon in sediment/soil, K_{AOC} is the amorphous organic carbon–water partitioning coefficient, and K_{BC} is the black carbon–water Freundlich coefficient described by the Freundlich isotherm and partitioning coefficient. Both K_{AOC} and K_{BC} can be approximated from the K_{ow}. The more accurate approach would be a direct measurement of freely dissolved concentration (C_{free}) by applying passive equilibrium samplers, like silicone-based polymer devices, where C_{free} is based on the contaminant-specific partitioning ratio between a polymer device and water ($K_{polymer,water}$) and contaminant concentration in a polymer ($C_{polymer}$)[166]:

$$C_{free} = \frac{C_{polymer}}{K_{polymer,water}}.$$

However, this approach still requires BCF for steady-state tissue residue, and as the actual measured estimates vary considerably too depending on the conditions, e.g., [110], some inconsistencies in results are expected. Recent developments in passive sampling techniques have advanced the equilibrium sampling approach a step further. Concentration in a polymer can be directly converted to equilibrium partitioning concentration in organisms' lipids ($C_{lipid,partitioning}$) when applying partitioning ratios between lipid and polymer ($K_{lipid,polymer}$) [124, 151]:

$$C_{lipid,equilibrum} = C_{polymer} \times K_{lipid,polymer}.$$

The success of the lipid–polymer partition ratio ($K_{lipid,polymer}$) is based on the assumption that model lipid (e.g., olive oil) has the same affinity (fugacity capacity) for a contaminant as organisms' lipids. Also, one needs to be sure that polymer has reached equilibrium in an exposure system.

The models can be tested simply by comparing modeled values to actual observations in a laboratory or field data. Deviations from modeled values can indicate biotransformation, biomagnification, or simply system disequilibrium. The concept of fugacity [168] or chemical activity [116] can be applied to investigate system equilibrium. Fugacity or activity difference, not the concentrations, runs the partitioning process until the activities are equal. Therefore, unequal activities would reveal disequilibrium. Passive samplers can also be used to detect activities of contaminants in environmental samples [132]:

$$a = C_{polymer} \times \gamma_{polymer},$$

where thermodynamic activity (a) is a result of concentration of contaminant in a polymer device ($C_{polymer}$) and activity coefficient ($\gamma_{polymer}$). Determination of chemical activity requires a calibration to a reference state or external activity standard by means of activity coefficient. This can be done by calculating coefficients in methanol standard using SPARC online calculator [123] or using contaminant partitioning and solubility data in methanol and polymer [132].

Application of chemical activity in distribution and fate modeling of organic contaminants is a new and still rarely used approach, e.g. [169], that improves understanding of bioaccumulation and fate of contaminants in environment. Besides detecting chemical (dis)equilibrium, the direction of net flux in a system would also be evident and the existence of biotransformation and biomagnification would be easier to evaluate.

5 Bioavailability and Risk Management

Contamination of surface waters and sediments is a common problem and poses risk to the environment and human health. Sediments have been shown to be sinks for contaminants, but they can also serve as sources of contamination for decades after the original source has ceased. Therefore, risk assessment and management of contaminated sediments is important, but at the same time challenging. Setting specific limit values for chemical concentrations would not necessarily improve the situation. One of the key issues in evaluation of risks of contaminants is exposure assessment. This chapter has presented that the bioavailability of contaminants is site specific, which means that local conditions together with the contaminant properties are drivers for the exposure of organisms and also possible harmful effects. This means that certain concentration can be safe on one site and highly harmful in another. And when one is not relying to certain absolute limit value concentrations, there is a possibility for a real site-specific assessment.

Since the total contaminant concentrations neither in sediment nor in water are giving the right picture about the potential risks of contaminants in aquatic systems, it is therefore important to include these aspects in the risk assessment and management. For example, US Environmental Protection Agency (USEPA) endorses an inclusion of bioavailability in the site-specific decisions for remediation and management of contaminated sediments [170]. Yet, there are still uncertainties and open questions, such as how to assess the bioavailability on a given site and to what extent will the biota be exposed in the long term. Selection of the best methods for bioavailability estimation can be still tricky and validation for contaminant mixtures is needed [171]. Another aspect is that ecologically different organisms may experience bioavailability differently; i.e., tissue concentrations may vary. In the end, the evidence is cumulating indicating the accurateness and usefulness of many chemical methods (e.g., passive sampling) to estimate bioavailability. Since zero tolerance for contaminants in the aquatic systems cannot be reached, bioavailability estimations would take us closer to reality and help with the management decisions. However, all this needs guidelines and standardized methods to estimate bioavailability, which do not exist yet.

References

1. Semple KT, Doick KJ, Jones KC et al (2004) Defining bioavailability and bioaccessibility of contaminated soil and sediment is complicated. Environ Sci Technol 38:228A–231A
2. Ehlers GAC, Loibner AP (2006) Linking organic pollutant (bio)availability with geosorbent properties and biomimetic methodology: a review of geosorbent characterisation and (bio) availability prediction. Environ Pollut 141:494–512
3. Kuster M, Lopez MJ, de Alda MJL et al (2004) Analysis and distribution of estrogens and progestogens in sewage sludge, soils and sediments. Trac Trends Anal Chem 23:790–798
4. Chiou CT, Peters LJ, Freed VH (1979) A physical concept of soil-water equilibria for nonionic organic compounds. Science 206:831–832
5. Karickhoff SW, Brown DS, Scott TA (1979) Sorption of hydrophobic pollutants on natural sediments. Water Res 13:241–248
6. Cuypers C, Grotenhuis T, Nierop KGJ et al (2002) Amorphous and condensed organic matter domains: the effect of persulfate oxidation on the composition of soil/sediment organic matter. Chemosphere 48:919–931
7. Chiou CT, Kile DE, Rutherford DW et al (2000) Sorption of selected organic compounds from water to a peat soil and its humic-acid and humin fractions: potential sources of the sorption nonlinearity. Environ Sci Technol 34:1254–1258
8. Weber WJ, McGinley PM, Katz LE (1992) A distributed reactivity model for sorption by soils and sediments. Conceptual basis and equilibrium assessments. Environ Sci Technol 26:1955–1962
9. Weber WJ, Huang WL, Leboeuf EJ (1999) Geosorbent organic matter and its relationship to the binding and sequestration of organic contaminants. Colloids Surf A Physicochem Eng Aspects 151:167–179
10. Xing B, Pignatello JJ (1997) Dual-mode sorption of low-polarity compounds in glassy poly (vinyl chloride) and soil organic matter. Environ Sci Technol 31:792–799
11. Pehkonen S, You J, Akkanen J et al (2010) Influence of black carbon and chemical planarity on bioavailability of sediment-associated contaminants. Environ Toxicol Chem 29: 1976–1983
12. Huang WL, Ping PA, Yu ZQ et al (2003) Effects of organic matter heterogeneity on sorption and desorption of organic contaminants by soils and sediments. Appl Geochem 18:955–972
13. Cornelissen G, Gustafsson O, Bucheli TD et al (2005) Extensive sorption of organic compounds to black carbon, coal, and kerogen in sediments and soils: mechanisms and consequences for distribution, bioaccumulation, and biodegradation. Environ Sci Technol 39:6881–6895
14. Pan B, Ghosh S, Xing BS (2008) Dissolved organic matter conformation and its interaction with pyrene as affected by water chemistry and concentration. Environ Sci Technol 42:1594–1599
15. Chin Y-P, Aiken GR, Danielsen KM (1997) Binding of pyrene to aquatic and commercial humic substances: the role of molecular weight and aromaticity. Environ Sci Technol 31:1630–1635
16. Kopinke FD, Georgi A, MacKenzie K (2001) Sorption of pyrene to dissolved humic substances and related model polymers. 1. Structure–property correlation. Environ Sci Technol 35:2536–2542
17. Hur J, Schlautman MA (2003) Using selected operational descriptors to examine the heterogeneity within a bulk humic substance. Environ Sci Technol 37:4020–4020
18. Hur J, Lee DH, Shin HS (2009) Comparison of the structural, spectroscopic and phenanthrene binding characteristics of humic acids from soils and lake sediments. Org Geochem 40:1091–1099
19. Huang W, Weber WJ Jr (1997) A distributed reactivity model for sorption by soil and sediments. 10. Relationships between desorption, hysteresis, and the chemical characteristics of organic domains. Environ Sci Technol 31:2562–2569

20. Hur J, Schlautman MA (2004) Influence of humic substance adsorptive fractionation on pyrene partitioning to dissolved and mineral-associated humic substances. Environ Sci Technol 38:5871–5877
21. Kang SH, Xing BS (2005) Phenanthrene sorption to sequentially extracted soil humic acids and humins. Environ Sci Technol 39:134–140
22. Stevenson FJ (1994) Humus chemistry: genesis, composition, reaction. Wiley, New Jersey
23. Schulten HR, Plage B, Schnitzer M (1991) A chemical-structure for humic substances. Naturwissenschaften 78:311–312
24. Lawrence MAM, Davies NA, Edwards PA et al (2000) Can adsorption isotherms predict sediment bioavailability? Chemosphere 41:1091–1100
25. Xing BS (2001) Sorption of anthropogenic organic compounds by soil organic matter: a mechanistic consideration. Can J Soil Sci 81:317–323
26. Pan B, Xing BS, Liu WX et al (2006) Distribution of sorbed phenanthrene and pyrene in different humic fractions of soils and importance of humin. Environ Pollut 143:24–33
27. Koelmans AA, Jonker MTO, Cornelissen G et al (2006) Black carbon: the reverse of its dark side. Chemosphere 63:365–377
28. Zhu DQ, Hyun SH, Pignatello JJ et al (2004) Evidence for pi-pi electron donor-acceptor interactions between pi-donor aromatic compounds and pi-acceptor sites in soil organic matter through pH effects on sorption. Environ Sci Technol 38:4361–4368
29. Jonker MTO, Koelmans AA (2002) Sorption of polycyclic aromatic hydrocarbons and polychlorinated biphenyls to soot and soot-like materials in the aqueous environment mechanistic considerations. Environ Sci Technol 36:3725–3734
30. Pignatello JJ, Xing B (1996) Mechanisms of slow sorption of organic chemicals to natural particles. Environ Sci Technol 30:1–11
31. Sanchez-Garcia L, Cato I, Gustafsson O (2010) Evaluation of the influence of black carbon on the distribution of PAHs in sediments from along the entire Swedish continental shelf. Mar Chem 119:44–51
32. Huang W, Weber WJ Jr (1998) A distributed reactivity model for sorption by soils and sediments.11. Slow concentration-dependent sorption rates. Environ Sci Technol 32: 3549–3555
33. Luthy RG, Aiken GR, Brusseau ML et al (1997) Sequestration of hydrophobic organic contaminants by geosorbents. Environ Sci Technol 31:3341–3347
34. Leppänen MT, Landrum PF, Kukkonen JVK et al (2003) Investigating the role of desorption on the bioavailability of sediment associated 3,4,3’,4’-tetrachlorobiphenyl in benthic invertebrates. Environ Toxicol Chem 22:2861–2871
35. Drori Y, Aizenshtat Z, Chefetz B (2005) Sorption–desorption behavior of atrazine in soils irrigated with reclaimed wastewater. Soil Sci 69:1703–1710
36. Alexander M (2000) Aging, bioavailability, and overestimation of risk from environmental pollutants. Environ Sci Technol 34:4259–4265
37. Nam K, Alexander M (1998) Role of nanoporosity and hydrophobicity in sequestration and bioavailability: tests with model solids. Environ Sci Technol 32:71–74
38. Weber WJ, Kim SH, Johnson MD (2002) Distributed reactivity model for sorption by soils and sediments. 15. High-concentration co-contaminant effects on phenanthrene sorption and Desorption. Environ Sci Technol 36:3625–3634
39. Lesan HM, Bhandari A (2003) Atrazine sorption on surface soils: time-dependent phase distribution and apparent desorption hysteresis. Water Res 37:1644–1654
40. Lu YF, Pignatello JJ (2004) Sorption of apolar aromatic compounds to soil humic acid particles affected by aluminum(III) ion cross-linking. J Environ Qual 33:1314–1321
41. Ju D, Young TM (2005) The influence of natural organic matter rigidity on the sorption, desorption, and competitive displacement rates of 1,2-dichlorobenzene. Environ Sci Technol 39:7956–7963
42. Kraaij RH, Ciarell S, Tolls J et al (2001) Bioavailability of lab-contaminated and native polycyclic aromatic hydrocarbons to the amphpod Corophium volutator relates to chemical desorption. Environ Toxicol Chem 20:1716–1724

43. Leppänen MT, Kukkonen JVK (2006) Evaluating the role of desorption in bioavailability of sediment associated contaminants using oligochaetes, semipermeable membrane devices and Tenax extraction. Environ Pollut 140:150–163
44. Schwarzenbach RP, Gschwend PM, Imboden DM (2003) Environmental organic chemistry, 2nd edn. Wiley, New Jersey
45. Van Noort PCM, Jonker MTO, Koelmans AA (2004) Modeling maximum adsorption capacities of soot and soot-like materials for PAHs and PCBs. Environ Sci Technol 38: 3305–3309
46. Thorsen WA, Cope WG, Shea D (2004) Bioavailability of PAHs: effects of soot carbon and PAH source. Environ Sci Technol 38:2029–2037
47. Moermond CTA, Zwolsman JJG, Koelmans AA (2005) Black carbon and ecological factors affect in situ biota to sediment accumulation factors for hydrophobic organic compounds in flood plain lakes. Environ Sci Technol 39:3101–3109
48. Neff JM, Stout SA, Gunster DG (2005) Ecological risk assesment of polycyclic aromatic hydrocarbons in sediments: identifying sources and ecological hazard. Integr Environ Assess Manag 1:22–33
49. Luo L, Lou L, Cui X et al (2011) Sorption and desorption of pentachlorophenol to black carbon of three different origins. J Hazard Mater 185:639–646
50. Åkerblom N, Goedkoop W, Nilsson T et al (2010) Particle-specific sorption/desorption properties determine test compound fate and bioavailability in toxicity tests with chironomus riparius-high-resolution studies with lindane. Environ Toxicol Chem 29:1520–1528
51. Cui XY, Hunter W, Yang Y et al (2010) Bioavailability of sorbed phenanthrene and permethrin in sediments to Chironomus tentans. Aquat Toxicol 98:83–90
52. Sijm D, Kraaij R, Belfroid A (2000) Bioavailability in soil or sediment: exposure of different organisms and approaches to study it. Environ Pollut 108:113–119
53. Mayer LM, Weston DP, Bock MJ (2001) Benzo[a]pyrene and zinc solubilization by digestive fluids of benthic invertebrates – a cross-phyletic study. Environ Toxicol Chem 20:1890–1900
54. Gobas FAPC, Zhang X, Wells R (1993) Gastrointestinal magnification – the mechanism of biomagnification and food-chain accumulation of organic-chemicals. Environ Sci Technol 27:2855–2863
55. Kortelainen P (1999) Occurence of humic waters. In: Keskitalo J, Eloranta P (eds) Limnology of humic waters. Backhuys, Leiden
56. Thurman EM (1985) Organic geochemistry of natural waters. Martinus Nijoff/ Dr. W Junk, Dordrecht
57. Kronberg L (1999) Content of humic substances in freshwater. In: Keskitalo J, Eloranta P (eds) Limnology of humic waters. Backhuys, Leiden
58. Leenheer JA, Croue JP (2003) Characterizing aquatic dissolved organic matter. Environ Sci Technol 37:18A–26A
59. Lam B, Baer A, Alaee M et al (2007) Major structural components in freshwater dissolved organic matter. Environ Sci Technol 41:8240–8247
60. Ma HZ, Allen HE, Yin YJ (2001) Characterization of isolated fractions of dissolved organic matter from natural waters and a wastewater effluent. Water Res 35:985–996
61. Pernet-Coudrier B, Clouzot L, Varrault G et al (2008) Dissolved organic matter from treated effluent of a major wastewater treatment plant: characterization and influence on copper toxicity. Chemosphere 73:593–599
62. Gustafsson Ö, Gschwend PM (1997) Aquatic colloids: concepts, definitions, and current challenges. Limnol Oceanogr 42:519–528
63. Akkanen J, Kukkonen JVK (2003) Measuring the bioavailability of two hydrophobic organic compounds in the presence of dissolved organic matter. Environ Toxicol Chem 22:518–524
64. Ohlenbusch G, Kumke MU, Frimmel FH (2000) Sorption of phenols to dissolved organic matter investigated by solid phase microextraction. Sci Total Environ 253:63–74
65. Burkhard LP (2000) Estimating dissolved organic carbon partition coefficients for nonionic organic chemicals. Environ Sci Technol 34:4663–4668

66. Kukkonen J, McCarthy JF, Oikari A (1990) Effects of XAD-8 fractions of dissolved organic carbon on the sorption and bioavailability of organic micropollutants. Arch Environ Contam Toxicol 19:551–557
67. Kukkonen J, Oikari A (1991) Bioavailability of organic pollutants in boreal waters with varying levels of dissolved organic material. Water Res 25:455–463
68. Cho H-H, Park J-W, Liu CCK (2002) Effect of molecular structures on the solubility enhancement of hydrophobic organic compounds by environmental amphiphiles. Environ Toxicol Chem 21:999–1003
69. Akkanen J, Vogt RD, Kukkonen JVK (2004) Essential characteristics of natural dissolved organic matter affecting the sorption of hydrophobic organic contaminants. Aquatic Sci 66:171–177
70. Haftka JJH, Govers HAJ, Parsons JR (2010) Influence of temperature and origin of dissolved organic matter on the partitioning behavior of polycyclic aromatic hydrocarbons. Environ Sci Pollut Res 17:1070–1079
71. Akkanen J, Tuikka A, Kukkonen JVK (2005) Comparative sorption and desorption of benzo [a]pyrene nad 3,4,3',4'-tetrachlorobiphenyl in natural lake water containing dissolved organic matter. Environ Sci Technol 39:7529–7534
72. Uhle ME, Chin Y-P, Aiken GR et al (1999) Binding of polychlorinated biphenyls to aquatic humic substances: the role of substrate and sorbate properties on partitioning. Environ Sci Technol 33:2715–2718
73. ter Laak TL, Van Eijkeren JCH, Busser FJM et al (2009) Facilitated transport of polychlorinated biphenyls and polybrominated diphenyl ethers by dissolved organic matter. Environ Sci Technol 43:1379–1385
74. Kuivikko M, Sorsa K, Kukkonen JVK et al (2010) Partitioning of tetra- and pentabromo diphenyl ether and benzo[a]pyrene among water and dissolved and particulate organic carbon along a salinity gradient in coastal waters. Environ Toxicol Chem 29:2443–2449
75. Yang WC, Spurlock F, Liu WP et al (2006) Effects of dissolved organic matter on permethrin bioavailability to Daphnia species. J Agric Food Chem 54:3967–3972
76. Yang WC, Hunter W, Spurlock F et al (2007) Bioavailability of permethrin and cyfluthrin in surface waters with low levels of dissolved organic matter. J Environ Qual 36:1678–1685
77. Kukkonen J (1991) Effects of pH and natural humic substances on the accumulation of organic pollutants in two freshwater invertebrates. In: Allard B, Borén H, Grimvall A (eds) Lecture notes in earth sciences: humic substances in the aquatic and terrestrial environment. Springer, Berlin
78. De Paolis F, Kukkonen J (1997) Binding of organic pollutants to humic and fulvic acids: influence of pH and the structure of humic material. Chemosphere 34:1693–1704
79. Haitzer M, Akkanen J, Steinberg C et al (2001) No enhancement in bioconcentration of organic contaminants by low levels of DOM. Chemosphere 44:165–171
80. Akkanen J, Penttinen S, Haitzer M, Kukkonen JVK (2001) Bioavailability of atrazine, pyrene and benzo[a]pyrene in European river waters. Chemosphere 45:453–462
81. Wiegand C, Pehkonen S, Akkanen J et al (2007) Bioaccumulation of paraquat by Lumbriculus variegatus in the presence of dissolved natural organic matter and impact on energy costs, biotransformation and antioxidative enzymes. Chemosphere 66:558–566
82. Kim HJ, Lee DS, Kwon JH (2010) Sorption of benzimidazole anthelmintics to dissolved organic matter surrogates and sewage sludge. Chemosphere 80:256–262
83. Maoz A, Chefetz B (2010) Sorption of the pharmaceuticals carbamazepine and naproxen to dissolved organic matter: role of structural fractions. Water Res 44:981–989
84. Landrum PF, Reinhold MD, Nihart SR et al (1985) Predicting the bioavailability of organic xenobiotics to Pontroreia hoyi in the presence of humic and fulvic materials and natural dissolved organic matter. Environ Toxicol Chem 4:459–467
85. Kukkonen J, Oikari A, Johnsen S et al (1989) Effects of humus concentrations on benzo[a] pyrene accumulation from water to Daphnia magna: comparison of natural waters and standard preparations. Sci Total Environ 79:197–207

86. Day KE (1991) Effects of dissolved organic carbon on accumulation and acute toxicity of fenvalerate, deltamethrin and cyhalothrin to Daphnia Magna (Straus). Environ Toxicol Chem 10:91–101
87. Haitzer M, Abbt-Braun G, Traunspurger W et al (1999) Effect of humic substances on the bioconcentration of polycyclic aromatic hydrocarbons: correlations with spectroscopic and chemical properties of humic substances. Environ Toxicol Chem 18:2782–2788
88. Gourlay C, Tusseau-Vuillemin MH, Garric J et al (2003) Effect of dissolved organic matter of various origins and biodegradabilities on the bioaccumulation of polycyclic aromatic hydrocarbons in Daphnia magna. Environ Toxicol Chem 22:1288–1294
89. Gourlay C, Tusseau-Vuillemin MH, Mouchel JM et al (2005) The ability of dissolved organic matter (DOM) to influence benzo[a]pyrene bioavailability increases with DOM biodegradation. Ecotoxicol Environ Saf 61:74–82
90. Kukkonen J (1991) Effects of lignin and chlorolignin in pulp mill effluents on the binding and bioavailability of hydrophobic organic pollutants. Water Res 26:1523–1532
91. Servos MR, Muir DCG, Webster GRB (1989) The effect of dissolved organic-matter on the bioavailability of polychlorinated dibenzo-para-dioxins. Aquat Toxicol 14:169–184
92. Traina SJ, Mcavoy DC, Versteeg DJ (1996) Association of linear alkylbenzenesulfonates with dissolved humic substances and its effect on bioavailability. Environ Sci Technol 30: 1300–1309
93. Haftka JJH, Parsons JR, Govers HAJ et al (2008) Enhanced kinetics of solid-phase microextraction and biodegradation of polycyclic aromatic hydrocarbons in the presence of dissolved organic matter. Environ Toxicol Chem 27:1526–1532
94. Smith KEC, Thullner M, Wick LY et al (2009) Sorption to humic acids enhances polycyclic aromatic hydrocarbon biodegradation. Environ Sci Technol 43:7205–7211
95. Mayer P, Karlson U, Christensen PS et al (2005) Quantifying the effect of medium composition on the diffusive mass transfer of hydrophobic organic chemicals through unstirred boundary layers. Environ Sci Technol 39:6123–6129
96. Mayer P, Fernqvist MM, Christensen PS et al (2007) Enhanced diffusion of polycyclic aromatic hydrocarhons in artificial and natural aqueous solutions. Environ Sci Technol 41: 6148–6155
97. ter Laak TL, ter Bekke MA, Hermens JLM (2009) Dissolved organic matter enhances transport of PAHs to aquatic organisms. Environ Sci Technol 43:7212–7217
98. Ghosh K, Schnitzer M (1980) Macromolecular structures of humic substances. Soil Sci 129:266–276
99. Aho J, Lehto O (1984) Effect of ionic strength on elution of aquatic humus in gel filtration chromatography. Arch Hydrobiol 101:21–38
100. Tsutsuki K, Kuwatsuka S (1984) Molecular size distribution of humic acids as affected by the ionic strength and the degree of humification. Soil Sci Plant Nutr 30:151–162
101. Engebretson RR, von Wandruszka R (1994) Microorganization in dissolved humic acids. Environ Sci Technol 28:1934–1941
102. Ephraim JH, Pettersson C, Nordén M, Allard B (1995) Potentiometric titrations of humic substances: do ionic strength effects depended on the molecular weight? Environ Sci Technol 29:622–628
103. Myneni SCB, Brown JT, Martinez GA, Meyer-Ilse W (1999) Imaging of humic substance macromolecular structures in water and soils. Science 286:1335–1337
104. Engebretson RR, Amos T, von Wandruszka R (1996) Quantitative approach to humic acid associations. Environ Sci Technol 30:990–997
105. Schlautman MA, Morgan JJ (1993) Effects of aqueous chemistry on the binding of polycyclic aromatic hydrocarbons by dissolved humic materials. Environ Sci Technol 27:961–969
106. Gauthier TD, Shane EC, Guerin WF et al (1986) Fluorescence quenching method for determining equilibrium constants for polycyclic aromatic hydrocarbons binding to dissolved humic materials. Environ Sci Technol 20:1162–1166

107. Jota MAT, Hassett JP (1991) Effects of environmental variables on binding of a PCB Congener by dissolved humic substances. Environ Toxicol Chem 10:483–491
108. Murphy EM, Zachara JM, Smith SC et al (1994) Interaction of hydrophobic organic compounds with mineral-bound humic substances. Environ Sci Technol 28:1291–1299
109. Jones KD, Tiller CL (1999) Effect of solution chemistry on the extent of binding of phenanthrene by a soil humic acid: a comparison of dissolved and clay bound humic. Environ Sci Technol 33:580–587
110. Akkanen J, Kukkonen JVK (2001) Effects of water hardness and dissolved organic material on bioavailability of selected organic chemicals. Environ Toxicol Chem 20:2303–2308
111. Verge C, Moreno A, Bravo J et al (2001) Influence of water hardness on the bioavailability and toxicity of linear alkylbenzene sulphonate (LAS). Chemosphere 44:1749–1757
112. Björk M (1995) Bioavailability and uptake of hydrophobic organic contaminants in bivalve filter feeders. Ann Zool Fennici 32:237–245
113. Bejarano AC, Widenfalk A, Decho AW et al (2003) Bioavailability of the organophosporous insecticide chlorpyrifos to the suspension-feeding bivalve, Mercenaria mercenaria, following exposure to dissolved and particulate matter. Environ Toxicol Chem 22:2100–2105
114. Gourlay C, Mouchel JM, Tusseau-Vuillemin MH et al (2005) Influence of algal and bacterial particulate organic matter on benzo[a]pyrene bioaccumulation in Daphnia magna. Sci Total Environ 346:220–230
115. Di Toro DM, Zabra CS, Hansen DJ et al (1991) Technical basis for establishing sediment quality criteria for nonionic organic chemicals using equilibrium partitioning. Environ Toxicol Chem 10:1541–1583
116. Reichenberg F, Mayer P (2006) Two complementary sides of bioavailability: accessibility and chemical activity of organic contaminants in sediments and soils. Environ Toxicol Chem 25:1239–1245
117. Huckins JN, Tubergen MW, Manuweera GK (1990) Semipermeable membrane devices containing model lipid: a new approach to monitoring the bioavailability of lipophilic contaminants and estimating their bioconcentration potential. Chemosphere 20:533–552
118. Booij K, Hofmans HE, Fischer CV et al (2003) Temperature-dependent uptake rates of nonpolar organic compounds by semipermeable membrane devices and low-density polyethylene membranes. Environ Sci Technol 37:361–366
119. McCarthy JF, Southworth GR, Ham KD et al (2000) Time-integrated, flux-based monitoring using semipermeable membrane devices to estimate the contribution of industrial facilities to regional polychlorinated biphenyl budgets. Environ Toxicol Chem 19:352–359
120. Lyytikäinen M, Rantalainen AL, Mikkelson P et al (2003) Similarities in bioaccumulation patterns of polychlorinated dibenzo-p-dioxins and furans and polychlorinated diphenyl ethers in laboratory-exposed oligochaetes and semipermeable membrane devices and in field-collected chironomids. Environ Toxicol Chem 22:2405–2415
121. Mayer P, Vaes WHJ, Wijnker F et al (2000) Sensing dissolved sediment porewater concentrations of persistent and bioaccumulative pollutants using disposable solid-phase microextraction fibers. Environ Sci Technol 34:5177–5183
122. Hawthorne SB, Grabanski CB, Miller DJ (2009) Solid-phase-microextraction measurement of 62 polychlorinated biphenyl congeners in milliliter sediment pore water samples and determination of K-DOC values. Anal Chem 81:6936–6943
123. Legind CN, Karlson U, Burken JG et al (2007) Determining chemical activity of (semi) volatile compounds by headspace solid-phase microextraction. Anal Chem 79:2869–2876
124. Mäenpää K, Leppänen MT, Reichenberg F et al (2011) Equilibrium sampling of persisten and bioaccumulative compounds in soil and sediment – comparison of two approaches to determine equilibrium partition concentration in lipids. Environ Sci Technol. doi:10.1021/es1029969
125. Zhou SN, Oakes KD, Servos MR et al (2008) Application of solid-phase microextraction for in vivo laboratory and field sampling of pharmaceuticals in fish. Environ Sci Technol 42:6073–6079

126. Zhang X, Oakes KD, Cui SF et al (2010) Tissue-specific in vivo bioconcentration of pharmaceuticals in rainbow trout (Oncorhynchus mykiss) using space-resolved solid-phase microextraction. Environ Sci Technol 44:3417–3422
127. Jahnke A, Mayer P (2010) Do complex matrices modify the sorptive properties of polydimethylsiloxane (PDMS) for non-polar organic chemicals? J Chromatogr A 1217: 4765–4770
128. Rusina TP, Smedes F, Klanova J et al (2007) Polymer selection for passive sampling: a comparison of critical properties. Chemosphere 68:1344–1351
129. Rusina TP, Smedes F, Klanova J (2010) Diffusion coefficients of polychlorinated biphenyls and polycyclic aromatic hydrocarbons in polydimethylsiloxane and low-density polylethylene polymers. J Appl Polymer Sci 116:1803–1810
130. Mayer P, Torang L, Glaesner N et al (2009) Silicone membrane equilibrator: measuring chemical activity of nonpolar chemicals with poly(dimethylsiloxane) microtubes immersed directly in tissue and lipids. Anal Chem 81:1536–1542
131. Jahnke A, Mayer P, Broman D et al (2009) Possibilities and limitations of equilibrium sampling using polydimethylsiloxane in fish tissue. Chemosphere 77:764–770
132. Reichenberg F, Smedes F, Jonsson JA et al (2008) Determining the chemical activity of hydrophobic organic compounds in soil using polymer coated vials. Chem Cent J 2:8
133. Meloche LM, deBruyn AMH, Otton SV et al (2009) Assessing exposure of sediment biota to organic contaminants by thin-film solid phase extraction. Environ Toxicol Chem 28:247–253
134. St George T, Vlahos P, Harner T et al (2011) A rapidly equilibrating, thin film, passive water sampler for organic contaminants; characterization and field testing. Environ Pollut 159: 481–486
135. Wilcockson JB, Gobas FAP (2001) Thin-film solid-phase extraction to measure fugacities of organic chemicals with low volatility in biological samples. Environ Sci Technol 35: 1425–1431
136. Hayatsu H (1992) Cellulose bearing covalently linked copper phthalocyamine trisulphonate as an adsorbent selective for polycyclic compounds and its use in studies of environmental mutagens and carcinogens. J Chromatogr 597:37–56
137. Sakamoto H, Hayatsu H (1990) A simple method for monitoring mutagenicity of river water – mutagens in Yodo river system, Kyoto Osaka. Bull Environ Contam Toxicol 44:521–528
138. Kummrow F, Rech CM, Coimbrao CA et al (2006) Blue rayon-anchored technique/ Salmonella microsome microsuspension assay as a tool to monitor for genotoxic polycyclic compounds in Santos estuary. Mutat Res-Genetic Toxicol Environ Mutagenesis 609:60–67
139. Watanabe T, Shiozawa T, Takahashi Y et al (2002) Mutagenicity of two 2-phenylbenzotriazole derivatives, 2-[2-(acetylamino)-4-(diethylamino)-5-methoxyphenyl]-5-amino-7-bromo-4-chloro-2 H-benzotriazole and 2-[2-(acetylamino)-4-(diallylamino)-5-methoxyphenyl]-5-amino-7-bromo-4-c hloro-2 H-benzotriazole and their detection in river water in Japan. Mutagenesis 17:293–299
140. Rico-Rico A, Droge STJ, Hermens JLM (2010) Predicting sediment sorption coefficients for linear alkylbenzenesulfonate congeners from polyacrylate-water partition coefficients at different salinities. Environ Sci Technol 44:941–947
141. Wu SF, Ding WH (2010) Fast determination of synthetic polycyclic musks in sewage sludge and sediments by microwave-assisted headspace solid-phase microextraction and gas chromatography–mass spectrometry. J Chromatogr A 1217:2776–2781
142. Hale SE, Meynet P, Davenport RJ et al (2010) Changes in polycyclic aromatic hydrocarbon availability in River Tyne sediment following bioremediation treatments or activated carbon amendment. Water Res 44:4529–4536
143. Friedman CL, Burgess RM, Perron MM et al (2009) Comparing polychaete and polyethylene uptake to assess sediment resuspension effects on PCB bioavailability. Environ Sci Technol 43:2865–2870

144. Sormunen AJ, Tuikka AI, Akkanen J et al (2010) Predicting the bioavailability of sediment-associated spiked compounds by using the polyoxymethylene passive sampling and Tenax® extraction methods in sediments from three river basins in Europe. Arch Environ Contam Toxicol 59:80–90
145. Jonker MTO, Koelmans AA (2001) Polyoxymethylene solid phase extraction as a partitioning method for hydrophobic organic chemicals in sediment and soot. Environ Sci Technol 35:3742–3748
146. van der Heijden SA, Jonker MTO (2009) PAH bioavailability in field sediments: comparing different methods for predicting in situ bioaccumulation. Environ Sci Technol 43:3757–3763
147. Cornelissen G, Rigterink H, Ten Hulscher DEM et al (2001) A simple Tenax® extraction method to determine the availability of sediment-sorbed organic compounds. Environ Toxicol Chem 20:706–711
148. Trimble TA, You J, Lydy MJ (2008) Bioavailability of PCBs from field-collected sediments: application of Tenax extraction and matrix-SPME techniques. Chemosphere 71:337–344
149. Dean JR, Scott WC (2004) Recent developments in assessing the bioavailability of persistent organic pollutants in the environment. Trac Trends Anal Chem 23:609–618
150. Mayer P, Tolls J, Hermens L et al (2003) Equilibrium sampling devices. Environ Sci Technol 37:184A–191A
151. Jahnke A, McLachlan MS, Mayer P (2008) Equilibrium sampling: partitioning of organochlorine compounds from lipids into polydimethylsiloxane. Chemosphere 73:1575–1581
152. Huckins JN, Manuweera GK, Petty JD et al (1993) Lipid-containing semipermeable-membrane devices for monitoring organic contaminants in water. Environ Sci Technol 27: 2489–2496
153. Arthur CL, Pawliszyn J (1990) Solid phase microextraction with thermal desorption using fused silica optical fibers. Anal Chem 62:2145–2148
154. Landrum PF, Lee HI, Lydy MJ (1992) Toxicokinetics in aquatic systems: model comparisons and use in hazard assessment. Environ Toxicol Chem 11:1709–1725
155. Arnot JA, Gobas FAPC (2004) A food web bioaccumulation model for organic chemicals in aquatic ecosystems. Environ Toxicol Chem 23:2343–2355
156. McKim JM, Nichols JW (1994) Use of physiologically based toxicokinetic models in a mechanistic approach to aquatic toxicology. In: Malins DC, Ostrander GK (eds) Aquatic toxicology: molecular, biochemical, and cellular perspectives. Lewis, Boca Raton, FL
157. Cahill TM, Cousins I, MacKay D (2003) Development and application of a generalized physiologically based pharmacokinetic model for multiple environmental contaminants. Environ Toxicol Chem 22:26–34
158. Landrum PF, Hayton WL, Lee H et al (1994) Synopsis of discussion session on the kinetics behind environmental bioavailability. In: Hamelink JL, Landrum PF, Bergman HL et al (eds) Bioavailability: physical, chemical and biological interactions. Lewis, Ann Arbor, MI
159. Sormunen AJ, Leppänen MT, Kukkonen JVK (2009) Examining the role of temperature and sediment-chemical contact time on desorption and bioavailability of sediment-associated tetrabromo diphenyl ether and benzo(a)pyrene. Ecotoxicol Environ Saf 72:1234–1241
160. Weston DP, Penry DL, Gulmann LK (2000) The role of ingestion as a route of contaminant bioaccumulation in a deposit-feeding polychaete. Arch Environ Contam Toxicol 38:446–454
161. Hamelink JL, Waybrant RC, Ball RC (1971) A proposal: exchange equilibria control the degree chlorinated hydrocarbons are biologically magnified in lentic environments. Trans Am Fish Soc 100:207–214
162. McFarland VA, Clarke JU (1986) Testing bioavailability of polychlorinated biphenyls from sediments using a two-level approach. Proceedings of the sixth USEPA committee on water quality. In: Wiler RG (ed) Proceedings of the US army engineer committee on water quality. Hydrologic Engineering Research Center, Davis, CA
163. Tracey GA, Hansen DJ (1996) Use of biota-sediment accumulation factors to assess similarity of nonionic organic chemical exposure to benthically-coupled organisms of differing trophic mode. Arch Environ Contam Toxicol 30:467–475

164. Wong CS, Capel PD, Nowell LH (2001) National-scale, field-based evaluation of the biota-sediment accumulation factor model. Environ Sci Technol 35:1709–1715
165. Webster E, Cowan-Ellsberry CE, McCarty L (2004) Putting science into persistence, bioaccumulation, and toxicity evaluations. Environ Toxicol Chem 23:2473–2482
166. Kraaij R, Mayer P, Busser FJM et al (2003) Measured pore-water concentrations make equilibrium partitioning work – a data analysis. Environ Sci Technol 37:268–274
167. Werner D, Hale SE, Ghosh U et al (2010) Polychlorinated biphenyl sorption and availability in field-contaminated sediments. Environ Sci Technol 44:2809–2815
168. MacKay D, Paterson S (1981) Calculating fugacity. Environ Sci Technol 15:1006–1014
169. Lohmann R, Burgess RM, Cantwell MG et al (2004) Dependency of polychlorinated biphenyl and polycyclic aromatic hydrocarbon bioaccumulation in Mya arenaria on both water column and sediment bed chemical activities. Environ Toxicol Chem 23:2551–2562
170. USEPA (2005) Contaminated sediment remediation: guidance for hazardous waste sites. http://www.epa.gov/superfund/health/conmedia/sediment/guidance.htm. Accessed 28 Jan 2011
171. Maruya KA, Landrum PF, Burgess RM et al. (2011) Incorporating contaminant bioavailability into sediment quality assessment frameworks. Integr Environ Assess Manage. doi:10.1002/ieam.135
172. McLeod PB, van den Heuvel-Greve MJ, Allen-King RM et al (2004) Effects of particulate carbonaceous matter on the bioavailability of benzo[a]pyrene and 2,2',5,5'-tetrachlorobiphenyl to the clam, Macoma baltica. Environ Sci Technol 38:4549–4556
173. Accardi-Dey A, Gschwend PM (2002) Assessing the combined roles of natural organic matter and black carbon as sorbents in sediments. Environ Sci Technol 36:21–29
174. Song JZ, Peng PA, Huang WL (2002) Black carbon and kerogen in soils and sediments. 1. Quantification and characterization. Environ Sci Technol 36:3960–3967
175. Prest HF, Richardson BJ, Jacobson LA et al. (1995) Monitoring organochlorines with semi-permeable membrane devices (SPMDs) and mussels (*Mytilus edulis*) in Corio Bay, Victoria, Australia. Mar Poll Bull 30:543–554

The Use of Attached Microbial Communities to Assess Ecological Risks of Pollutants in River Ecosystems: The Role of Heterotrophs

Lorenzo Proia, Fernanda Cassió, Claudia Pascoal, Ahmed Tlili, and Anna M. Romaní

Abstract The aim of this chapter is to highlight the importance of microbial attached communities in the assessment of the effects of pollutants on freshwater ecosystems. We particularly focus on the role of heterotrophs in biofilms developing on different substrata. Firstly, an overview of the importance of microbial communities for the whole ecosystem processes is given, focusing on bacteria and fungi either living in consortia with autotrophs or as the microbial decomposing community on plant litter in river ecosystems. A series of detailed examples of direct effects of priority and emerging pollutants on bacteria in epilithic biofilms and on attached decomposers are included. Microbial ecological interactions between organisms in heterogeneous complex communities are highlighted describing the indirect effects observed in a series of study cases. A collection of laboratory and field study data is used to demonstrate the relevance of natural heterogeneous communities to obtain a more realistic approach to ecosystem processes. Finally, an upscaling from the effects observed at the microbial scale to the potential implication for ecosystems health and risk is included.

Keywords Biofilms • Decomposers • Ecological risk • Heterotrophs • Pollutants effects • Rivers

L. Proia (✉) • A.M. Romaní
Institut d'Ecologia Aquàtica, Universitat de Girona, Girona, Espanya
e-mail: lorenzo.proia@udg.edu

F. Cassió • C. Pascoal
Departamento de Biologia, Universidade do Minho, Braga, Portugal

A. Tlili
CEMAGREF, Lyon, France

H. Guasch et al. (eds.), *Emerging and Priority Pollutants in Rivers*,
Hdb Env Chem (2012) 19: 55–84, DOI 10.1007/978-3-642-25722-3_3,

Contents

1 Relevance of the Heterotrophic Microbial Processes in River Ecosystems and Their Sensitivity to Water Pollutants

River and stream ecosystems are extremely heterogeneous and dynamic environments continuously changing because of physical and biological processes. Physical processes are mainly driven by geomorphology and hydrology characteristics, determining the quantity and quality inputs of organic and inorganic materials in the flowing ecosystem as well as the arrangement of habitat patchiness (hyporheic zones, wetlands, rapid pools, floodplains, [1]). At the catchment scale, climate influences these processes. The biological processes, interacting with the river physical dynamics, are the main responsible for the matter cycling—release, uptake, and transformations—throughout the river system. Key biological processes are occurring at the microbial scale where the microbially driven biogeochemical reactions might be then relevant at the ecosystem scale [2].

The study of microbial ecology of river ecosystems is a relatively recent discipline that became important in the last decades. From the first description of the microbial loop by Azam et al. [3] in marine systems numerous studies were performed on marine, lentic, and lotic environments (i.e., [4–6]), highlighting relevant differences in the structure of the microbial communities. In lakes and marine systems, the importance of the planktonic food webs for organic matter re-mineralization and carbon fluxes has been widely described [7, 8]. In contrast, a crucial role of the benthic microbial community in river energy flow has been highlighted [9]. Moreover, in river and streams, considerable differences are denoted in biotic interactions and processes between suspended and attached microbial communities such as a greater bacterial activity and biomass and extracellular enzyme activities in the benthic biofilms than in the flowing water [8, 10]. Attached microbial communities are in general structured by autotrophs (diatoms, green algae, and cyanobacteria) and heterotrophs (bacteria, fungi, and protozoa) embedded in extracellular polymeric matrix adhered on substrata conferring a biofilm [11]. Attached bacteria start colonization process by covering mineral surfaces with polysaccharide glycocalyx [12]. This extracellular matrix facilitates the adherence of micro-autotrophs and traps organic matter particles permitting the development of more complex biofilms [13]. In spite of their mobility, micro-metazoans can also be found interstitially as well as on the substrate surface and they must be considered as part of the attached community because of their trophic relationship with biofilm microorganisms [14]. This close spatial relation between so different life-strategy organisms results in important interactions (i.e., bacterial utilization of algal exudates, [15]), that allow us to consider attached communities as micro-ecosystems with a complex structure where important processes take place. In freshwater systems different attached communities develop on different streambed substrata [16]. Communities developed on submerged coarse particulate organic matter (leaves and wood) are dominated by fungi [17–19], while biofilms developed on mineral surfaces (fine sediment) and fine particulate organic matter are dominated by bacteria and protozoa [13, 20]. Conversely, inorganic hard and relative inert substrata (rocks and cobbles) are colonized by communities dominated by autotrophs (diatoms mainly; [21]).

The river microbial attached communities play an important role in freshwater ecosystems for organic matter re-mineralization and inorganic nutrient fluxes through hydrographic web. In the case of allochthonous energy inputs, i.e., plant material from riparian vegetation, the initial leaves breakdown is mainly carried out by aquatic hyphomycetes [22] while attached bacteria play an important role as decomposers of fine particulate organic matter accumulating in sediments and in the overlying water [23]. The epilithic biofilms developing on rocks and cobbles have been also described as relevant sites for organic and inorganic nutrient uptake and retention [24]. One of the key roles of bacteria and fungi within the attached communities is their capability to decompose and transform organic molecules to finally mineralize them, thanks to their extracellular enzyme capabilities. The size of molecules limits its transport across biological membranes, making the activity of extracellular enzymes the first biotic step of organic matter turnover in ecosystems [13]. Extracellular enzymes produced by fungi and bacteria exhibit substrate specificity. In general, fungal communities produce enzymes capable of degrading complex and recalcitrant compounds such as cellulose, hemicelluloses, and lignin (by means of β-xylosidase, β-glucosidase, peroxidase, and phenoloxidase, [25]), while bacteria mostly synthesize enzymes involved in degrading more simple polymers such as polysaccharides, peptides, or organic phosphorus compounds (β-glucosidase, peptidase, and phosphatase). However, these enzyme capabilities could be modulated by microbial interactions within the attached community, such as the positive interaction between algae and bacteria in epilithic biofilms [26, 27] or the synergistic/competitive relationship between fungi and bacteria in leaf litter breakdown [28, 29]. Overall, microbial attached communities dominate the ecosystem metabolism in many aquatic systems being a major component for the uptake, storage, and cycling of carbon, nitrogen, and phosphorous [9, 13]; therefore they are an important compartment in water purification processes [126].

Rivers and streams are the most important source of water for human use (drinking water, industry, agriculture, etc.) even if less than 1% of the total water in the earth is stored in freshwater systems [30]. At present, most rivers and streams are affected by any anthropogenic disturbance. Most pollutants have the potential to enter aquatic habitats from direct application, terrestrial runoff, or wind-borne drift. Input of pollutants from wastewater treatment plants (WWTP) and industrial sewage are widespread along river networks strongly changing quantitatively and qualitatively the timing and location of solute inputs, including some well-known toxicants but new emerging pollutants as well (i.e., pharmaceuticals).

Environmental stressors, including emergent and priority pollutants, cause a variety of responses in heterotrophic microbes at different biological levels (from molecules and cells to communities and ecosystems) that can provide measurable endpoints and are useful in ecological risk assessment and environmental management. Because there are thousands of different contaminants and their potential toxicity may vary with water chemistry and physical characteristics of the habitat, there is an increasing need of developing reliable bioindicators and early warning biomarkers of stress. Over the last few years, the use of microbial communities as model systems in ecology and ecotoxicology has been increasing. As described above, microorganisms play a crucial role in several environmental services, like nutrient recycling, water purification, and carbon sequestration [31]; they are ubiquitous, have short generation times, and are easy to manipulate under controlled conditions, constituting ideal systems for assessing the impacts of stressors that are not feasible to manipulate in the field [32–34].

The microbial attached communities have been defined as interfaces that integrate a variety of responses to environmental changes and chemical stressors; their rapid interaction with dissolved substances results in functional (short-term) and structural (long-term) responses, making them useful as "early warning systems" of disturbances [24]. As attached communities are micro-ecosystems, responses to multiple stressors can be different depending on target organisms and thus both direct and indirect effects can be inferred (Fig. 1). Thus, the study of the potential effects of priority and new emerging pollutants found in continental water bodies on the structure and function of microbial attached communities might be relevant to assist management strategies of freshwater ecosystems. Several examples of both direct and indirect effects of priority and emerging pollutants on heterotrophs of microbial attached communities are presented in the following sections.

2 Assessing the Direct Effects of Pollutants on Heterotrophs

2.1 *Bacteria in Epilithic Biofilms*

Biofilm communities are one of the first victims of pollutants which can affect their structure and function. With increasing number of toxic molecules found in freshwaters (e.g., pharmaceuticals, nanomaterials, PAH, and metals), which can directly target heterotrophic organisms, recent research has focused on the bacterial biofilm responses to pollutants ([35, 36], Table 1).

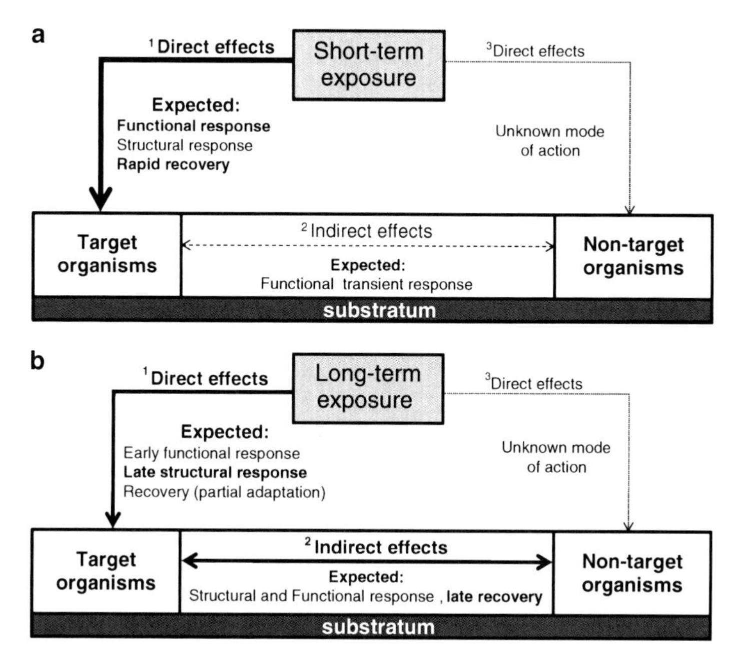

Fig. 1 Schematic representation of the expected direct and indirect effects of sublethal concentrations of toxic compound/s on target and nontarget organisms of freshwater microbial attached communities in case of (**a**) short-term and (**b**) long-term exposure. In the case of short-term exposure to toxic compound/s (**a**) direct effects of target organisms ([1]) are expected to be the most important ones. In particular, rapid effect and recovery of functional response would be expected. Some structural response could also occur as well as some direct effect on nontarget organisms ([3]) due to unknown mode of action. These effects can also generate an indirect effect on target organisms. The magnitude of these effects is expected to be less relevant than direct ones. Indirect effects ([2]) are expected to be transient and mainly on function. In the case of long-term exposure to toxic compound/s (**b**) direct ([1]) and indirect ([2]) effects are expected to occur. In particular, target organisms would respond quickly in terms of function and later at structural level ([1]). Recovery of these effects is expected to be partial depending on the magnitude of the response. For example, the exposure to some bactericide could result in an initial negative effect on some function sustained by bacteria (i.e., extracellular enzymatic activity). Nevertheless if exposure persisted, some resistant species are expected to be selected. This selection will result in the shift of community composition (structural response). The structural response may therefore restore previous functional levels resulting in a general recovery of functional parameters. Nevertheless, the structural response (shift in community composition) could not be considered recovered until the original community will not be restored or even an adapted community will establish. Occurrence of indirect effects ([2]) is expected to be delayed respect to direct ones. Thus, the structural and functional response of nontarget organisms will occur after the target organisms responded. The magnitude and delay of indirect effects depend on the interaction with the target organisms directly affected and on physiology, metabolism, and life cycle of the nontarget organism. Recovery of these indirect effects mainly depends on resilience potential of nontarget organisms, the magnitude, and the duration of the direct effect observed that generated it. Direct effects on nontarget organisms ([3]) could also occur in consequence of some unknown mode of action. These effects can also generate an indirect effect on target organisms. The magnitude of these effects is expected to be less relevant than direct ones

Table 1 Review of the effects of priority and emerging contaminants on autotrophs and bacteria of fluvial biofilms and interaction between them

Reference	Compound	Effects on autotrophs	Effects on bacteria	Interaction
Pesce et al. [96]	Diuron	Low Chl *a* concentration low algal density	↓ of bacterial density and productivity	Direct inhibition of primary producers generate indirect negative effect on bacterial community
Ricart et al. [94]	Diuron	↑ Chl *a* density ↓ photosynthesis ↓ diatoms biovolume Diatoms community composition	↓ live bacteria ↑ peptidase activity	Direct diuron damage on autotrophs affects indirectly bacteria. Algal cell lysis provide organic compounds to bacteria resulting in ↑ of peptidase activity
Ricart et al. [35]	Triclosan	↓ photosynthetic efficiency ↓ non-photochemical quenching ↑ diatoms mortality	↑ bacteria mortality	Direct effect on bacteria generates indirect effect on autotrophs (mainly diatoms) Direct effect on algae photosystems is not excluded
Lopez-Doval et al. [95]	Diuron + grazing	↓ photosynthesis	↑ bacterial mortality that recovered	Direct inhibition of photosynthesis ↓ organic compounds available for bacteria inducing mortality
Boninneau et al. [40]	Atenolol	↓ photosynthetic efficiency	↑ bacteria mortality ↓ Peptidase activity	↓of algal exudates result in ↓ of peptidase activity and ↑ of bacteria mortality Direct effects on bacteria
	Propranolol	↓ Photosynthetic efficiency and capacity	↓ Peptidase activity (transitory)	Not described
Lawrence et al. [84]	Nikel	↓ Photosynthetic biomass and changes in specie composition Cyanobacteria eliminated	↓ biomass and cell viability ↓ denitrification Community shift	↓ EPS ↓ carbon utilization Inhibition of many catabolic pathways
Lawrence et al. (2005)	Carbamazepine	↓cyanobacteria biomass	↓ bacteria biomass ↑ L/D ratio	↓ EPS Significantly different communities in composition and function
	Caffeine	↓cyanobacteria biomass	↑ Bacteria biomass. ↓ L/D ratio Shift from γ- to β-proteobacteria	↑ thickness Changes in EPS composition (+) ↑ of activity (C utilization)

	Furosemide	↓ cyanobacteria biomass ↑ algal biomass	↑ bacteria biomass ↓ L/D ratio Shift from γ- to α-proteobacteria	↑ of activity (C utilization) ↑ thickness Changes in EPS composition (+)
	Ibuprofen	↓ cyanobacteria biomass	↑ L/D ratio Shift from γ- to β-proteobacteria	Changes in EPS composition (–) Toxic effect on cyanobacteria impacts linkages between phototrophic biomass and bacteria
Lawrence et al. [127]	Chlorhexidine	↑ of algal and cyanobacterial biomass	Changes in bacterial community	Changes in carbon utilization
Lawrence et al. [100]	Triclosan and triclocarban	Negative effect on biomass mainly filamentous forms	Shift from autotrophic to heterotrophic community	Negative impact on carbon utilization
Tlili et al. [97]	Diuron	Direct effects on biomass, pigments composition and photosynthesis	Late changes in bacterial community composition	Long-term exposure affects autotrophs, which changes the quantity and quality of algal products available to bacteria leading to a shift in the composition of the bacterial community
Barranguet et al. [102]	Copper	↓ of biomass Shift from diatoms to cyanobacteria	↓ of substrata metabolized Late ↑ of Cu-tolerant bacteria	Change in the algal community composition induced by Cu results in a different composition of algal products available for heterotrophs causing the change in the heterotrophic community composition

↑ = increase, ↓ = decrease

Direct effects of toxicants on bacterial community could lead to two biofilm responses, which may differ in their temporal pattern: short-term biochemical and physiological alterations and long-term changes in community structure (Fig. 1). Short-term effects include increases in bacterial mortality and reduction in bacterial growth, production, respiration, and extracellular enzyme activities. In an experimental study, Ricart et al. [35] assessed the mortality of biofilm bacteria after their acute exposure to the bactericide triclosan, which inhibits bacterial fatty acid synthesis [37]. The authors showed that environmentally relevant concentrations of triclosan caused an increase of bacterial mortality with a non-effect concentration (NEC) of 0.21 μg L^{-1}, after 48 h of exposure. Lawrence et al. [38] exposed biofilms to the anti-inflammatory drug diclofenac, demonstrating a rapid and negative impact on the growth of bacteria at 100 μg L^{-1}.

Physiological effects of toxicants on bacteria have been also shown such as the significant decrease in bacterial production (as measured by thymidine and leucine incorporation) after the biofilms being exposed to metals [39]. On the other hand, recent studies on the effect of pharmaceuticals showed significant effects on biofilm extracellular enzyme activities, thus compromising the biofilm organic matter cycling role in the ecosystem. Boninneau et al. [40] showed that leucine-aminopeptidase activity was significantly inhibited after 6-h exposure of biofilm bacteria to the β-blockers propranolol and atenolol, reflecting the negative effect of these pharmaceuticals on the bacterial ability to hydrolyze peptides. Moreover, Chenier et al. [41] focused on the effect of the aliphatic hydrocarbons hexadecane on the anaerobic denitrification activity of biofilm bacteria, showing that 1 μg L^{-1} of hexadecane inhibited denitrification and reduced the N_2/N_2O ratio. These results indicate that petroleum hydrocarbons, even when present at low concentrations, can alter the flux of N in river ecosystems. The authors concluded that, since petroleum hydrocarbons are hydrophobic, they can accumulate in lipid biolayers of bacterial cytoplasmic membranes or within the external membranes of gram-negative bacteria, causing alterations of membrane structure and function, which have a negative impact on cellular activity.

In addition to the effects on heterotrophic metabolism, the presence of pollutants in lotic systems may lead to a change in microbial biomass or to a shift in community structure. It is widely accepted that chemical pollution could reduce bacterial diversity [42]. Nevertheless, the effects of pollutants on bacteria may also favor the selection of certain species. Bacterial species composing the biofilm may have ranging sensitivities and responses toward various anthropogenic pressures [39]. During exposure to toxic agents such as metals or antibiotics, the most sensitive organisms may be overtaken by the more resistant or more tolerant ones. An experimental study conducted by Tlili et al. [36] showed that the pattern of biofilm bacterial diversity obtained by molecular fingerprinting was modified under long-term exposure to 30 μg L^{-1} of copper, indicating that this metal was a strong driver of bacterial structural changes. In the same study, authors highlighted heterotrophic biofilm sensitivity to copper by performing short-term inhibition tests based on extracellular enzyme activity (β-glucosidase and leucine-aminopeptidase) and substrate-induced respiration activity. Whatever, the results indicated that the observed shift in community structure due to copper exposure was accompanied by an enhanced tolerance of bacterial communities to the metal. Lawrence et al. [38] also assessed the structural responses of biofilm bacterial communities after their exposure to 100 $\mu g\ L^{-1}$ of diclofenac. In this study, the community analyses by FISH probes indicated that diclofenac induces significant alterations of community composition with significant increases in overall Eubacterial, β-, γ-, and Cytophaga-Flavobacterium probe-positive populations. Furthermore, denaturing gradient gel electrophoresis analyses confirmed changes of bacterial community composition of diclofenac-treated biofilms, in comparison with control.

On the other hand, in case of prolonged exposure to pollutants, the processes of bacterial adaptation may also lead to stimulation of biodegradation capacities of these compounds ([43], Fig. 1 long-term exposure). Indeed, because of their species-rich communities and the possibility to adapt their metabolisms, biofilm bacteria provide a great potential for the removal of contaminants from water systems by using the pollutants as a source of organic matter. Some studies have explored the degradation pathways of pollutants by microbial communities taken from rivers and confirmed their important role in elimination process of pharmaceutical or other chemical residues [38, 44]. Pieper et al. [45] investigated the degradation of the pharmaceuticals phenazone, a pyrazolone derivate in widespread use, by bacterial isolates from natural biofilms. Results indicated that some bacterial strains were able to metabolize phenazone in its metabolite 1,5-dimethyl-1,2-dehydro-3-pyrazolone over the sampling period of 8 weeks. Authors concluded that the microorganisms need a reasonably long time to adapt their metabolisms to enable the removal of phenazone from water samples. Even if it is not yet well developed in aquatic environments, the functional approach based on the bacterial biodegradation of pollutants offers promising prospects for ecosystem biomonitoring as these activities are generally focused on specific molecules or groups of molecules. This could be enriched by including in situ measurements of expression of genes encoding enzymes involved in degradation mechanisms, and diversity studies to identify the organisms involved in these processes.

In conclusion, two complementary approaches coexist to assess direct effects of pollutants on bacterial communities. The first one is based on structural analysis of communities (biomass, taxonomy, and diversity), and the second one is a functional approach generally based on the metabolic activity measurements (respiration, enzymatic activities, and potential for biodegradation). The use of bacterial communities in epilithic biofilms as bioindicator might therefore allow an important choice of descriptors, adapted to different types of contaminants [36].

2.2 *Attached Community on Plant Material: Decomposers*

In freshwaters, bacteria and fungi play a key role in organic matter decomposition and convert plant litter from the surrounding vegetation into a more suitable food source for invertebrates [46, 47]. Fungi, mainly aquatic hyphomycetes, are often dominant over bacteria during earlier stages of plant litter decomposition and contribute to up to 39% to leaf carbon loss [19]. Even though the contribution of bacteria to organic matter turnover may increase in polluted streams, fungi accounted for 89–99% of the total microbial production [19, 48, 49].

Pollution by mine drainage is among the best-documented stressors affecting microbial decomposers and litter breakdown in streams with most studies documenting a reduction in taxonomic diversity and fungal reproductive ability (Table 2). Slower leaf breakdown and reduced microbial biomass accumulation occur in streams severely impacted by mining or metallurgic activities [50–54]. Also, stream acidification through atmospheric inputs and related impacts (e.g., high levels of aluminum) is generally accompanied by reduced litter breakdown rates (e.g., [55, 56]). However, moderate metal pollution affects fungal diversity more than fungal biomass and decomposition [57]. In a low-order stream impacted by metals and eutrophication, the decline of aquatic hyphomycete diversity and sporulation was not accompanied by decreased fungal biomass or leaf decomposition [58]. These findings suggest that aquatic hyphomycete communities may respond to stress according to the redundancy model, in which processes are stable because increasing biomass of tolerant species compensates for the loss of sensitive ones. Hence, it is predicted that biodiversity has a lower threshold of response to anthropogenic stress, whereas biomass and function are stable or increase under low to moderate stress and decrease only under high stress conditions.

In the last few years, the extensive use of engineered nanoparticles increases the chance of their release into aquatic environments, raising the question whether they can pose risk to aquatic biota and the associated ecological processes. The exposure of microbial decomposers to nano-Cu oxide and nano-silver led to a reduction in fungal sporulation and leaf decomposition rates [59]. These effects were accompanied by shifts in the structure of fungal and bacterial communities based on DNA fingerprints and fungal spore morphology. Generally, the negative effects of nano- and ionic metals were stronger on bacterial biomass than on fungal biomass [33, 59]. However, the impacts of metal nanoparticles on leaf decomposition by aquatic microbes were less pronounced compared to their ionic forms.

A prerequisite for using any type of indicator of ecosystem integrity is that it responds unequivocally to anthropogenic stresses. The cited studies support that leaf breakdown rate and microbial activity are potential indicators of functional integrity of streams (Table 2). Among several fungi-based metrics, spore production emerged as the most sensitive indicator of stress. This is supported by a meta-analysis comprising 23 studies [60] and by microcosm experiments using stream dwelling microbial assemblages with stressors added alone or in combination [33, 59, 61].

The combined effects of metals on aquatic fungi show that Zn alleviates the effect of Cd toxicity on the growth of five aquatic hyphomycete species [62]. However, this study analyzed the effects of metal mixtures at the species level. The observed effects at the community level were mostly additive, particularly on fungal biomass and leaf decomposition [33], suggesting no interaction between Cu and Zn or that the magnitude of synergistic effects in some species was offset by

Table 2 Effects of priority and emerging contaminants on microbial decomposers and implications to stream ecosystem functioning

Reference	Stressor	Endpoints	Biological level and set up	Results
Duarte et al. [61]	Zn	Leaf mass loss, fungal productivity, reproduction and diversity	Stream-dwelling microbes, *Alnus glutinosa* (alder) leaves, 25 days in microcosms, two times, four Zn levels up to 150 μM	Reduction of fungal production and litter decomposition; shifts in fungal assemblage
Pascoal et al. [107]	Zn	Leaf mass loss, fungal biomass and reproduction	Up to four fungal species, 21 days in microcosms, three levels up to 100 μM	Positive diversity effects on decomposition were reduced under stress; diversity increased functional stability mainly under stress
Duarte et al. [33]	Zn, Cu	Leaf mass loss, fungal and bacterial biomass and diversity, and fungal reproduction	Stream-dwelling microbes on alder leaves, 40 days in microcosms, three times, three metal levels up to 100 μM in all combinations	Reduction in fungal reproduction, bacterial biomass, and litter decomposition; shifts in microbial assemblages; stronger effects for Cu alone or in mixtures; mostly additive effects
Duarte et al. [34]	Zn, Cu	Leaf mass loss, fungal and bacterial biomass, and diversity and fungal reproduction	Stream-dwelling microbes on alder leaves, 30 days in microcosms, 50 μM of metals added alone or in mixtures, together or sequentially in time (0 and 10 days); release from stressors after 20 days	Reduction of fungal reproduction, bacterial biomass, and litter decomposition; shifts in fungal and bacterial assemblages; recovery of microbial activity after metal release
Roussel et al. [108]	Cu, habitat	Leaf mass loss, fungal biomass and reproduction, and macroinvertebrates	Stream mesocosms, two habitats (shallow with pebbles and deep fine sediments), four Cu levels up to 75 μg/L, mixtures of alder, oak, and maple leaves, 7 weeks	Reduction in leaf decomposition and abundance of dominant shredder in pebbles at the highest metal level; no differences in fungal biomass and sporulation
Pradhan et al. [59]	Ionic and nano-copper and silver	Leaf mass loss, fungal reproduction, fungal and bacterial biomass, and diversity	Stream-dwelling microbes on alder leaves, 21 days in microcosms, four levels for copper up to 500 ppm nano-CuO or 30 ppm $CuCl_2$, three	Reduction in decomposition; shifts in microbial assemblages; bacteria more sensitive than fungi; stronger effects of ionic metals than

(continued)

Table 2 (continued)

Reference	Stressor	Endpoints	Biological level and set up	Results
			levels for silver up to 300 ppm nano-Ag or 20 ppm $AgNO_3$	nanometals; nano-copper more toxic than nano-silver
Fernandes et al. [68]	Cd	Leaf mass loss and fungal biomass	Up to three fungal species on alder leaves, two functional types, 35 days in microcosms; 1.5 mg/L Cd	Positive diversity effects were only kept in assemblages with the Cd-resistant type with shifts from complementarity to dominance
Moreirinha et al. [63]	Cd, PAH	Leaf mass loss, fungal biomass, reproduction, and diversity	Stream-dwelling microbes on alder leaves, 14 days in microcosms, six levels up to 4.5 mg/L Cd and 0.2 mg/L phenanthrene	Reduction in fungal diversity, reproduction, and decomposition; phenanthrene potentiated Cd effects
Baudoin et al. [56]	Al and low pH	Leaf mass loss, fungal diversity, and biomass	Six headwater streams, Al up to 960 μg/L, pH 3.8–7.7, *Fagus sylvatica* (beech), leaf bags	Fungal richness correlates negatively with Al and positively with pH; reduction in leaf breakdown
Medeiros et al. [64]	Zn, Mn, or Fe	Leaf mass loss, fungal biomass, and reproduction	Microbes on alder leaves from reference and impacted sites, 16 days in microcosms, six levels, up to 9.6 ppm Zn or Mn and up to 20 ppm Fe	Metals decreased fungal sporulation and changed fungal species composition; effects were less pronounced in metal-adapted assemblages
Solé et al. [123]	Metals, sulfate, nitrate and hypoxia	Diversity and biomass of fungi	In situ experiment, alder leaf bags, 11 sites, streams, and groundwater wells, 2 weeks	Reduction in fungal biomass and diversity; shifts in species composition; fungi as potential bioindicators of stress
Sridhar et al. [51]	Mining	Diversity, sporulation and biomass of fungi, leaf mass loss, invertebrate feeding	In situ experiment, alder leaf bags, two sites, six sampling times up to 42 days, Zn up to 2.6 g/L and Cu, Pb, Cs, Cd, As up to 13 mg/L	Reduction in sporulation, low fungal biomass; decomposition stopped after 20 days in the highly polluted site; invertebrate preferred conditioned leaves
Niyogi et al. [57]	Mining	Litter breakdown, microbial respiration and invertebrate biomass	27 sites (8 pristine, 19 varying in the degree of impact), Salix sp.	Reduction in breakdown, mainly when dissolved Zn and metal oxides were high, microbial activity correlates

			(willow) leaves in bags, Zn up to 80 mg/L, pH 2.7–7.9	negatively with metal oxide deposition on leaves, shredder biomass correlates negatively with dissolved Zn
Niyogi et al. [54]	Mining	Litter decomposition, fungal respiration biomass, and diversity	20 stream sites, willow leaves in bags	Fungal diversity decreased with high Zn (>1 mg/L) and low pH (<6); fungal biomass and activity not affect by dissolved Zn but decreased with metal oxide deposition
Medeiros et al. [53]	Abandoned gold mine (As, Cd, Cu and Zn)	Litter breakdown, fungal sporulation, biomass and diversity, microbial respiration, invertebrate diversity, abundance, and feeding	Three sites (one reference, one intermediate and one polluted), leaf bags with alder, up 64 days	Reduction in breakdown, but not when only microbes were present, reduction in fungal sporulation; lower feeding invertebrate activity
Lecerf and Chauvet [60]	Mining, eutrophication, altered riparian vegetation	Litter breakdown, fungal sporulation, biomass and diversity, meta-analysis	Three pairs of impact-control stream sites; up to 3,256 μg/L N-NO3, 100.7 μg/L P-PO4, 387.6 μg/L NH_4, As 122 μg/L, Mn 45 μg/L; 4–12	Biomass and sporulation were higher in eutrophied stream; spore production depressed with mining and slightly enhanced in the stream affected by forestry; reduction in fungal diversity at impacted sites
Pascoal et al. [58]	Urbanization, agriculture and industry	Leaf mass loss, microbial biomass, fungal diversity, and reproduction	In situ experiment, alder leaf bags, reference versus impacted site, seven sampling times up to 54 days	Reduction in fungal diversity and sporulation, stimulation of bacterial biomass but no effect on fungal biomass or decomposition
Dangles et al. [55]	Acidification gradient and Al	Leaf breakdown rates, microbial respiration, fungal biomass, shredder diversity, abundance, and biomass	25 woodland headwater streams, Vosges Mountains, France	Litter breakdown responds to stream acidification and Al; *Gammarus* abundance and microbial respiration accounted for 85% of differences in breakdown among streams
Augustin et al. [75]	1-Naphthol	Analysis of degradation products	Axenic cultures of *H. lugdunensis* grown 7 days in malt extract with	*H. lugdunensis* metabolized 74% of 1-naphthol in 5 days

(continued)

Table 2 (continued)

Reference	Stressor	Endpoints	Biological level and set up	Results
			50 μM Cd (added after 2 days growth)	
Bundschuh et al. [110]	Antibiotics in mixtures	Bacterial and fungal biomass, invertebrate feeding	Microcosm with alder leaves, 2 or 200 μg/L total antibiotic concentration	Increase fungal biomass and invertebrate feeding
Bermingham et al. [109]	Mecoprop	Fungal biomass and feeding experiments	Alder leaves inoculated with spore suspensions of one fungus, five levels up to 1 g/L 5 days, *G. pseudolimnaeus* allowed to feed on leaves	Reduction in fungal biomass but not in a dose-dependent manner; animals discriminated between exposed and control leaves
Cheng et al. [124]	Herbicides and low metal levels	Leaf breakdown, fungal diversity and sporulation, and shredder diversity and density	Bags with beech and *Populus nigra* (poplar) leaves; atrazine 0.45 ppb, simazine 0.45 ppb and diuron 1.43 ppb	Reduction in leaf breakdown, fungal sporulation, and shredder density

antagonistic effects in others. On the other hand, the exposure to the polycyclic aromatic hydrocarbon phenanthrene potentiated the negative effects of Cd on fungal diversity and activity [63], suggesting that the co-occurrence of these stressors may pose additional risk to aquatic biodiversity and stream ecosystem functioning.

Some contaminants have a long history in aquatic environments, and such pressure on communities can select for resistant genotypes leading to changes in community structure and ecosystem functions. Indeed, pollution from an abandoned gold mine in Portugal affected leaf-associated fungi, but effects appeared to be less pronounced in metal-adapted communities [64]. Certain species of aquatic hyphomycetes, such as *Heliscus lugdunensis* and *Tetracladium marchalianum*, have been found among the top ranked species in highly polluted streams in Germany [65]. The species spectrum in Portugal was different, with *Flagellospora curta* and *Heliscus submersus* mainly associated with polluted streams [58, 66]. Such species are obvious candidates when (1) searching for biomarkers of stress, (2) analyzing the physiological mechanisms underlying fungal survival in polluted environments, and (3) exploring the potential of fungi for bioremediation of polluted sites. Recent studies based on ITS rRNA barcodes show that aquatic hyphomycete species have intraspecific diversity [67]. This suggests the existence of different populations, some of which might be adapted to pollution. Indeed, a manipulative experiment using two functional types (Cd-resistant and Cd-sensitive phenotypes) of *Articulospora tetracladia* showed that intraspecific traits altered biodiversity effects under stress: the positive effects of fungal diversity on leaf decomposition and fungal biomass production were kept under Cd stress only when the assemblage had a Cd-resistant functional type [68].

The effects of organic xenobiotics have been also addressed on freshwater microbial decomposers. Aquatic hyphomycetes accumulate dichloro-diphenyl-trichloroethane (DDT) and low concentrations of this insecticide enhanced the growth of several fungal species [69]. The fungicides mancozeb and carbendazim did not inhibit fungal growth or reproduction up to 5 mg/L or conidial germination up to 1 mg/L [70]. In contrast, Bärlocher and Premdas [71] reported a linear negative relationship between conidia production and concentration of the fungicide pentachlorophenol (0.0001–10 mg/L). Some freshwater fungi, including the aquatic hyphomycete *Clavariopsis aquatica*, have shown ability to degrade some xenobiotics, such as xenoestrogen nonylphenol [72], polycyclic musks [73], and synthetic azo and anthraquinone dyes [74]. This has been related to the activity of extracellular laccases, an oxidoreductase enzyme, which appears to be involved in lignin biodegradation. Also, *H. lugdunensis* showed ability to metabolize polycyclic aromatic hydrocarbons such as naphthol [75]. These findings support a role of fungi in affecting the environmental fate of pollutants in aquatic ecosystems.

3 Microbial Interactions and Indirect Effects of Pollutants

3.1 Bacterial–Algal Interactions

The close spatial relation and isolation from the overlying water column facilitate complex interactions within stream microbial assemblages [76, 77]. Probably the most important interactions occurring in attached microbial communities are between autotrophs and heterotrophs [78].

Autotroph–heterotroph trophic and metabolic interactions include (1) bacterial use of organic compounds from algal exudates released during photosynthesis (e.g., [15, 79, 80]), and heterotrophic use of O_2 released during photosynthesis [81], (2) algal use of inorganic nutrients released during mineralization activities by heterotrophic assemblages [77] and CO_2 released during heterotrophic respiration (reviewed by [82]), and (3) algal use of vitamins provided by bacteria [83]. Physical interaction where algae provide substrata for bacterial colonization may also be significant [11]. It has been suggested that the importance of these linkages may be controlled by the abundance of algae in the system [84]. Bacteria may be tightly linked to algae when the algal biomass is high and not when it is low [85], or there may be threshold levels before this coupling occurs [86]. Several studies demonstrated the stimulation of bacterial extracellular enzymatic activities by algal photosynthesis in freshwater biofilms [27, 78, 87–89]. At the same time, nutrient enrichment can decouple algal–bacterial production and it is hypothesized that competition occurs in nutrient-poor conditions [90]. Decoupling of algal–bacterial metabolism has been also observed when adding labile dissolved organic matter (glucose) in laboratory grown biofilms [91].

When biofilms are exposed to pollutants we therefore might expect both direct and indirect effects on the autotrophic and/or heterotrophic compartments. This might depend on the mode of action of the compound and on the target organisms. Effects can be acute or chronic depending on the time of exposure and on the concentration levels of the pollutant. An appropriate set of endpoints should be selected in order to detect the wide range of possible responses of such complex microbial communities. As a general rule, the more functional and physiological descriptors should be used when acute responses are expected because of short-time exposure or transient perturbation in natural systems, while the more structural descriptors should be selected as endpoints in studies aiming the investigation of potential chronic effects (Fig. 1). However, due to the potential interaction between the biofilm compartments, both functional and structural parameters might be analyzed, enhancing the relevance of a multibiomarker approach in ecotoxicological studies [40]. In Table 1 several examples of direct and indirect effects of pollutants on autotrophs and bacteria as well as evidences of interactions between them in fluvial biofilms are reported.

In general, target organisms are affected directly and the effects being shown after relative short-time exposure to the toxic compound. Direct effects are normally dose-dependent and can be recovered if the toxic concentration is sublethal, the exposure does not persist in time, and the organism has a good resilience potential. Thus, in short-time exposure indirect effects on nontarget organisms are less expected (Fig. 1). However, the chronic long-term exposure of biofilm to pollutants, in spite of being at sublethal concentrations, could result in the persistence of direct effects and the appearance of indirect effects due to interactions between target and nontarget organisms within the microbial biofilm. As an example, this short- and long-term exposure effect in the biofilm photosynthetic response after diuron exposure is shown in Fig. 2. Diuron (herbicide) blocks the electron transport in the photosystem II [92] causing loss of photosynthetic efficiency of autotrophic compartment of biofilms, which is recovered after 1 week at the short-term but persists at the long-term exposure ([35, 93], Fig. 2). The metabolic changes but also structural changes at community level to the target organisms after a chronic long-term exposure might determine greater effects to potential nontarget organisms. Thus, the long-term exposure of fluvial biofilms to low diuron concentrations resulted in significant direct effects on diatom community (shift in species composition and biovolume reduction), increase in Chl *a* density, persistent decrease of photosynthetic parameters (Fig. 2), and also in a delayed indirect effect on bacterial viability and peptidase activity ([94], Table 1, Figs. 3 and 4). This study

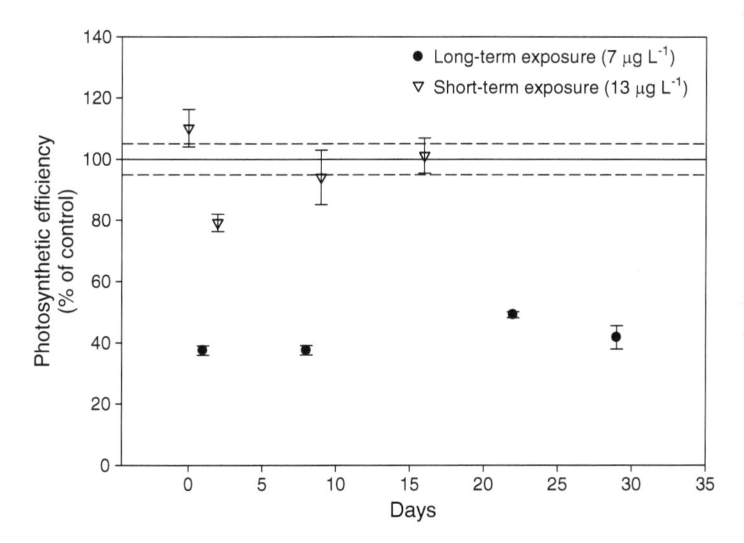

Fig. 2 Direct effect of the herbicide diuron on photosynthetic efficiency of river biofilms. Results from a short-term exposure (*empty triangles*, [93]) and long-term exposure (*black points*, [94]) studies are plotted together in order to compare responses in case of acute and chronic contamination scenario. Values are mean ± standard error ($n = 4$). The *horizontal continuous line* represents controls ± 95% confidence intervals (*dashed lines*). In the short-term experiment biofilms were exposed during 48 h to pulses (13 μg L^{-1}) of diuron and recovery process was followed during 2 weeks after the end of exposure. Long-term experiment consisted in 29 days of exposure to 7 μg L^{-1} of diuron. In particular, photosynthesis is inhibited and rapidly recovered after short-term exposure while persistent inhibition of photosynthesis was measured in long-term exposure experiment

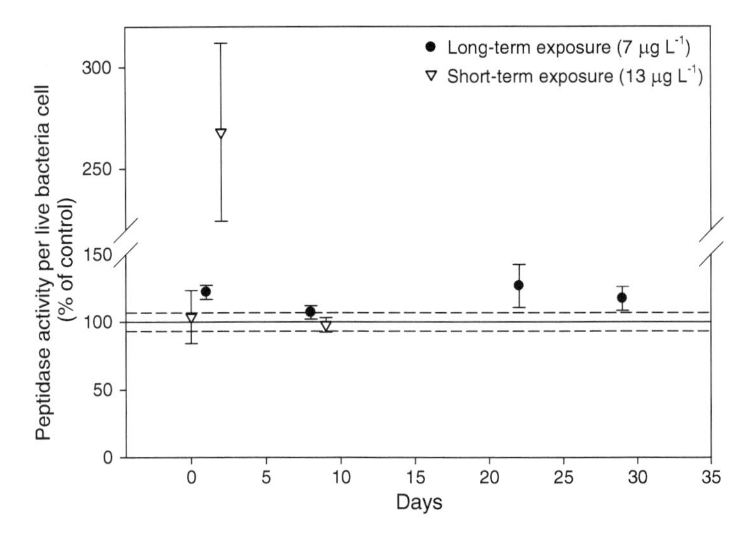

Fig. 3 Indirect effect of the herbicide diuron on specific extracellular peptidase activity per bacterial cell. Results from a short-term exposure (*empty triangles*, [93]) and long-term exposure (*black points*, [94]) studies are plotted together in order to compare responses in case of acute and chronic contamination scenario. Values are mean ± standard error ($n = 4$). The *horizontal continuous line* represents controls ± 95% confidence intervals (*dashed lines*). In the short-term experiment biofilms were exposed during 48 h to pulses of diuron (13 μg L^{-1}) and recovery process was followed during 2 weeks after the end of exposure. Long-term experiment consisted in 29 days of exposure to 7 μg L^{-1} of diuron. In particular, transient increase in specific peptidase activity per cell was measured after the short-term exposure while the same effect was observed later in long-term exposure experiment

suggests direct diuron damage on autotrophs affecting indirectly the structure and functioning of the biofilm bacterial community. Other long-term diuron exposure studies have also shown indirect effects on bacteria (nontarget organisms) after direct effects on autotrophs in fluvial biofilms ([95–97], Table 1).

Nontarget organisms may be affected both directly (unknown mode of action) and indirectly because of interaction with target organisms. As an example, biofilm autotrophs were affected (diatom mortality) after a direct effect of triclosan on bacteria [35, 93]. In this case, direct effects of triclosan on bacteria viability might generate indirect effects on autotrophs (mainly diatoms), but, at the same time, a direct effect of triclosan on autotrophic organisms as suggested by other authors cannot be ruled out ([98–101, 125], Table 1).

Indirect effects are strictly dependent on the magnitude and timing of direct effect observed. Obviously they appear after the direct response occurred and the delay depends on the interaction with the target organism directly affected and on physiology, metabolism, and life cycle of the nontarget organism. For example, Boninneau et al. [40] described a rapid increase of bacterial mortality and decrease of extracellular peptidase activity after 24 h of exposure to high concentration of the β-blocker atenolol. The authors concluded that direct negative effect observed on photosynthesis could indirectly (reduction of algal exudates available for bacteria)

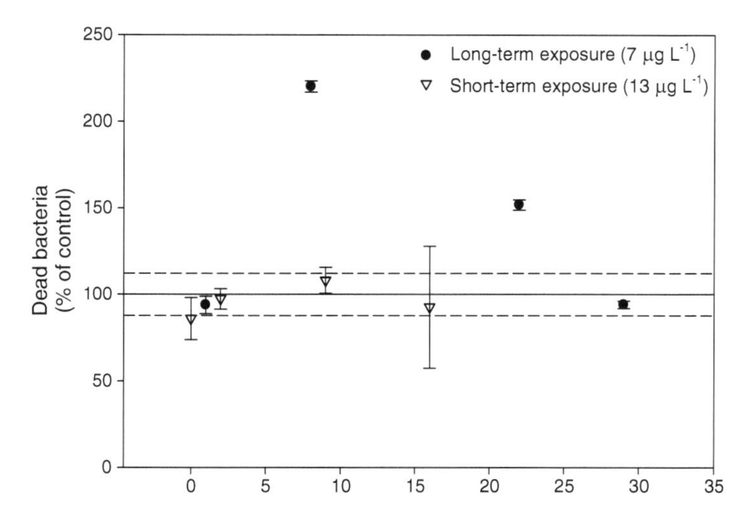

Fig. 4 Indirect effect of the herbicide diuron on bacteria. Results from a short-term exposure (*empty triangles*, [93]) and long-term exposure (*black points*, [94]) studies are plotted together in order to compare responses in case of acute and chronic contamination scenario. Values are mean ± standard error ($n = 4$). The *horizontal line* represents controls ± 95% confidence intervals (*dashed lines*). In the short-term experiment biofilms were exposed during 48 h to pulses of diuron (13 μg L^{-1}) and recovery process was followed during 2 weeks after the end of exposure. Long-term experiment consisted in 29 days of exposure to 7 μg L^{-1} of diuron. In particular, no response was observed after short-term exposure while late increase of dead bacterial cells, finally recovered, is evidenced in long-term exposure experiment

enhance the stress provoked directly by the toxic on bacteria (Table 1). Moreover, Barranguet et al. [102] highlighted that the change in the algal community composition induced by Cu results in a different composition of algal products available for heterotrophs causing changes of the heterotrophic community composition in a long-term study (Table 1). These are two examples demonstrating as the indirect effects timescale could vary depending on the direct effects' magnitude and timing.

Recovery of indirect effects is therefore a more complex process with high ecological concern. Indirect effects' recovery is also dependent on resilience potential of affected organisms, time of exposure, and toxicant concentration but is moreover mainly dependent on the nature of the interaction with target organisms that generated it. In particular, it strictly depends on the behavior and the duration of the direct effect observed. Not many ecotoxicological studies assessed resistance and recovery of complex attached microbial communities exposed to priority and emerging pollutants. Lopez-Doval et al. [95] found significant increase of dead bacteria (nontarget organisms) in fluvial biofilms after 1 day of exposure to 2 μg L^{-1} of the herbicide diuron, but bacteria were completely recovered 28 day later. The low concentration applied in this study and the probable shift of bacterial community ([96, 97, 103], Table 1) might determine the recovery of bacteria. Another example showing biofilm recovery after exposure to pollutants was that shown by Proia et al. [93] who specifically followed recovery of fluvial biofilms during 2 weeks after 48 h exposure to pulses of triclosan and diuron. The recovery of diatoms' viability (indirect effect) occurred 2 weeks after the end of triclosan exposure, exactly 1 week after bacteria viability totally recovered, highlighting the different time needed for recovery of indirect effects on the different biofilm compartments.

In conclusion it is important to remark the importance of studies on microbial complex communities and the interactions occurring within them in order to evaluate possible effects of pollutants on freshwater ecosystem. As stated by Barranguet et al. [102] "...the effects of toxicants on each biofilm compartment studied separately will not give an accurate picture of its sensitivity under natural conditions..." However, the known tight microbial interactions within the biofilms suggest that pollutant effects to the biofilm might be buffered when compared to single species, as suggested by their resistance to life in extreme environments [104].

3.2 Fungal–Bacterial Interaction in Decomposition Process

Interactions among microbes on plant litter can result from differences in sequestering resources, direct interference between cells, and production of inhibitory substances. Contribution of bacteria to carbon flow from decaying leaf litter is lower than that of fungi, regardless of the presence of fungi and nutrient availability [18, 19]. However, both synergistic and antagonistic interactions between litter-associated fungi and bacteria have been found [18, 29, 105]. Some authors found negative effects of bacteria on fungal growth and biomass accrual [105]. On the contrary, bacteria appear to grow better together with fungi than alone [105], probably because fungi provide bacteria with resources that they are not able to acquire on their own. On the other hand, culture filtrates of several aquatic hyphomycetes inhibited bacterial growth and showed antifungal activities [106]. However, little is known on the factors controlling the interactions between microbes (e.g., fungus–fungus, fungi–bacteria) on decomposing leaves. Moreover, the presence of contaminants is expected to modulate such interactions. For instance, the complementarily relationships among aquatic hyphomycetes on decomposing leaves were weakened under metal stress [68, 107]. Some authors found that metal ions and metal nanoparticles reduce litter decomposition, fungal sporulation, and bacterial biomass, but appear to have little effect on fungal biomass [33, 59]. If metal stress affects bacteria more than fungi, the negative effects of bacteria on fungi are expected to decrease. This, together with the carbon released from dead bacteria, might override the expected negative impacts of metals on fungal biomass.

Due to the complex trophic interactions between organisms that govern organic matter decomposition in streams, indirect effects of contaminants on detritivores that feed on microbially colonized litter have also to be considered. In an in situ mesocosm study, low levels of Cu did not affect aquatic hyphomycetes but reduced the abundance of the dominant shredder *Gammarus pulex* on decomposing leaves [108]. Other studies report that invertebrate shredders are able to discriminate between leaves exposed to reference and metal contaminated streams [53], probably because considerable amounts of metals can accumulate or adsorb on fungi or leaves [51]. Therefore, toxicity for invertebrates can result from direct effects of metals via water or indirectly via food (i.e., leaves and/or microbes growing on leaves).

The contamination of alder leaves with the herbicide Mecoprop reduced the biomass of *H. lugdunensis* and inhibited leaf consumption by the shredder *G. pseudolimnaeus*, but no evidence of indirect effects of Mecoprop was found [109]. On the other hand, antibiotic mixtures, at concentrations typical of streams receiving wastewater treatment plant effluents, had no effects on bacterial assemblages but led to a stimulation of fungal biomass on leaves; this increased leaf palatability for shredders which preferred to feed on conditioned leaves in the presence of antibiotics [110].

Some studies have demonstrated that plant detritus, particularly when colonized by aquatic fungi, contribute to the removal of pollutants from the water column (e.g., metals, [51]), which would be beneficial to aquatic biota. However, because conditioned leaves serve as a food resource to invertebrate detritivores, pollutants can still enter in the food web. On the other hand, as referred above, some fungi show ability to degrade recalcitrant organic pollutants (e.g., [72, 75]) consistent with the known complex fungal enzymatic machinery. Hence, preserving fungal biodiversity would be beneficial to maintain stream water quality and ecosystem functions.

Generally, impacts of pollutants on microbial community functions depend on duration and type of disturbance. One may expect that after a brief disturbance, ecosystem functions eventually return to its former state, whereas sustained disturbance may result in a new state. Although some evidence of fungal acclimation to stressors has been found, the resistance of microbial decomposers to one stressor did not increase when communities were previously acclimated to another stressor [34]. It is conceivable that some species are more affected by certain stressors than others, explaining the shifts in species composition and dominance often found. It is plausible that microbial decomposers divert energy toward defense mechanisms to cope with stressors, and away from growth or reproduction. One should point out that fungal communities under stress consistently respond with a reduction of sporulation. Hence, when communities are exposed to stressors for long periods of time, species with reduced dispersal will tend to disappear.

3.3 *Role of Protozoa*

Protozoa (ciliate and flagellate) are also a relevant biofilm compartment, playing a key role in the biofilm microbial food web [13]. Protozoa might feed on bacteria, thus controlling their biomass or either use organic molecules from the matrix EPS

[111]. Within a biofilm, effects of protozoa grazing on bacterial community structure have been also reported [112]. Thus, when the whole biofilm is exposed to any pollutant which affects a biofilm compartment (i.e., algae or bacteria) changes in the protozoa community might also occur. In this sense, in a laboratory mesocosm experiment where the whole microbial food web was considered, a direct effect of atrazine on algae was measured, as expected, but a significant increase in ciliates and flagellates abundance was measured after 48 h of exposure [128]. Similarly, an increase in biofilm protozoa density was observed after its exposure to triclosan [100]. Friberg-Jensen et al. [113], when analyzing the effects of cypermethrin (a pyrethroid insecticide) on a freshwater community by a multispecies field experiment considering different trophic levels, observed the major effect of the insecticide on crustaceans (cypermethrin $\geq 0.13\ \mu g\ L^{-1}$). The proliferation of rotifers, protozoans, bacteria, and algae (planktonic and periphytic) but the decrease of crustaceans suggests a reduced grazer control from crustaceans due to the cypermethrin exposure.

Few studies address the direct impact of pollutants on ciliates and heterotrophic flagellates in biofilms. Most studies on natural communities focus on planktonic communities or either use one single species, commonly the ciliate *Tetrahymena thermophila*. Effects of metal nanoparticles on *T. thermophila* showed a greater effect of nano-ZnO than nano-CuO (EC_{50} values of 5 mg metal/l versus 128 mg metal/l, [114]). Generation of reactive oxygen species (ROS) induced by exposure to heavy metals (Cd, Cu, or Zn) in ciliated protozoa (*Tetrahymena* sp. and three strains of *Colpoda steinii*) was also detected, indicating a physiologic response of these organisms to metal exposure [115]. The effect of salty urban waters might also determine reduction in ciliate bacterivory as it was shown for *T. thermophila* [116]. On the other hand, the ciliate *Euplotes mutabilis* isolated from industrial effluent has been found to adapt and then uptake metals [117].

Within the biofilm, due to the close contact between their components (algae, bacteria, fungi, and protozoa) the effects of one toxicant probably cannot be simplified as their direct effect, but in contrast the whole response of the biofilm would be either enhanced or buffered. In the case of protozoa, an increase of bacterivory might be expected when their resource (bacteria or EPS) is increasing. This protozoa increase might lead to changes in the biofilm's three-dimensional structure and thus influencing higher trophic levels (micrometazoans feeding on the biofilm).

4 From Microbial Scale to Ecological Risks Assessment

Normally, in freshwater ecosystems priority and emerging compounds are detected at low concentration and can chronically enter by WWTP (e.g., pharmaceuticals and personal care products) or reach the system by pulses from punctual or diffuse sources (i.e., herbicides). Climate and hydrology of the systems strongly influence levels of pollutants detected in freshwater bodies. For example, flood events after rain could result in pulses of herbicides in agricultural zones as described by Rabiet et al. [118] as well as dry periods could enhance the relative importance of discharge from WWTP effluents resulting in higher concentration of compounds entering by this source. This situation is mainly occurring in the Mediterranean area of Europe and in all regions included in water scarcity scenarios by the climate change prevision [119]. In this context, it appears fundamental to study the potential effects of unknown number of new emerging compounds on ecosystem function and services in freshwaters.

We already discussed in the introduction of this chapter about the importance of microbial processes in freshwaters systems at ecosystem scale. Moreover, the positive relation between biodiversity and ecosystem functioning and services has been widely demonstrated (i.e., [120, 121]). Freshwaters provide several ecosystem services that are directly related to microbial processes. The main ecosystems services provided by rivers and streams and regulated by microbial processes are related to nutrient recycling (self-depuration capacity) and organic matter processing and transport. Is it possible to assess how pollutants may influence these ecosystem services by affecting microbial community structure and function? All results of the studies presented in this chapter may be used in some way to evaluate risks at ecosystem level. Shifts of community structure in response to exposure to some pollutants, reported in several studies presented in this chapter, are an example of effects of emerging compounds that could affect ecosystem processes as well as indirect effects due to microbial interactions within attached communities. Studying direct and indirect effects of pollutants on natural microbial communities provides a more realistic ecological approach than monospecific toxicology tests. Although this is a first step trying to link laboratory studies to ecological relevant evidence, the limitations in upscaling from laboratory experiments to ecosystems are well known as well as the scarce cause–effects relationships identified in field studies [122]. The best choice would be probably to combine specific endpoints linked to target and nontarget organisms and on short and long expected effects and include as well some whole community functioning measurement more directly related to ecological services. As an example, Proia et al. [93] used the measurement of inorganic phosphorus uptake capacity by the biofilm which was significantly reduced after triclosan exposure. Although this effect is also highlighted by the triclosan damage on biofilm structure and function, the use of phosphorus uptake results evidenced risks for self-depuration capacity of running waters associated with triclosan pulses. The use of such kind of whole community function descriptors related to ecosystem services should be implemented in ecotoxicological studies as they are able to integrate direct and indirect effects observed at different trophic levels within microbial communities.

In conclusion, we believe that the arguments discussed and the examples provided in this chapter highlighted and demonstrated the importance of using complex microbial communities, and interactions occurring within them, to assess ecological risks associated with the entrance of new emerging nonpriority pollutants in freshwater ecosystems.

References

1. Allan JD (1995) Stream ecology. Structure and functioning of running waters. Chapman & Hall, London
2. Zehr J (2010) Microbes in Earth's aqueous environments. Front Microbiol, Aquat Microbiol, Volume 1, Article 4. doi:10.3389/fmicb.2010.00004
3. Azam F, Fenche T, Field JG et al (1983) The ecological role of water-column microbes in the sea. Mar Ecol Prog Ser 10:257–263
4. Descy J-P, Leporq B, Viroux L et al (2002) Phytoplankton production, exudation and bacterial reassimilation in the River Meuse (Belgium). J Plankton Res 24:161–166
5. Hart DR, Stone L, Berman T (2000) Seasonal dynamics of the Lake Kinneret food web: the importance of the microbial loop. Limnol Oceanogr 45:350–361
6. Fenchel T (2008) The microbial loop – 25 years later. J Exp Mar Biol Ecol 366:99–103
7. Pomeroy LR, Wiebe WJ (1988) Energetics of microbial food webs. Hydrobiologia 159:7–18
8. Edwards RT, Meyer JL, Findlay SEG (1990) The relative contribution of benthic and suspended bacteria to system biomass, production, and metabolism in a low-gradient blackwater river. J N Am Benthol Soc 9:216–228
9. Battin T, Butturini A, Sabater F (1999) Immobilization and metabolism of dissolved organic carbon by natural sediment biofilms in two climatically contrasting streams. Aquat Microb Ecol 19:297–305
10. Romaní AM, Sabater S (1999) Epilithic ectoenzyme activity in a nutrient-rich Mediterranean river. Aquat Sci 61:122–132
11. Lock MA (1993) Attached microbial communities in rivers. In: Ford TE (ed) Aquatic microbiology: an ecological approach. Blackwell, Oxford
12. Bärlocher F, Murdoch JH (1989) Hyporheic biofilms—a potential food source for interstitial animals. Hydrobiologia 184:61–67
13. Pusch M, Fiebig D, Brettar I et al (1998) The role of micro-organisms in the ecological connectivity of running waters. Freshwat Biol 40:453–495
14. Lamberti GA (1996) The role of periphyton in benthic food webs. In: Stevenson RJ, Bothwell ML, Lowe RL (eds) Algal ecology. Freshwater benthic ecosystems. Academic, San Diego, CA
15. Murray RE, Cooksey KE, Priscu JC (1986) Stimulation of bacterial DNA synthesis by algal exudates in attached algal-bacterial consortia. Appl Environ Microbiol 52:1177–1182
16. Artigas J (2008) The role of fungi and bacteria on the organic matter decomposition process in streams: interaction and relevance in biofilms. Ph.D. thesis
17. Diez J, Elosegi A, Chauvet E (2002) Breakdown of wood in the Agüera stream. Freshwat Biol 47:2205–2215
18. Gulis V, Suberkropp K (2003) Effect of inorganic nutrients on relative contributions of fungi and bacteria to carbon flow from submerged decomposing leaf litter. Microb Ecol 45:11–19
19. Pascoal C, Cássio F (2004) Contribution of fungi and bacteria to leaf litter decomposition in a polluted river. Appl Environ Microbiol 70:5266–5273
20. Findlay S, Tank J, Dye S et al (2002) A cross-system comparison of bacterial and fungal biomass in detritus pools of headwater streams. Microb Ecol 43:55–66
21. Stevenson RJ (1996) In: Stevenson RJ, Bothwell ML, Lowe RL (eds) Algal ecology, freshwater benthic ecosystems. Academic, San Diego, CA
22. Mathuriau C, Chauvet E (2002) Breakdown of litter in a neotropical stream. J N Am Benthol Soc 21:384–396
23. Findlay S, Strayer D, Goumbala C et al (1993) Metabolism of streamwater dissolved organic carbon in the shallow hyporheic zone. Limnol Oceanogr 38:1493–1499
24. Sabater S, Guasch H, Ricart M et al (2007) Monitoring the effect of chemicals on biological communities. The biofilm as an interface. Anal Bioanal Chem 387:1425–1434
25. Romaní AM (2010) Freshwater biofilms. In: Dürr S, Thomason JC (eds) Biofouling, 1st edn. Wiley-Blackwell, Oxford

26. Rier ST, Stevenson RJ (2002) Effects of light, dissolved organic carbon, and inorganic nutrients on the relationship between algae and heterotrophic bacteria in stream periphyton. Hydrobiologia 489:179–184
27. Francoeur SN, Wetzel RG (2003) Regulation of periphytic leucine-aminopeptidase activity. Aquat Microb Ecol 31:249–258
28. Bengtsson G (1992) Interactions between fungi, bacteria and beech leaves in a stream mesocosm. Oecologia 89:542–549
29. Mille-Lindblom C, Tranvik LJ (2003) Antagonism between bacteria and fungi on decomposing aquatic plant litter. Microb Ecol 45:173–182
30. Sabater S, Elosegi A (2009) Conceptos y técnicas en ecológia fluvial. Fundación BBVA, Bilbao
31. Ducklow H (2008) Microbial services: challenges for microbial ecologists in a changing world. Aquat Microb Ecol 53:13–19
32. Jessup CM, Kassen R, Forde SE et al (2004) Big questions, small worlds: microbial model systems in ecology. Trends Ecol Evol 19:189–197
33. Duarte S, Pascoal C, Alves A et al (2008) Copper and zinc mixtures induce shifts in microbial communities and reduce leaf litter decomposition in streams. Freshw Biol 53:91–102
34. Duarte S, Pascoal C, Cássio F (2009) Functional stability of stream-dwelling microbial decomposers exposed to copper and zinc stress. Freshw Biol 54:1638–1691
35. Ricart M, Guasch H, Alberch M et al (2010) Triclosan persistence through wastewater treatment plants and its potential toxic effects on river biofilms. Aquat Toxicol 100:346–353
36. Tlili A, Bérard A, Roulier JL et al (2010) PO_4^{3-} dependence of the tolerance of autotrophic and heterotrophic biofilm communities to copper and diuron. Aquat Toxicol 98:165–177
37. McMurry LM, Oethinger M, Levy SB (1998) Triclosan targets lipid synthesis. Nature 394:531–532
38. Lawrence JR, Swerhone GDW, Topp E et al (2007) Structural and functional responses of river biofilm communities to the nonsteroidal anti-inflammatory diclofenac. Environ Toxicol Chem 26:573–582
39. Blanck H, Admiraal W, Cleven RFMJ et al (2003) Variability in zinc tolerance, measured as incorporation of radio-labeled carbon dioxide and thymidine, in periphyton communities sampled from 15 European river stretches. Arch Environ Contam Toxicol 44:17–29
40. Boninneau C, Guasch H, Proia L et al (2010) Fluvial biofilms: a pertinent tool to assess β-blockers toxicity. Aquat Toxicol 96:225–233
41. Chenier MR, Beaumier D, Fortin N et al (2006) Influence of nutrient inputs, hexadecane and temporal variations on denitrification and community composition of river biofilms. Appl Environ Microbiol 72:575–584
42. Mahmoud HMA, Goulder R, Carvalho GR (2005) The response of epilithic bacteria to different metals regime in two upland streams: assessed by conventional microbiological methods and PCR-DGGE. Arch Hydrobiol 163:405–427
43. Watanabe K, Baker PW (2000) Environmentally relevant microorganisms. J Biosci Bioeng 89:1–11
44. Paje MLF, Kuhlicke U, Winkler M et al (2002) Inhibition of lotic biofilms by Diclofenac. Appl Microbiol Biotechnol 59:488–492
45. Pieper C, Risse D, Schmidt B et al (2010) Investigation of the microbial degradation of phenazone-type drugs and their metabolites by natural biofilms derived from river water using liquid chromatography/tandem mass spectrometry (LC-MS/MS). Water Res 44:4559–4569
46. Bärlocher F (2005) Freshwater fungal communities. In: Dighton J, Oudemans P, White J (eds) The fungal community, 3rd edn. CRC, Boca Raton, FL
47. Pascoal C, Cássio F (2008) Linking fungal diversity to the functioning of freshwater ecosystems. In: Sridhar KR, Bärlocher F, Hyde KD (eds) Novel techniques and ideas in mycology. Fungal Diversity Press, Hong Kong

48. Pascoal C, Cássio F, Marcotegui A et al (2005) The role of fungi, bacteria, and invertebrates in leaf litter breakdown in a polluted river. J N Am Benthol Soc 24:784–797
49. Baldy V, Gobert V, Guérold F et al (2007) Leaf litter breakdown budgets in streams of various trophic status: effects of dissolved inorganic nutrients on microorganisms and invertebrates. Freshw Biol 52:1322–1335
50. Krauss G-J, Wesenberg D, Ehrman J et al (2008) Fungal responses to heavy metals. In: Sridhar S, Bärlocher F, Hyde K (eds) Novel techniques and ideas in mycology. Fungal Diversity Press, Hong Kong
51. Sridhar KR, Krauss G, Bärlocher F et al (2001) Decomposition of alder leaves in two heavy metal-polluted streams in central Germany. Aquat Microb Ecol 26:73–80
52. Bermingham S, Maltby L, Cooke RC (1996) Effects of a coal mine effluent on aquatic hyphomycetes. II. Laboratory toxicity experiment. J Appl Ecol 33:1311–1321
53. Medeiros AO, Rocha P, Rosa CA et al (2008) Litter breakdown in a stream affected by drainage from a gold mine. Fund Appl Limnol 172:59–70
54. Niyogi DK, McKnight DM, Lewis WM Jr (2002) Fungal communities and biomass in mountain streams affected by mine drainage. Arch Hydrobiol 155:255–271
55. Dangles O, Gessner MO, Guerold F et al (2004) Impacts of stream acidification on litter breakdown: implications for assessing ecosystem functioning. J Appl Ecol 41:365–378
56. Baudoin JM, Guérold F, Felten V et al (2008) Elevated aluminium concentration in acidified headwater streams lowers aquatic hyphomycete diversity and impairs leaf-litter breakdown. Microb Ecol 56:260–269
57. Niyogi DK, Lewis WM Jr, McKnight DM (2001) Litter breakdown in mountain streams affected by mine drainage: biotic mediation of abiotic controls. Ecol Appl 11:506–516
58. Pascoal C, Cássio F, Marvanová L (2005) Anthropogenic stress may affect aquatic hyphomycete diversity more than leaf decomposition in a low order stream. Arch Hydrobiol 162:481–496
59. Pradhan A, Seena S, Pascoal C et al (2011) Can increased production and usage of metal nanoparticles be a threat to freshwater microbial decomposers? Microb Ecol 62:58–68
60. Lecerf A, Chauvet E (2008) Diversity and functions of leaf-decaying fungi in human-altered streams. Freshw Biol 53:1658–1672
61. Duarte S, Pascoal C, Cássio F (2004) Effects of zinc on leaf decomposition by fungi in streams: studies in microcosms. Microb Ecol 48:366–374
62. Abel TH, Bärlocher F (1984) Effects of cadmium on aquatic hyphomycetes. Appl Environ Microbiol 48:245–251
63. Moreirinha C, Duarte S, Pascoal C et al (2010) Effects of cadmium and phenanthrene mixtures on leaf-litter decomposition and associated aquatic fungi. Arch Environ Contam Toxicol 61(2):211–219
64. Medeiros AO, Duarte S, Pascoal C et al (2010) Effects of Zn, Fe and Mn on leaf litter breakdown by aquatic fungi: a microcosm study. Int Rev Hydrobiol 95:12–26
65. Sridhar KR, Krauss G, Bärlocher F et al (2000) Fungal diversity in heavy metal polluted waters in central Germany. In: Hyde KD, Ho WH, Pointing SB (eds) Aquatic mycology across the Millenium. Fungal Diversity Press, Hong Kong
66. Pascoal C, Marvanová L, Cássio F (2005) Aquatic hyphomycete diversity in streams of Northwest Portugal. Fungal Divers 19:109–128
67. Seena S, Pascoal C, Marvanová L et al (2010) DNA barcoding of fungi: a case study using ITS sequences for identifying aquatic hyphomycete species. Fungal Divers 44:77–87
68. Fernandes I, Pascoal C, Cássio F (2011) Intraspecific traits change biodiversity effects on ecosystem functioning under metal stress. Oecologia 164(4):1019–1028
69. Hodkinson M (1976) Interactions between aquatic fungi and DDT. In: Jones EBG (ed) Recent advances in aquatic mycology. Wiley, New York
70. Chandrashekar KR, Kaveriappa KM (1994) Effects of pesticides on sporulation and germination of conidia of aquatic hyphomycetes. J Environ Biol 15:315–324

71. Bärlocher F, Premdas PD (1988) Effects of pentachlorophenol on aquatic hyphomycetes. Mycologia 80:135–137
72. Junghanns C, Möder M, Krauss G et al (2005) Degradation of the xenoestrogen nonylphenol by aquatic fungi and their laccases. Microbiology 151:45–57
73. Martin C, Moeder M, Daniel X et al (2007) Biotransformation of the polycyclic musks HHCB and AHTN and metabolite formation by fungi occurring in freshwater environments. Environ Sci Technol 41:5395–5402
74. Junghanns C, Krauss G, Schlosser D (2008) Potential of fungi derived from diverse freshwater environments to decolourise synthetic azo and anthraquinone dyes. Bioresour Technol 99:1225–1235
75. Augustin T, Schlosser D, Baumbach R et al (2006) Biotransformation of 1-naphthol by a strictly aquatic fungus. Curr Microbiol 52:216–220
76. Freeman C, Lock MA (1995) The biofilm polysaccharide matrix: a buffer against changing organic substrate supply? Limnol Oceanogr 40:273–278
77. Wetzel RG (1993) Microcommunities and microgradients: linking nutrient regeneration, microbial mutualism, and high sustained aquatic primary production. Neth J Aquat Ecol 27:3–9
78. Rier ST, Kuehn KA, Francoeur SN (2007) Algal regulation of extracellular enzyme activity in stream microbial communities associated with inert substrata and detritus. J N Am Benthol Soc 26:439–449
79. Haack TK, McFeters GA (1982) Microbial dynamics of an epilithic mat community in a high alpine stream. Appl Environ Microbiol 43:702–707
80. Kaplan LA, Bott TL (1989) Diel fluctuations in bacterial activity on streambed substrata during vernal algal blooms: effects of temperature, water chemistry, and habitat. Limnol Oceanogr 34:718–733
81. Kühl M, Glud RN, Ploug H et al (1996) Microenvironmental control of photosynthesis and photosynthesis-coupled respiration in an epilithic syanobacterial biofilms. J Phycol 32:799–812
82. Cole JJ (1982) Interactions between bacteria and algae in aquatic ecosystems. Annu Rev Ecol Systemat 13:291–314
83. Croft MT, Lawrence AD, Raux-Deery E et al (2005) Algae acquire vitamin B12 through a symbiotic relationship with bacteria. Nature 438:90–93
84. Lawrence JR, Chenier MR, Roy R et al (2004) Microscale and molecular assessment of impacts of nickel, nutrients, and oxygen level on structure and function of river biofilm communities. Appl Environ Microbiol 70:4326–4339
85. Sobczak WV, Burton TM (1996) Epilithic bacterial and algal colonization in a stream run, riffle, and pool: a test of biomass covariation. Hydrobiologia 332:159–166
86. Findlay S, Howe K (1993) Bacterial-algal relationships in streams of the Hubbard brook experimental forest. Ecology 74:2326–2336
87. Neely RK (1994) Evidence for positive interactions between epiphytic algae and heterotrophic decomposers during the decomposition of *Typha latifolia*. Arch Hydrobiol 129:443–457
88. Espeland EM, Wetzel RG (2001) Complexation, stabilization, and UV photolysis of extracellular and surface-bound glucosidase and alkaline phosphatase: implications for biofilm microbiota. Microb Ecol 42:572–585
89. Romaní AM, Guasch H, Muñoz I et al (2004) Biofilm structure and function and possible implications for riverine DOC dynamics. Microb Ecol 47:316–328
90. Scott JT, Back JA, Taylor JM et al (2008) Does nutrient enrichment decouple algal–bacterial production in periphyton? J N Am Benthol Soc 27:332–344
91. Ylla I, Borrego C, Romaní AM et al (2009) Availability of glucose and light modulates the structure and function of a microbial biofilms. FEMS Microbiol Ecol 69:27–42
92. Van Rensen JJS (1989) Herbicides interacting with photosystem II. In: Dodge AD (ed) Herbicides and plant metabolism. Cambridge University Press, Cambridge

93. Proia L, Morin S, Peipoch M et al (2011) Resistance and recovery of stream biofilms to Triclosan and Diuron pulses. Sci Total Environ 409:3129–3137
94. Ricart M, Barceló D, Geiszinger A et al (2009) Effects of low concentrations of the phenylurea herbicide diuron on biofilm algae and bacteria. Chemosphere 76:1392–1401
95. Lopez-Doval JC, Ricart M, Guasch H et al (2010) Does grazing pressure modify diuron toxicity in a biofilm community? Arch Environ Contam Toxicol 58:955–962
96. Pesce S, Fajon C, Bardot C et al (2006) Effects of the phenylurea herbicide diuron on natural riverine microbial communities in an experimental study. Aquat Toxicol 78:303–314
97. Tlili A, Dorigo U, Montuelle B et al (2008) Responses of chronically contaminated biofilms to short pulses of diuron. An experimental study simulating flooding events in a small river. Aquat Toxicol 87:252–263
98. Capdevielle M, Van Egmond R, Whelan M et al (2008) Consideration of exposure and species sensitivity of triclosan in the freshwater environment. Integr Environ Assess Manag 4:15–23
99. Franz S, Altenburger R, Heilmeier H et al (2008) What contributes to the sensitivity of microalgae to triclosan? Aquat Toxicol 90:102–108
100. Lawrence JR, Zhu B, Swerhone GDW et al (2009) Comparative microscale analysis of the effects of triclosan and triclocarban on the structure and function of river biofilm communities. Sci Total Environ 407:3307–3316
101. Wilson BA, Smith V, Denoyelles F Jr et al (2003) Effects of three pharmaceutical and personal care products on natural freshwater algal assemblages. Environ Sci Technol 37:1713–1719
102. Barranguet C, Van den Ende FP, Rutgers M et al (2003) Copper-induced modifications of the trophic relations in riverine algal-bacterial biofilms. Environ Toxicol Chem 22:1340–1349
103. Dorigo U, Leboulanger C, Bèrard A et al (2007) Lotic biofilm community structure and pesticide tolerance along a contamination gradient in a vineyard area. Aquat Microb Ecol 50:91–102
104. Brake SS, Hasiotis ST (2010) Eukaryote-dominated biofilms and their significance in acidic environments. Geomicrobiol J 27:534–558
105. Romaní AM, Fischer H, Mille-Lindblom C et al (2006) Interactions of bacteria and fungi on decomposing litter: differential extracellular enzyme activities. Ecology 87:2559–2569
106. Gulis V, Stephanovich AI (1999) Antibiotic effects of some aquatic hyphomycetes. Mycol Res 103:111–115
107. Pascoal C, Cássio F, Nikolcheva LG et al (2010) Realized fungal diversity increases functional stability of leaf-litter decomposition under zinc stress. Microb Ecol 59:84–93
108. Roussel H, Chauvet E, Bonzom JM (2008) Alteration of leaf decomposition in copper-contaminated freshwater mesocosms. Environ Toxicol Chem 27:637–644
109. Bermingham S, Fisher PJ, Martin A et al (1998) The effect of the herbicide mecoprop on *Heliscus lugdunensis* and its influence on the preferential feeding of *Gammarus pseudolimnaeus*. Microb Ecol 35:199–204
110. Bundschuh M, Hahn T, Gessner MO et al (2009) Antibiotics as a chemical stressor affecting an aquatic decomposer-detritivore system. Environ Toxicol Chem 28:197–203
111. Joubert LM, Wolfaardt GM, Botha A (2006) Microbial exopolymers link predator and prey in a model yeast biofilm system. Microb Ecol 52:187–197
112. Matz C, Kjelleberg S (2005) Off the hook-how bacteria survive protozoan grazing. Trends Microbiol 13:302–307
113. Friberg-Jensen UL, Wendt-Rasch WP et al (2003) Effects of the pyrethroid insecticide, cypermethrin, on a freshwater community studied under field conditions. I. Direct and indirect effects on abundance measures of organisms at different trophic levels. Aquat Toxicol 63:357–371
114. Mortimera M, Kasemets K, Kahrua A (2010) Toxicity of ZnO and CuO nanoparticles to ciliated protozoa *Tetrahymena thermophila*. Toxicology 269:182–189

115. Rico D, Martín-González A, Díaz S et al (2009) Heavy metals generate reactive oxygen species in terrestrial and aquatic ciliated protozoa. Comp Biochem Physiol C Toxicol Pharmacol 149:90–96
116. St. Denis CH, Pinheiro MDO, Power ME et al (2010) Effect of salt and urban water samples on bacterivory by the ciliate, *Tetrahymena thermophila*. Environ Pollut 158:502–507
117. Rehman A, Shakoori FR, Shakoori AR (2008) Heavy metal resistant freshwater ciliate, *Euplotes mutabilis*, isolated from industrial effluents has potential to decontaminate wastewater of toxic metals. Bioresour Technol 99:3890–3895
118. Rabiet M, Margoum C, Gouy V et al (2010) Assessing pesticide concentrations and fluxes in the stream of a small vineyard catchment—effect of sampling frequency. Environ Pollut 158:737–748
119. Sabater S, Tockner K (2010) Effects of hydrologic alterations on the ecological quality of river ecosystems. In: Sabater S, Barceló D (eds) Water scarcity in the Mediterranean: perspectives under global change. Springer, Berlin
120. Balvanera P, Pfisterer AB, Buchmann N et al (2006) Quantifying the evidence for biodiversity effects on ecosystem functioning and services. Ecol Lett 9:1146–1156
121. Hector A, Bagchi R (2007) Biodiversity and ecosystem multifunctionality. Nature 448:188–191
122. Clements WH, Newman MC (2002) Community ecotoxicology. Wiley, Chichester
123. Solé M, Fetzer I, Wennrich R et al (2008) Aquatic hyphomycete communities as potential bioindicators for assessing anthropogenic stress. Sci Total Environ 389:557–565
124. Cheng ZL, Andre P, Chiang C (1997) Hyphomycetes and macroinvertebrates colonizing leaf litter in two belgian streams with contrasting water quality. Limnetica 13:57–63
125. Morin S, Proia L, Ricart M et al (2010) Effects of a bactericide on the structure and survival of benthic diatom communities. Vie Milieu 60:107–114
126. Cazelles B, Fontvieille D, Chau NP (1991) Self-purification in a lotic ecosystem: a model of dissolved organic carbon and benthic microorganisms dynamics. Ecol Model 58:91–117
127. Lawrence JR, Zhu B, Swerhone GDW et al. (2008) Community-Level Assessment of the Effects of the Broad-Spectrum Antimicrobial Chlorhexidine on the Outcome of River Microbial Biofilm Development. Appl Environ Microb 74:3541–3550
128. DeLorenzo ME, Lauth J, Pennington PL et al. (1999) Atrazine effects on the microbial food web in tidal creek. Aquat Toxicol 46:241–251

The Use of Photosynthetic Fluorescence Parameters from Autotrophic Biofilms for Monitoring the Effect of Chemicals in River Ecosystems

Natàlia Corcoll, Marta Ricart, Stephanie Franz, Frédéric Sans-Piché, Mechthild Schmitt-Jansen, and Helena Guasch

Abstract Photosynthetic processes play a key role in aquatic ecosystems. These processes are highly sensitive to the presence of toxicants, leading to an increase in their use as ecotoxicological endpoints. The use of chlorophyll-*a* fluorescence techniques to assess the impact of toxicants on the photosynthesis of the autotrophic component of fluvial biofilms has increased in the last decades. However, these photosynthetic endpoints are not currently used in water quality monitoring programs.

A review of the currently available literature—including studies dealing with toxicity assessment of both priority and emerging compounds—allowed the discussion of the pros and cons of their use as ecotoxicological endpoints in fluvial systems as well as their inclusion in regular monitoring programs.

Chlorophyll-*a* fluorescence measurements have the ability to detect effects of a large panel of chemical substances on the photosynthetic processes of fluvial biofilms, covering both functional and structural aspects of the biofilm community. Moreover, they might provide early warning signals of toxic effects.

Thus, the application of the chlorophyll-*a* fluorescence measurement is recommended as a complementary measurement of toxic stress in aquatic ecosystems.

N. Corcoll (✉) • H. Guasch
Department of Environmental Sciences, Institute of Aquatic Ecology, University of Girona, Campus de Montilivi, 17071 Girona, Spain
e-mail: natalia.corcoll@udg.edu

M. Ricart
Department of Environmental Sciences, Institute of Aquatic Ecology, University of Girona, Campus de Montilivi, 17071 Girona, Spain

Catalan Institute for Water Research (ICRA), Scientific and Technologic Park of the University of Girona, 17003 Girona, Spain

S. Franz • F. Sans-Piché • M. Schmitt-Jansen
Department of Bioanalytical Ecotoxicology, Helmholtz-Centre for Environmental Research-UFZ, Permoserstr. 15, 04318 Leipzig, Germany

H. Guasch et al. (eds.), *Emerging and Priority Pollutants in Rivers*,
Hdb Env Chem (2012) 19: 85–116, DOI 10.1007/978-3-642-25722-3_4,

Their application is of special interest in the context of the Water Framework Directive (WFD, Directive 2000/60/EC), where the development of new structural and functional endpoints of the biological quality elements (e.g., biofilms) is required.

Keywords Biofilms • Chl-*a* fluorescence parameters • Emerging substances • Priority substances • Rivers

Contents

Abbreviations

AL	Actinic light
BQE	Biological quality element
F	Fluorescence yield at the maximal reduced state
Fe	Ferrodoxin
Fm	Maximal fluorescence yield
Fm′	Fluorescence yield at actinic light steady state
Fo	Minimal fluorescence yield
Fo(Bl)	Fluorescence signal linked to cyanobacteria group
Fo(Br)	Fluorescence signal linked to diatoms' algal group
Fo(Gr)	Fluorescence signal linked to green-algae algal group
Fo/Fv	Efficiency of the water-splitting apparatus of PSII
Fo′	Fluorescence yield when actinic light is omitted
Fv	Variable fluorescence yield
Fv/2	Fluorescence measurment of plastoquinone pool
ML	Measuring light
NPQ	Non-photochemical quenching without measuring Fo′
PAM	Pulse amplitude modulated
PEA	Plant efficiency analyzer
Pheo	Pheophytin
PQ	Plastoquinone pool
PQ_A	Plastoquinone A

PQ_B	Plastoquinone B
PS	Photosystem
PSI	Photosystem I
PSII	Photosystem II
qN	Non-photochemical quenching
qP	Photochemical quenching
SP	Saturation pulse
UQD_{rel}	Relative unquenched fluorescence
WFD	Water framework directive
Φ'_{PSII}	Effective quantum yield of PSII
Φ_{PSII}	Maximal quantum yield of PSII

1 Introduction

Photosynthesis is a major process for all illuminated ecosystems as it provides the main source of organic material for the food chain. Autotrophic organisms carry out this physiological process by converting the light energy into chemical energy, building up organic molecules out of CO_2 and water [1]. Derived organic matter serves as food of the heterotrophic organisms. Understanding the physiological/photochemical processes supporting this key function was focus of research for hundreds of years, and several measuring techniques were developed based on chemical analysis of pigments, ^{14}C fixation, oxygen production, or chl-*a* fluorescence-based methods, among others. The latter methods provide a good basis for application in monitoring algal density or photosynthetic processes in terms of electron transport activity and energy dissipation processes [2].

The use of chl-*a* fluorescence techniques to assess photosynthesis performance under changing environmental conditions has widely been proved as a rapid, noninvasive, reliable method [3, 4]. Since the first acknowledgement of the analytical potential of chl-a fluorescence techniques [5, 6] until now, extensive research has been carried out to apply this technique in different research fields. In the case of ecotoxicology, chl-*a* fluorescence techniques have been used to evaluate the toxicity of different pollutants and to locate their primary sites of damage in photosynthetic organisms.

This review focuses on the use of chl-*a* fluorescence techniques to assess chemical effects on the autotrophic component of fluvial biofilms, a nontarget community present in all aquatic systems. Fluvial biofilms are made up mainly of algae, bacteria, fungi, and other micro- and meiofauna organisms, embedded in a matrix of extracellular polymeric substances [7]. In contrast to other autotrophic groups, such as phytoplankton, biofilms have the particularity to live immobile; therefore they are suitable for long-term monitoring and allow chemical toxicity to be assessed at community level. This community approach is much closer to the processes of an ecosystem than the use of single species tests and uses a biological quality element (BQE) [8, 9], which is regularly monitored within the Water Framework Directive (WFD, Directive 2000/60/EC) for the assessment of the ecological status of water systems. Biofilms have the capacity to modify the transport and accumulation of substances such as nutrients [10] as well as organic toxicants and heavy metals [11]. Several studies have highlighted the sensitivity of these communities to a large panel of toxicants such as heavy metals [12–14], herbicides [15–19], pharmaceuticals, and personal care products [20–23]. Due to this sensitivity, fluvial biofilms can be used as early warning systems for the detection of the effects of toxicants on aquatic systems [24]. The pertinence of the use of chl-*a* fluorescence techniques for toxicity assessment on fluvial biofilms is attributed to the basis that if a chemical produces effects on the photosynthesis processes (in a direct or indirect way) the chl-*a* fluorescence parameters will reflect it. In the last decade several investigations followed this approach [12, 15, 18, 20, 25–31, 33].

Based on the chl-*a* fluorescence techniques mainly two types of analysis of fluorescence have been developed to assess photosynthesis performance: (a) the fast fluorescence induction kinetics, measured by plant efficiency analyzer (PEA) instruments, and (b) the slow fluorescence induction kinetics, measured by pulse amplitude modulated (PAM) fluorometers. Although the first provides interesting information on various steps of photosynthetic electron transport [32], it has not been applied to complex samples such as fluvial biofilms.

Several PAM fluorescence instruments are provided to assess the photosynthetic performance in biofilms; each one presenting a specific characteristic: (1) Standard PAM fluorometers excite chlorophyll fluorescence at one wavelength (665 nm, excitation maximum of the chlorophyll-*a* molecule) and have been applied on biofilms to evaluate the global photosynthetic response of the autotrophic compartment (e.g., [15, 25]). PAM fluorometers are available in different technical settings (Fa. Walz, Effeltrich, Germany). Next to standard applications working with cuvettes, a microscopy PAM fluorometer is available, suitable for the assessment of selected cells within a biofilm on a microscopic scale [33]. The Maxi-Imaging PAM fluorometer was developed to assess the photosynthetic capacity of large surfaces, e.g., leaves or multiwell plates and is also suitable for measuring biofilms [34, 35]. (2) Multiwavelength excitation PAM fluorometers (e.g., Phyto-PAM) present the singularity to work with several excitation wavelengths, exciting pigments with different absorption spectra, e.g., carotenoids, which are characteristic for defined algal classes. After deconvolution of the fluorescent signals from mixed algal samples the multiwavelength PAM fluorometry has the potential to reveal the contribution of algal groups with different absorption spectra [36].

Different sources of chemical contamination (industrial, agricultural, or urban activities) discharge a wide variety of compounds with different modes of action (MoA), toxic concentrations, persistence into the ecosystem (accumulation, degradation), etc.

The WFD (2000/60/EC) defines a strategy for protecting and restoring clean waters across Europe. As a first step of this strategy, a list of priority substances was adopted in the Directive 2008/105/EC, in which 33 substances of priority concern were identified and regulated. The list includes mainly organic contaminants, such as pesticides, and four toxic metals [37]. Besides this recognized contaminants, more substances are being detected in the environment. The so-called emerging contaminants are compounds that are not currently covered by existing water quality regulations and are thought to be potential threats to environmental ecosystems [38]. The WFD follows two different assessment strategies: the chemical status evaluation and the ecological status. The "ecological status" represents the "quality of the structure and functioning of aquatic ecosystems associated with surface waters" (Directive 2000/60/EC). In order to establish the "ecological status," the WFD requires the sampling and interpretation of data on a broad suite of "BQEs." The WFD required that observed metric values for BQEs in a water body undergoing monitoring were mathematically compared with expected values for reference condition sites based on predictive modeling, hind casting, or expert judgment [39]. Biofilms, referred to as periphyton in the WFD, have been included as one major BQE. Concerning this BQE, the WFD recognized the diatom index within biofilms to evaluate structural effects, which has been widely applied for water managers. For non-diatom species from periphyton there is not a recognized parameter to be evaluated.

This chapter introduces the measuring principles of chl-*a* fluorescence techniques and summarizes studies that assess the toxicity of priority and emerging substances on autotrophic biofilms by using chl-*a* fluorescence techniques. The photosynthesis mechanism has been shown to be very sensitive to several toxic substances such as heavy metals or herbicides [40, 41]. So the photosynthesis process which is essential for the overall survival of phototrophs can be used as an ecotoxicological endpoint to assess the impact of many toxicants [42].

The main aim of this chapter is to analyze the pros and cons of the use of chl-*a* fluorescence techniques for water quality monitoring programs in the context of the WFD. While not being an exhaustive review, 29 different investigations have been analysed, including only field and laboratory investigations dealing about the application of chl-a fluorescence parameters on biofilm communities.

2 Physiological Basis of Photosynthesis

Photoautotrophic organisms have the ability to synthesize organic compounds from CO_2 and water by converting light energy to molecular energy during photosynthesis. Oxygenic photosynthesis is catalyzed in two photosystems (PS): PSI and PSII containing chlorophyll molecules that are embedded in the thylakoid membrane of the chloroplasts as integral membrane protein complexes. Starting point of photosynthesis is the absorption of photons by the pigment molecules (chl *a*, chl *b*, phycobiliproteins, and carotenoids) of the antennae systems and the transfer of energy of excited molecules to the reaction centers of PSI and PSII (Fig. 1). The central chlorophyll molecule of the PSII, the first of the two photosystems activated during this process, is excited by a previously excited molecule, in a type of energy transfer that is called resonance energy transfer or excitation transfer. Promoted by a Mn complex (an enzyme complex), the reaction center of the PSII gets electrons from the cleavage of a water molecule, a reaction that produces O_2 and protons H^+.

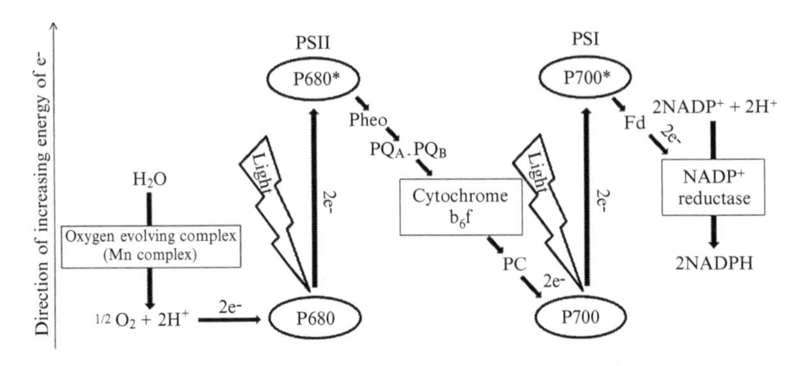

Fig. 1 Schematic overview of the *z*-scheme of electron transport between the two photosystems (PSII and PSI) in the photosynthesis processes

The electrons from the water cleavage are subsequently transferred by a complex system of acceptor and donor molecules: pheophytin (Pheo), plastoquinone A (PQ_A) and B (PQ_B), the plastoquinone pool (PQ), and the cytochrome complex to the PSI (Fig. 1). Finally, the excited molecules are transferred from the PSI to ferredoxin (Fd), another acceptor molecule, which catalyzes the reduction of $NADP^+$, a phosphorylated derivate that carries reducing electrons. These energy-rich products from photosynthesis are later used in the Calvin cycle to build up hexoses and other organic matter [1]. The chl-*a* fluorescence techniques allow to evaluate different photosynthetic processes occurring under light excitation.

3 Chl-*a* Fluorescence Analysis and Derived Parameters

Two types of chl-*a* fluorescence analysis have been developed and applied in ecotoxicological studies: the "fast fluorescence induction kinetics" [32] and the "slow fluorescence induction kinetics" [43].

3.1 *The Fast Fluorescence Induction Kinetics: PEA Fluorometry*

The rapid rise of fluorescence is measured with PEA and provides information on various steps of the photosynthetic electron transport [32]. In the fast fluorescence induction kinetics, the chl-*a* fluorescence transient follows a polyphasic pattern of O–J–I–P electrons transients from the initial fluorescence level (Fo or O) to the maximum fluorescence level (Fm or P) (Fig. 2). The rise from O (at 0.05 ms) to the J phase (at 2 ms) is due to the net photochemical reduction of plastoquinone (PQ_A–PQ_A) (photochemical phase). The J–I phase (at 30 ms) is due to the closure of the remaining photosynthetic centers, and the I–P (ends about at 500 ms) is due to the removal of plastoquinone quenching due to the reduction of PQ (non-photochemical phase) [32]. The decrease in Fm and fluorescence levels at phases

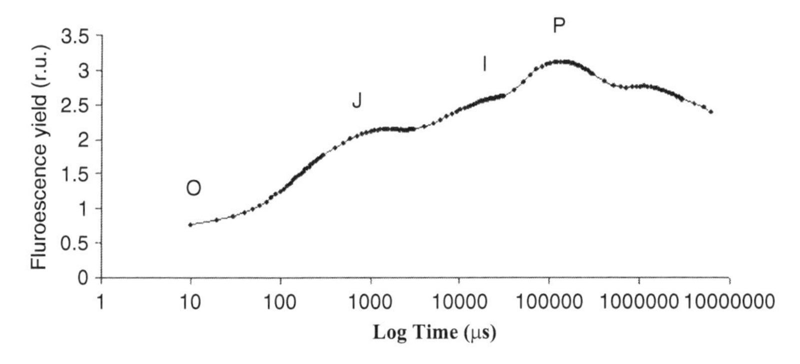

Fig. 2 Typical rapid fluorescence kinetics measured with PEA fluorometer

J and I is normally explained by the inhibition of the electron transport at the electron donor side of PSII, which results in the accumulation of the excited reaction center of PS II (P680*), a strong fluorescence quencher [44]. Thus the fluorescence rise provides information on various steps of photosynthetic electron transport [32]. The kinetics of OJIP transients obtained allows evaluating the toxic effect of chemicals on specific characteristics of PSII as energy trapping processes or antenna size. Besides, by using this fast fluorescence kinetics it is also possible to calculate different photosynthetic parameters and the complementary area (CA) [45]. The CA is a measure of the kinetics of fluorescence induction up to the P level and it has been reported as a direct indicator of the PSII photochemistry [46]. CA has been used for many years to assess the phytotoxicity of pollutants [40]. It is reported that fast fluorescence induction kinetics can also allow the location of the primary site of damage induced by environmental stress [47].

3.2 The Slow Fluorescence Induction Kinetics: Standard PAM Fluorometry

The slow fluorescence induction analyses are carried out by using PAM fluorometry. The so-called saturation pulse quenching analysis based on the principle that light energy absorbed by PSII pigments can drive the photochemical energy conversion at PSII reaction centers, be dissipated into heat, or be emitted in the form of chl *a* fluorescence. As these three pathways of energy conversion are complementary, the fluorescence yield may serve as a convenient indicator of time- and state-dependent changes in the relative rates of photosynthesis and heat dissipation. The PAM fluorescence method employs a combination of three different types of light: modulated or measuring light (ML) = 0.05 μmols photons m^{-2} s^{-1} in μsec pulses; actinic light (AL) = from 1 to 600 μmols photons m^{-2} s^{-1}; and strong saturation pulses (SP) = 8,000 μmols photons m^{-2} s^{-1} (Fig. 3), which allow the adequate analysis of the fluorescence induction kinetics of photosynthetic organisms.

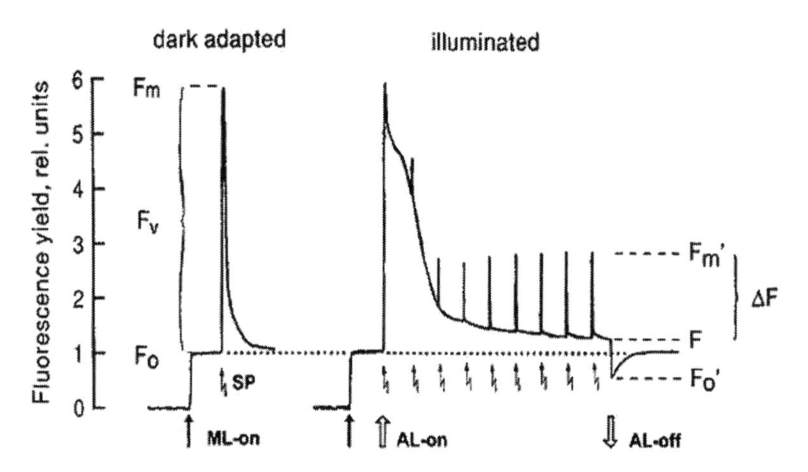

Fig. 3 Schematic representation of the slow fluorescence kinetics analysis by using the PAM fluorometry. The different types of light are indicated (*ML* measuring light, *AL* actinic light, *SP* saturation pulse). Modified from Schreiber [89]

By PAM fluorometry we can get two different types of measures: a biomass-related ones (based on basal fluorescence emission) and a functional ones (based on changes in fluorescence emission caused by strong saturation pulses).

During PAM fluorescence analysis, different information can be obtained from a dark- or light-adapted sample.

3.2.1 Chlorophyll Fluorescence Characteristics in the Dark-Adapted State

In a dark-adapted phototropic organism modulated light with low energy is unable to trigger electron transport; then the small fluorescence yield corresponds to the constant fluorescence (Fo), which represents the light dissipation by excited antennae of Chl-*a* molecules before the excitation energy is transferred to the reaction centers of PSII [48–50].

The maximal fluorescence yield (Fm) is observed, when all PSII reaction centers are closed after the application of a strong saturation pulse. The difference between Fm and Fo is the maximal variable fluorescence yield (Fv). Fv/Fm is used as a measure of the maximal photochemical efficiency of PSII (Fig. 3).

3.2.2 Chlorophyll Fluorescence Characteristics in the Light-Adapted State

Actinic light maintains the photosynthetic process in an active state, providing the appropriate conditions to analyze fluorescence kinetics. The maximum fluorescence yield, reached under this condition, will determine the fluorescence transient (F), where electron transport carriers are at a maximum reduced state. The maximum variable fluorescence induced by a set of saturating flashes at a steady state of fluorescence is represented by Fm′. The difference between Fm′ and F indicates the variable fluorescence induced by actinic light at a steady state of electron transport (ΔF). The Fo′ is the constant fluorescence when actinic light is omitted (Fig. 3).

3.2.3 Parameters Used and Their Interpretation

Deriving from these dark and light chl-*a* fluorescence measurements several parameters can be obtained (Table 1) and applied in ecotoxicological studies, to assess the "health" status of the photosynthetic organisms through photosynthetic activity measurements.

The *minimal fluorescence yield (Fo)* reflects the chl-*a* fluorescence emission of all open reaction centers in a non-excited status. Fo can be used as a surrogate of algal biomass since chlorophyll fluorescence is proportional to total chlorophyll content [51, 52]. It is expected that Fo will decrease if toxic exposure causes a reduction in the number of cells due to cell death (structural damage) or the chlorophyll content of a sample. Biomass estimations based on Fo are not always possible, losing linearity above a given biomass level if the thickness of the biofilms is excessive producing self-shading [15, 53].

Table 1 Fluorescence parameters obtained by using PAM fluorometry

Parameter	Name	Equation	References
Φ_{PSII}	Maximal quantum yield	$\Phi_{PSII} = (Fm - Fo)/Fm$	Schreiber et al. [49] and Genty et al. [54]
Φ'_{PSII}	Effective quantum yield	$\Phi'_{PSII} = (Fm' - F)/Fm'$	Schreiber et al. [49] and Genty et al. [54]
qP	Photochemical quenching	$qP = (Fm' - F)/(Fm' - Fo')$	Schreiber et al. [49], Horton et al. [57], and Müller et al. [90]
qN	Non-photochemical quenching	$qN = 1 - [(Fm' - Fo')/(Fm - Fo)$	Schreiber et al. [49], Horton et al. [94], and Müller et al. [90]
NPQ	Non-photochemical quenching	$NPQ = (Fm - Fm')/Fm'$	Bilger and Björkman [91]
UQF_{rel}	Relative unquenched fluorescence	$UQF_{rel} = (F - Fo')/(Fm' - Fo')$	Juneau et al. [60]
Fv/2	Plastoquinone pool	$Fv/2 = (Fm - Fo)/2$	Bolhàr-Nordenkampt and Öquist [92]
Fo/Fv	Efficiency of the water-splitting apparatus of PSII	$Fo/Fv = Fo/(Fm - Fo)$	Kriedemann et al. [93]
Fo	Minimal fluorescence yield		Serôdio et al. [52] and Rysgaard et al. [51]

The *maximal or optimal quantum yield* (Φ_{PSII}) measures the quantum yield of PSII electron transport in a dark-adapted state (Table 1). It has been shown in many studies that Φ_{PSII} can also be a measure of the quantum yield of photosynthesis [4, 54]. It is an estimate of the potential maximal photosynthetic activity and it is expected to be effected if a chemical produces alterations in the structure of the photosynthetic apparatus (e.g., shade-adapted chloroplasts); in general this situation occurs when biofilms are exposed to chemicals at high concentrations or during a long-term exposure. The Φ_{PSII} parameter could be named with other nomenclatures or names in the bibliography: Φ_o [55], Fv/Fmax [18], photosynthetic capacity [20], or photosynthetic activity [21]. In order to standardize nomenclatures we recommend the use of Φ_{PSII}.

The *effective or operation quantum yield* (Φ'_{PSII}) measures the efficiency of excitation energy capture by the open PSII reaction centers under light conditions [49, 54], proportional to photosynthetic efficiency. A reduction of the Φ'_{PSII} indicates that the toxicant is reducing electron flow in the PSII. This parameter is very sensitive to PSII-inhibiting herbicides, if they block the electron transport flow (Table 1). The Φ'_{PSII} parameter could be named with other nomenclatures or names in the bibliography: Φ_{II} [25], Φ_{PSII} [56], yield or Y [28], photon yield [26], photosynthetic efficiency at PSII or Yeff [29], or Yield II [53]. In order to standardize nomenclatures the use of Φ'_{PSII} is recommended.

The *photochemical quenching (qP)*, which is determined by the redox state of Q_A, the primary electron acceptor of PSII [4], represents the proportion of excitation energy "trapped" by open PSII reaction centers used for electron transport [57] (Table 1).

The *non-photochemical quenching (qN)* reflects the amount of light energy dissipation inducing fluorescence quenching that involves nonradiative energy process [57] (Table 1).

The *non-photochemical quenching without measuring Fo′ (NPQ)* is a simplified non-photochemical quenching value, which assumed that NPQ is caused only by one quenching factor [58], omitting other energy-consuming processes, not directly involved in the PSII activity [59] (Table 1).

The *relative unquenched fluorescence* (UQD_{rel}) is a parameter proposed by Juneau et al. [60]. It is complementary to the relative quenching components qP_{rel} and qN_{rel} proposed by Buschmann [58] that take into account the fraction of the non-quenched fluorescence yield related to the proportion of closed PSI reactive centers present under continuous irradiation (Table 1).

The *plastoquinone pool (Fv/2)* is a measure of the state of the pool of plastoquinones. Its reduction was linked with inhibitory effects of metals on the photosynthetic efficiency [61] (Table 1).

The *efficiency of the water-splitting apparatus of PSII (Fo/Fv)* reflects the state of water-splitting this complex. Metal exposure may damage this apparatus [61] (Table 1).

3.2.4 Multiwavelengths PAM Fluorometry

The PAM fluorescence methodology is now available and applied in several technical settings to monitor the influence of stress factors on microalgae photosynthesis (e.g., [95, 33, 36, 62]). In principle, this technique allows the researcher to address different levels of biological complexity using the same fluorescence parameters and was used in a comparative study by Schmitt-Jansen and Altenburger [33]. For uniform suspensions of unicellular algae measurement of variable chl-*a* fluorescence is a relatively straightforward exercise. Schreiber et al. [36] used an ultrasensitive dual-channel chlorophyll fluorometer for the assessment of diuron, deriving detection limits, sufficient for the demands of the European Commission drinking water regulation (0.1 μg/L for each individual substance).

Observation of variable fluorescence in biofilm communities, however, requires distinction between the contributions from the different algal components with respect to their differences in pigmentation and photosynthetic properties. Recently, a microscopic setup became available enabling PAM chl-*a* fluorescence measurements at a microscopic scale that allows excitation and detection of fluorescence from single algal cells [95]. Therefore, the noninvasive assessment of individual stress responses of cells in biofilms became possible. Another way to address this challenge is to simultaneously use excitation light of different wavelengths for excitation of the algal class-specific light harvesting complexes. For instance, Chl *b*, which characterizes chlorophytes as a key pigment, shows absorption peaks at 470 and 645 nm. Diatoms which are characterized by the key pigments Chl *c* and carotenoids, especially fucoxanthin, can be studied by fluorescence excitation in the blue and green spectral range (400–665 nm). On the other hand, in the case of cyanobacteria, fluorescence excitation in the blue and green is weak, while strong excitation is observed around 620–640 nm, in the absorption range of phycocyanin/allophycocyanin.

Using a four-wavelength excitation method, a differentiation of photosynthetic parameters of microalgal communities between different spectral groups, e.g., chlorophytes, cyanophytes, and diatoms, seems possible [62]. This application employs an array of light-emitting diodes (LED) to excite chlorophyll fluorescence at different wavelengths (470, 520, 645, and 665 nm). This particularity gives the opportunity to deconvolute the overall fluorescence signal into the contributions of algal groups [cyanobacteria-Fo(Bl), diatoms-Fo(Br), and green algae-Fo(Gr)] from community samples, e.g., phytoplankton or biofilms, based on the internal "reference excitation spectra" of a pure culture [36]. This application has been validated in fluvial biofilms by Schmitt-Jansen and Altenburger [53]. Besides the provision of a relative measure of the fluorescence abundance of each algal group it is possible to obtain values from fluorescence parameters specific of each phototrophic group (green algae, diatoms, or cyanobacteria) present in the community.

Beutler et al. [63] successfully applied a five-wavelength fluorometer to characterize a depth profile of the plankton algal community of Lake Plußsee, Germany.

4 The Use of Fluorescence Parameters to Assess the Effects of Toxicants on Biofilms

Chl-*a* fluorescence techniques based on PAM fluorometry have commonly been used to assess the effects of chemicals on biofilms. More precisely, photosystems I and II were recognized to be a good target for pesticides to control weed in agriculture; therefore several herbicides were developed, blocking photosystems, like the triazines or the phenylurea herbicides. The usefulness of fluorescence parameters derived after multiturnover flashes has been demonstrated for herbicides specifically inhibiting PSII activity in several studies [18, 19, 64–66]. On the other hand, fast fluorescence techniques measured by plant efficient analysis (PEA) have never been applied to assess chemical toxicity on biofilms. However, its use on algal monocultures is well reported [40, 45, 67, 68]. The information provided by PEA flurometers is different to that obtained by PAM fluorometry; the first allows evaluating the toxic effects on PSII energy trapping and PSII antenna size. For instance, Dewez et al. [68] assessed Cu and fludioxonil effects on *Selenastrum obliquus* by using PAM and PEA photosynthetic methods as well as by evaluating different antioxidant enzymatic activities. They observed that Cu may alter the energy storage in algae during photosynthesis but fludioxonil inhibitory effect appeared not to be directly associated with photosynthetic electron transport as seen for copper inhibition. By using the PEA method it was observed that the fluorescence yields at J, I, and P transients were quenched under Cu exposure (1–3 mg/L) and the authors attributed this disturbance on the energy trapping to an inhibition of the PSII electrons transport via PQ_A, PQ_B, and the plastoquinone pool. Also, they observed that copper altered the energy dissipation via the non-photochemical energy dissipation (by PAM method) and the change of the structural organization of the antenna size via PSII light harvesting complex (by PEA method). So, by using PEA and PAM fluorometers different but complementary information was obtained to evaluate deeply Cu toxicity on photosynthesis processes of *S. obliquus*.

Several studies are presented in order to illustrate the pros and cons of the application of PAM fluorometry technique on biofilm communities' growth in the laboratory to assess chemical toxicity of compounds with different MoA: PSII inhibitors (Sect. 4.1), other photosynthetic inhibitors (Sect. 4.2), and toxicants with unknown MoA on algae (Sect. 4.3). The use of PAM fluorometry to assess chemical toxicity in field studies is presented in Sect. 4.4. PAM fluorometry gives the opportunity to evaluate both functional and structural alterations in autotrophic organisms. These alterations are related with the MoA of the toxicants as well as the dose and the time of exposure.

4.1 PSII Inhibitors

Pollutants with MoA directly interfering with the PSII are mainly represented by organic herbicides (e.g., diuron, atrazine). Due to their specific MoA, studies dealing with their toxicity commonly use photosynthetic endpoints. The inhibition of the maximal and the effective quantum yield (Φ_{PSII}, Φ'_{PSII}) in biofilms exposed to PSII inhibitors for short periods of time has been already demonstrated. McClellan et al. [34] detected a reduction of the Φ_{PSII} in biofilms exposed to diuron (Table 2). Similar results have been obtained with atrazine, prometryn, and isoproturon ([19, 33, 53]; see Table 2 for detailed information). The sensitivity of biofilms to isoproturon under different light intensities was determined by Laviale et al. [69], showing a clear reduction of photosynthesis (Φ'_{PSII} and Φ_{PSII}), and, under dynamic light conditions, a clear reduction of the non-photochemical quenching (NPQ) mechanisms was observed (Table 2). Effects of PSII inhibitors on the biofilm structure are less reported, especially in the short-term toxicity assessment. Schmitt-Jansen and Altenburger [18] found a 90% reduction of algal biomass after 1 h of exposure to isoproturon (Table 2).

The chl-*a* fluorescence methods have also been applied to long-term toxicity studies. Ricart et al. [29] analyzed the long-term effect of the herbicide diuron on fluvial biofilms and detected a clear decrease of the Φ'_{PSII} effective quantum (Table 2). Ricart et al. [29] analyzed the long-term effect of the herbicide diuron on fluvial biofilms and detected a clear decrease of the effective quantum (Table 2) with increasing concentrations of diuron (Fig. 4), indicating a clear dose-dependent effect, probably attributable to the MoA of the herbicide. A reduction of the Φ_{PSII} of biofilms was also observed by Tlili et al. [70] after 3 weeks of exposure to diuron (Table 2). Schmitt-Jansen and Altenburger [18, 53] used these techniques to evaluate the long-term effects of isoproturon to biofilm structure. They detected a reduction in the algal biomass (measured as Fo) and a shift of the algal classes to a dominance of green algae (Table 2). Effects of diuron on a simple food chain (biofilm and grazers (snail *Physella [Costatella] acuta*)) have also been evaluated using PAM fluorometry techniques. In a long-term experiment, López-Doval et al. [71] used Φ'_{PSII} to monitor the physiological state of the biofilm, as well as to detect the effect of diuron, grazers, and their interaction on the biofilm community. A significant reduction in Φ'_{PSII} was detected in both diuron and diuron + grazers microcosms.

Chl-*a* fluorescence parameters have been used as a physiological endpoint to investigate community adaptation following the pollution-induced community tolerance (PICT) approach [72]. Briefly, the replacement of sensitive species by tolerant ones driven by toxicant's selection pressure is expected to increase the EC_{50} of the community measured in acute dose–response tests using physiological endpoints such as Φ'_{PSII}. PICT of periphyton to isoproturon [18, 53], atrazine, prometryn [19], and diuron [70] has been determined by testing inhibition of Φ'_{PSII} on pre-exposed and non-pre-exposed biofilm communities.

Table 2 Summary of studies applying chl-*a* fluorescence techniques to assess the toxicity of PSII inhibitors on fluvial biofilm communities

Toxicant	Biofilm community	Exposure	Conc.	Φ_{PSII}	Φ'_{PSII}	NPQ	Fo	Fo(Bl)	Fo(Gr)	Fo(Br)	Reference
Diuron	Microcosms	Short-term	0.4–100 μg/L		– – –						McClellan et al. [34]
Diuron	Microcosms	29 days	0.07–7 μg/L		– – –						Ricart et al. [29]
Diuron	Microcosms	3 weeks	10 μg/L	– –							Tlili et al. [70]
			10 μg/L + PO_4^{3-}	n.s.							
Diuron	Microcosms	29 days	2 μg/L		– –						López-Doval et al. [71]
Atrazine	Microcosms	1 h	7.5–2,000 μg/L		– – –						Schmitt-Jansen and Altenburger [19]
Prometryn			5–5,110 μg/L		– – –						
Isoproturon			2.4–3,120 μg/L		– – –						
Isoproturon	Microcosms	1 h	2.4–312 μg/L		– – –				+	– – –	Schmitt-Jansen and Altenburger [18]
Atrazine	Microcosms	1 h	7.5–2,000 μg/L		– – –						Schmitt-Jansen and Altenburger [33]
Isoproturon		1 h	2.4–312 μg/L		– – –						
Atrazine	Microcosms	1 h	7.5–2,000 μg/L		– – –						Schmitt-Jansen and Altenburger [53]
Prometryn			2.5–320 μg/L		– – –						
Isoproturon			2.4–312 μg/L		– – –						
Isoproturon	Microcosms	26 days	2.4–156 μg/L				– –		+ +	– –	
Isoproturon	Microcosms (constant light)	7 h	0–2,000 μg/L	– –	– – –	n.s.					Laviale et al. [69]
	Microscosms (dynamic light)	7 h	2, 6 and 20 μg/L	– –	– – –	+ + +					

The significant responses observed are indicated by "–" for inhibition or by "+" for increase and its approximate magnitude by: "–" or "+" if it was slight (10–30% of control), by: "– –" or "+ +" if it was moderate (30–50% of control) and by "– – –" or "+ + +" if it was high (>50% of control)
n.s. not significant ($p > 0.05$)

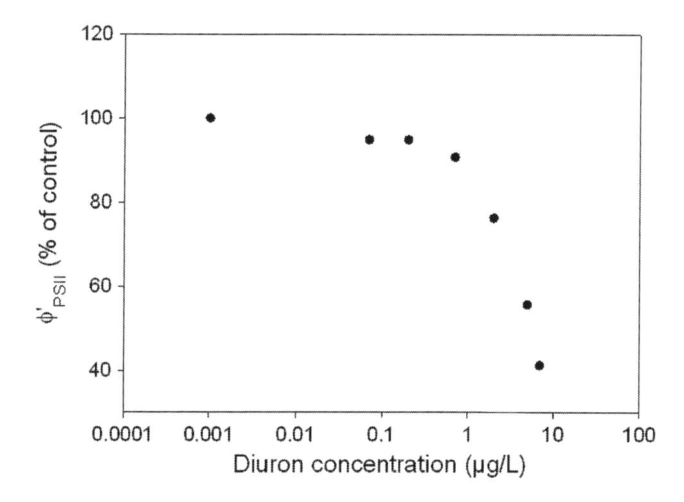

Fig. 4 Diuron effects on effective quantum yield of biofilm communities exposed to increasing concentrations of the herbicide. Modified from Ricart et al. [29]

4.2 Other Photosynthetic Inhibitors

Chl-*a* fluorescence parameters have commonly been used to assess the toxicity of chemicals affecting other processes than PSII electron transport. In these cases, the obtained parameters may indicate secondary effects on the performance of the photosynthetic apparatus or on algal biomass. For instance copper is a phytotoxic chemical affecting the activity of photosystem II and electron transfer rates [61]. However, this metal is also known to cause the oxidation of sulphydryl groups of enzymes leading to their inhibition [73] or to generate reactive oxygen species like superoxide and hydroxyl radicals causing oxidative stress (for a review see [14]) reducing growth as well as photosynthetic and respiratory activities [74]. Several biofilm studies have used copper as model toxic compound (Table 3), applying in many cases chl-*a* fluorescence parameters to assess biological responses. Serra et al. [28] measured Φ'_{PSII} to check the photosynthetic performance of biofilms during the Cu retention experiment that lasted for few hours. Transient inhibition of Φ'_{PSII} was also observed by Corcoll et al. [31] after few hours of Zn exposure; however this parameter recovered after longer exposure (Table 3).

Physiological chl-*a* parameters such as Φ'_{PSII} may not consistently show the effects caused by non-PSII inhibitors after longer exposure due to community adaptation. Barranguet et al. [56] observed that biofilm exposed during 2 weeks to Cu presented alterations in the Fo and that Φ'_{PSII} was less affected. Barranguet et al. [25] observed that the main factor regulating the sensitivity of biofilms to Cu toxicity (based on the Φ'_{PSII} endpoint) during short-term exposures was the physical structure of the biofilm (package of cells and thickness), and not to the species composition. These endpoints were used to show differences in sensitivity between suspended cells and biofilm communities. Fo and Φ'_{PSII} were also used

Table 3 Summary of studies applying chl-*a* fluorescence techniques to assess toxicity on fluvial biofilm communities of compounds with other effects on photosynthesis (different from PSII inhibition)

Toxicant	Biofilm community	Exposure	Conc.	Φ_{PSII}	Φ'_{PSII}	NPQ	Fo	Fo(Bl)	Fo(Gr)	Fo(Br)	References
Cu	Microcosms	24 h; 3 days	445–1,905; 126 µg/L		–/– –		– – –				Barranguet et al. [25, 56]
Zn	Microcosms	72 h	320 µg/L	n.s.	–	n.s	n.s.	n.s.	n.s.	n.s.	Corcoll et al. [31]
		5 weeks	320 µg/L	–	n.s.	– – –	– – –	n.s.	n.s.	–	
Cu	Microcosms	6; 16 days	0–100 µg/L		–		– – –				Guasch et al. [12, 27]
Cu, Zn, Cu + Zn	Microcosms	5 days	636; 6,356 µg/L	–	n.s.			– – –	– – –		García-Meza et al. [55]
Cu	Microcosms	4 weeks	30 µg/L					+ + +	n.s.	– – –	Serra et al. [28]
	no Cu vs. Cu pre-exposure	6 h	100 µg/L		–/n.s.		n.s				
Cu	Microcosms	3 weeks	30 µg/L	–							Tlili et al. [70]
			30 µg/L + PO_4^{3-}	n.s.							

Significant responses observed are indicated by "–" for inhibition or by "+" for increase and its approximate magnitude by "–" or "+" if it was slight (10–30% of control), by "– –" or "+ +" if it was moderate (30–50% of control) and by "– – –" or "+ + +" if it was high (>50% of control)
n.s. not significant ($p > 0.05$)

by Guasch et al. [12] and [55] to assess Cu and Cu plus Zn toxicity, respectively. In both cases Fo appeared to be the most sensitive endpoint, whereas effective concentrations based on Φ'_{PSII} were always much higher (Table 3). In the second example, multiwavelength fluorometers allowed following effects on the biomass of the cyanobacteria and green-algae groups of the biofilm as well as their respective Φ_{PSII}. Similarly, Serra et al. [28] demonstrated that community composition differed among biofilms with different Cu-exposure history. Community changes were identified using multiwavelength fluorometry, whereas Cu adaptation was assessed by comparing the Φ'_{PSII} responses to short Cu exposure (Table 3). Tlili et al. [70] used the Φ_{PSII} to evaluate the role of phosphorus on Cu toxicity on biofilms. They observed that biofilms pre-exposed to Cu presented a lower Φ_{PSII} than biofilms non-pre-exposed to Cu or pre-exposed to Cu plus phosphorus (Table 3).

4.3 Toxicants with Unknown Mode of Action on Algae

The impacts of personal care products and pharmaceuticals on the environment are not generally well known [75] and its effects on fluvial biofilms have poorly been investigated. Chemicals that damage membranes or proteins associated with photosynthetic electron transport or that inhibit any cellular process downstream of PSII, such as carbon assimilation or respiration, will lead to excitation pressure on PSII [76]. Therefore, chl-*a* fluorescence methods can also be used to assess the effects of chemicals affecting metabolic processes not directly linked to photosynthetic electron transport. This approach has recently been used to provide complementary endpoints to evaluate direct and indirect effects of these compounds on biofilm [20, 21, 23].

The acute toxicity of three β-blockers: metoprolol, propranolol, and atenolol on fluvial biofilms was assessed by using several biomarkers, including chl-*a* fluorescence parameters measured with a multiwavelength PAM to evaluate toxic effects on the autotrophic biofilm compartments (Fig. 5) [20]. They observed that propranolol was the most toxic for algae, causing 85% inhibition of Φ'_{PSII} (photosynthetic efficiency). Metoprolol was particularly toxic for bacteria and atenolol affected similarly bacterial and algal compartments of the biofilm but to a lesser extent (Table 4). Moreover, the use of chl-*a* fluorescence parameters for three algal groups gave an interesting insight into the algal compartment (Fig. 5). The relative contribution of the different algal groups was also used to obtain the effective photosynthetic efficiency for each of them, using the fluorescence signal linked to green algae, cyanobacteria, and diatoms. This approach allowed the detection of the earlier sensitivity of cyanobacteria-Fo(Bl) compared to green-Fo(Gr) and brown algae-Fo(Br) to propranolol. On the other hand, atenolol affected green algae and cyanobacteria photosynthetic efficiencies while diatoms seemed resistant to this toxic compound. Consequently, they concluded that the chl-*a* fluorescence parameters are powerful tools within a multibiomarker approach to detect effects on the phototrophic compartment of biofilms indicating potential effects on the community structure.

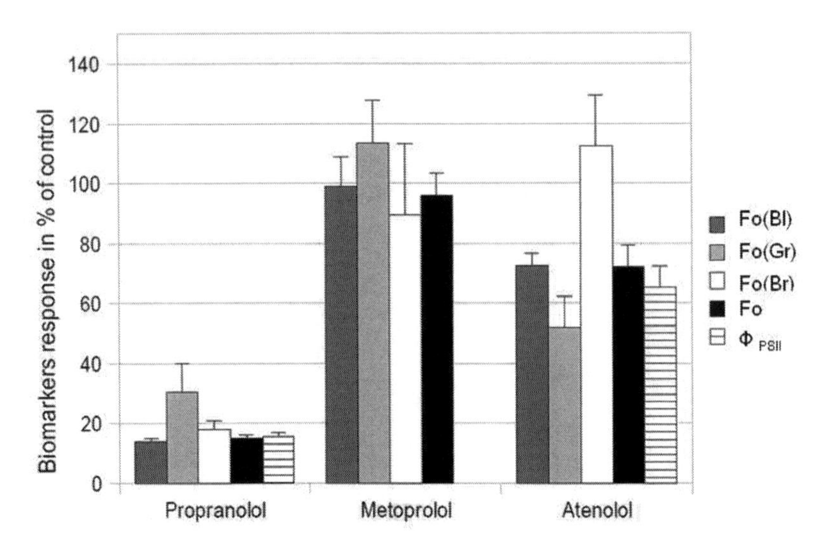

Fig. 5 Response of the minimal fluorescence yield of cyanobacteria [Fo(Bl), *dark gray*], green algae [Fo(Gr), *gray*], brown algae [Fo(Br), *white*] and the whole biofilm (Fo, *black*) and maximal quantum yield (Φ_{PSII}) of all biofilm (horizontal hatches), expressed in percentage of control, after 24 h of exposure to 531 μg/L of propranolol, 522 μg/L of metoprolol and 707,000 μg/L of atenolol. Error bars depict the standard error ($n = 3$). Modified from Bonnineau et al. [20]

Ricart et al. [23] studied the effects of the antimicrobial agent triclosan on fluvial communities. In this case, the PAM techniques were used to assess the effects of this antimicrobial agent on the biotic interaction between algae and bacteria within the biofilm community. The effective quantum yield (Φ'_{PSII}) was progressively reduced with increasing concentrations of toxicant up to 25% (Table 4). The NPQ parameter showed effects up to 70% (Table 4 and Fig. 6), indicating that the NPQ parameter was more sensitive to triclosan toxicity on biofilm than the Φ'_{PSII}. Triclosan effects on algae were also studied by Franz et al. [21] (Table 4). They used the Φ_{PSII} to compare differences in sensitivity to triclosan between algal monocultures and biofilm communities.

4.4 Field Studies

Pollution in fluvial ecosystems is related with the main activities that are carried out in their catchment. As a general rule, agricultural areas are more affected by organic chemicals (herbicides, insecticides, or fungicides), industrial and mining areas by metals, and urban areas by personal care products and pharmaceuticals via sewage treatment plant effluents [75, 77]. However, the predominant scenario found in most polluted rivers is a mixture of different chemicals. From an ecotoxicological point of view, mixture pollution implies that compounds with different chemical

Table 4 Summary of studies applying chl-*a* fluorescence techniques to assess the toxicity of toxicants with an unknown mode of action on fluvial biofilm communities

Toxicant	Biofilm community	Exposure (h)	Conc.	Φ_{PSII}	Φ'_{PSII}	NPQ	Fo	Φ'_{PSII} Fo (Bl)	Φ'_{PSII} Fo (Gr)	Φ'_{PSII} Fo (Br)	Reference
Atenolol	Experimental channels	24	707 mg/L	–	–			–	– –	n.s.	Bonnineau et al. [20]
Propranolol		24	531 μg/L	– – –	– – –			– –	– – –	– – –	
Metoprolol		24	522 μg/L		n.s.			n.s.	n.s.	n.s.	
Triclosan	Aquaria	24	270–17,300 μg/L	– – –							Franz et al. [21]
Triclosan	Experimental channels	48 h	0.05–500 μg/L		–	– – –					Ricart et al. [23]

The significant responses observed are indicated by "–" for inhibition or by "+" for increase and its approximate magnitude by "–" or "+" if it was slight (10–30% of control), by "– –" or "+ +" if it was moderate (30–50% of control) and by "– – –" or "+ + +" if it was high (>50% of control)
n.s. not significant ($p>0.05$)

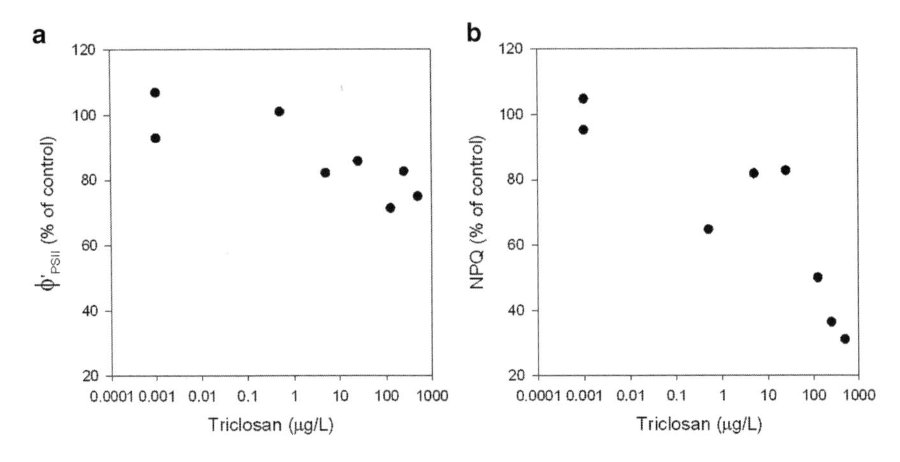

Fig. 6 Effects of triclosan on (**a**) effective quantum yield (Φ'_{PSII}) and (**b**) non-photochemical quenching (NPQ) of biofilms. Modified from Ricart et al. [23]

properties, MoA, and concentrations will coexist. Moreover, environmental factors such as light, nutrients, or flow regime may also influence the fate and effects of chemicals on fluvial communities. Due to its complexity, the effects of toxicants on biofilm communities have been addressed by different methodological approaches including periodic monitoring sampling, translocation experiments, or PICT approaches. In addition, different biofilm targeting endpoints have been used, including, in many cases, chl-*a* fluorescence parameters.

Ricart et al. [30] examined the presence of pesticides in the Llobregat river basin (Barcelona, Spain) and their effects on benthic communities (invertebrates and natural biofilms) (Table 5). Several biofilm metrics, including PAM parameters, were used as response variables to identify possible cause–effect relationships between pesticide pollution and biotic responses. Certain effects of organophosphates and phenylureas in both structural (chl-*a* content and Φ_{PSII}) and functional aspects (Φ'_{PSII}) of the biofilm community were suggested. The authors concluded that complemented with laboratory experiments, which are needed to confirm causality, this approach could be successfully incorporated into environmental risk assessments to better summarize biotic integrity and improve the ecological management. In the Morcille River, located in a vineyard area of France, several studies have been performed to evaluate pesticide effects on biofilms [78–80] (Table 5). Morin et al. [78] studied the structural changes in biofilm assemblages induced by the transfer of biofilm communities from two contaminated sites to a reference site, expecting a recovery of the translocated communities, either in structure or in diversity. The proportions of the different algal groups in biofilms were estimated by in vivo chl-*a* fluorescence measurements, observing that biofilm was mainly composed by diatoms before and after translocation (Table 5). In the same river Pesce et al. [79] used the Φ_{PSII} to evaluate individual and mixture effects of diuron and its main metabolites

Table 5 Review of field studies where chl-*a* fluorescence techniques have been applied to assess the effects of toxicants on photosynthesis performance of fluvial biofilm communities

Toxicant	Mode of action	River	Growth substrate	Experimental conditions	Φ_{PSII}	Φ'_{PSII}	NPQ	Fo	Fo (Bl)	Fo (Gr)	Fo (Br)	References
Cu	b	Ter River, Spain	Artificial substrata	Biofilm from different polluted sites (spring and summer) was transported to the laboratory for Cu and atrazine short-term toxicity tests		– – –						Navarro et al. [26]
Atrazine	a					– – –						
22 pesticides (mixture)	a and c	Llobregat River, Spain	Natural biofilm	Field monitoring of biofilm from different polluted sites during 2 years—natural exposure	–	–						Ricart et al. [30]
Atrazine	a	Seven streams from Catalonia Area, Spain	Artificial substrata	Biofilm from different polluted sites and during different seasons was transported to the laboratory for atrazine and Zn short-term toxicity tests		– – –						Guasch et al. [15]
Zinc	b					– –						
Atrazine	a	Ozanne River, France	Artificial substrata	Biofilm from different polluted sites and during different seasons was transported to the laboratory for atrazine and isoproturon toxicity tests—PICT concept	–							Dorigo and Leboulanger [65]
Isoproturon	a				–							
Zn, Cd, Ni, Fe (metal pollution)	b	Mort River, France	Natural biofilm	Biofilm from different polluted sites was evaluated—natural exposure		–						Bonnineau et al. [83]
Diuron, DCPMU and 3,4-DCA	a	Morcille River, France	Artificial substrata	Biofilm from different polluted sites, presenting a natural exposure, were sampled	–				– –			Pesce et al. [79, 80]
				Biofilm from different polluted sites was transported to the laboratory for short-term toxicity tests—PICT concept	– – –							

(continued)

Table 5 (continued)

Toxicant	Mode of action	River	Growth substrate	Experimental conditions	Φ_{PSII}	Φ'_{PSII}	NPQ	Fo	Fo (Bl)	Fo (Gr)	Fo (Br)	References
Pesticides	a and b	Morcille River, France	Artificial substrata	Biofilm from each sampling site was sampled before translocation to be structurally characterized—natural exposure					n.s.	n.s.	n.s.	Morin et al. [78]
Prometryn	a	Elbe River, Germany	Artificial substrata	Translocation experiment. Local and translocated biofilm was transported to the laboratory for short-term prometryn toxicity tests—PICT concept, structural characterization	− − −				+	+	− −	Rotter et al. [81]

The significant responses observed are indicated by "−" for inhibition or by "+" for increase and its approximate magnitude by "−" or "+" if it was slight (10–30% of control), by "− −" or "+ +" if it was moderate (30–50% of control) and by "− − −" or "+ + +" if it was high (>50% of control). The terminology used for mentioned the mode of action of each compound was: "a" for PSII inhibitors, "b" for photosynthesis inhibitors not directly targeting PSII and "c" for unknown mode of action on algae

n.s. not significant ($p > 0.05$)

(DCPMU and 3,4-DCA) on biofilms. They observed that diuron was the most toxic of the evaluated compounds and that biofilms from contaminated sites presented a higher tolerance to diuron and DCPMU than biofilms from a reference site. In another study, it was investigated how closely diuron tolerance acquisition by photoautrophic biofilm communities could reflect their previous in situ exposure to this herbicide. For this propose, the use of PICT (using the Φ_{PSII} as endpoint) together with multivariate statistical analyses was combined. A spatiotemporal variation in diuron tolerance capacities of photoautotrophic communities was observed, the biofilm from the most polluted site being the most tolerant [80] (Table 5). Similar tolerance results were found by Rotter et al. [81] in a study in the Elbe River where prometryn toxicity on biofilm was evaluated by combining a translocation experiment and the PICT approach (Table 5). They observed that biofilms from a polluted site presented a higher EC_{50} (based on Φ'_{PSII} measures) than biofilms from a reference site, suggesting a prometryn induction tolerance. The proportion of each group (cyanobacteria, green algae, and diatoms) in transferred and control biofilms was also measured by PAM fluorometry. In other studies [65, 82]) also used the PICT approach to investigate the toxicity of atrazine and isoproturon on biofilm from an herbicide polluted river (Ozanne River, France) (Table 5). Sampling was performed at different polluted sites and at different moments during the year. Using Φ_{PSII} as endpoint in short-term toxicity tests, differences in the EC_{50} between biofilms naturally exposed to different levels of atrazine and isoproturon could be observed.

Other field studies presenting organic and inorganic pollution were based on short-term toxicity tests to evaluate biofilms' sensitivity to toxic exposure [26] in the Ter River (NE, Spain) (Table 5). It was concluded that the short-term toxicity tests using the Φ'_{PSII} as a physiological endpoint may provide an early prospective quantification of the transient effects of a toxic compound on separate communities, predict which of these effects are not reversible, and determine their intensity. Guasch et al. [15] investigated the ecological and the structural parameters influencing Zn and atrazine toxicity on biofilm communities by testing the validity of the use of a short-term physiological method (based on the Φ'_{PSII} as microalgae endpoint) (Table 5). They concluded that short-term toxicity tests seem to be pertinent to assess atrazine toxicity on biofilms. In contrast, to assess Zn toxicity on biofilm the use of longer term toxicity tests was proposed to overcome the influence that biofilm thickness exerts on Zn diffusion and toxicity. Bonnineau et al. [83] assessed in situ the effects of metal pollution on Φ'_{PSII} in natural communities from the Riou-Mort river (France) by comparing reference and polluted sites. Slight effects were observed based on Φ'_{PSII} and marked effects were observed based on antioxidant enzymatic activities, supporting the low sensitivity of Φ'_{PSII} for the assessment in situ of the chronic effects of metals.

Based on the field studies reported here, it can be concluded that in field studies the responses of the chl-*a* fluorescence parameter could serve as an early warning signal of biological effects after acute exposure to photosystem II inhibitors, but in regular monitoring PAM fluorometry may show low sensitivity or give false-negative signals [84] if the community is already adapted to the prevailing toxic exposure conditions. Tolerance induction evaluations and multibiomarker approaches, together with the use of appropriate multivariate statistical analyses, may partially overcome this limitation [85].

5 General Discussion and Perspectives

The main aim of this chapter was to analyze the pros and cons of the use of chl-*a* fluorescence techniques for water quality monitoring programs, based on biofilm communities, in the context of the WFD.

The use chl-*a* fluorescence parameters as biomarkers of toxicity on biofilm communities has increased in the last few years, illustrating their applicability to show early and long-term effects of toxicants at community level using functional and structural descriptors.

Differences in sensitivity are reported between chl-*a* fluorescence parameters. Focusing on functional parameters, the effective quantum yield (Φ'_{PSII}) seems to detect the effects on algae of toxic compounds targeting the PSII. Its use is well reported in both laboratory and field studies (Tables 2 and 5), supporting its use for these types of compounds. On the other hand, its sensitivity to chemicals with a mode of action different to the PSII is highly influenced by the time of exposure of the toxicant (short-term vs. long-term exposure) (Tables 3–5). In short-term studies, the Φ'_{PSII} seems to be a sensitive endpoint of toxicity but in long-term studies its sensitivity is often less obvious (Tables 3 and 5). This lack of sensitivity during long-term exposures could be related with a development of biofilm tolerance. Therefore, in long-term studies both in field and in laboratory conditions, supplementary methodological approaches, such as PICT tests based on Φ'_{PSII} may be used to show community adaptation.

The maximal photosynthetic capacity of biofilms (Φ_{PSII}) showed a similar sensitivity to chemicals as the Φ'_{PSII}. This parameter has been used more in ecotoxicological studies than the Φ'_{PSII}, probably due to its simplicity for being measured. The Φ_{PSII}, which is measured on dark-adapted samples, is independent of the light conditions prevailing during the incubation period. Therefore, it is more suitable than Φ'_{PSII} for comparing results of experiments using different light conditions [69]. The use of both photosynthetic parameters, Φ'_{PSII} and Φ_{PSII}, provides more information than only using one of these parameters, since both functional and structural alterations are evaluated.

The use of the NPQ parameter to assess chemical toxicity on biofilm is less documented than those reported above and applied only in microcosm studies. However, its use is promising due to its sensitivity to different types of toxicants, and also due to its complementarity to the more classical photosynthetic related parameters (Φ'_{PSII} and Φ_{PSII}) [23, 31, 69]. However, as the activation of the NPQ processes is restricted to a very specific period of time (from minutes to few hours after the stress started), its use seems to be restricted to evaluate short-term toxicity. On the other hand, its sensitivity has also been observed in long-term studies. In these cases, it was linked to structural damage in the photosynthetic apparatus, specifically a NPQ reduction as a result of a modification of accessory pigments where NPQ processes occur [23, 31].

Finally, the use of the fast fluorescence techniques by PEA to assess chemical toxicity on biofilms is also recommended, indeed not yet applied. As this technique gives information about the kinetics of OJIP transients (Fig. 2), its application on biofilm could contribute to detect toxic effect of the PSII energy trapping or PSII antenna size and by this way complement the information obtained by PAM fluorometry. Its application on future ecotoxicological biofilm studies could be of special interest in studies focusing on the confirmation of the MoA of a determinate chemical on the photosynthetic processes of species composing biofilm community.

Concerning more structural chl-*a* fluorescence parameters, the minimal fluorescence parameter (Fo) has been applied for monitoring algal biomass (Tables 2–5) and biofilm growth rate [28, 31] in microcosms and field studies. It can be considered as a global indicator, which integrates effects of chemicals on different cellular metabolisms causing pigment damage and cell death. It has been successfully applied in ecotoxicological studies focusing on compounds that damage photosynthetic processes without targeting the PSII (Table 3). Since Fo is influenced by measuring conditions: gain, distance between sensor and the biofilm sample, or the measuring light intensity, absolute values are difficult to compare between different studies without using the same calibration. However, toxic effects on Fo, in ecotoxicological studies, are usually reported in comparison to a non-exposed community (in relative terms) to avoid this limitation. The measurements of Fo may present other limitations not related to the algal biomass or biofilm thickness. Fo may increase if the addition of a herbicide causes a transfer of the electron on PQ_B to the primary quinine acceptor (PQ_A) and displacement of PQ_B by the herbicide (Fig. 1); the reduced PQ_A leads to a higher Fo [86]. Consequently an overestimation of Fo could occur. Also, it is known that some chemicals emit fluorescence under excitation light (e.g., DCMU) that could not be distinguished from the fluorescence emitted from excited chl-*a* (Fo measure). In these cases, the use of Fo to assess algal biomass is not recommended. In field applications, this phenomenon could play an important role, since pollution is often due to a mixture of compounds of unknown origin. On the other hand, the greening effects could also contribute to limit the use of Fo for measuring algal biomass or algal growth. This phenomenon may occur due to an increase in the number of chl-*a* molecules per unit of cell, without increasing the number of cells [29, 87]. This situation could produce an increase of Fo values with no increase in the number of cells.

A significant portion of the ecotoxicological studies included in this chapter has incorporated the deconvolution of the chl-*a* fluorescence signal into the contribution of the main autotrophic groups (Tables 2–5). These measurements allow the main autotrophic groups of biofilms (structural approach) to be characterized.

Nowadays, PAM fluorometers which incorporate a four-wavelength excitation method to deconvolute these photosynthetic parameters of microalgal communities [Fo(Bl), Fo(Gr), and Fo(Br)] by incorporating spectral groups based on a single species pigment spectra. This calibration could be improved in the future if spectra used for calibrating the fluorometer are based on a mix of species from each group, based on the key species of each autotrophic group. This procedure could probably help to decrease uncertainty in these measurements.

6 Conclusions

Biofilms or periphyton communities are recognized for the Water Framework Directive (WFD, Directive 2000/60/EC) as a BQE. The use of chl-*a* fluorescence parameters as biofilm biomarkers allows to obtain a community approach which is much closer to the processes of an ecosystem than the use of single species tests. In fluvial ecosystems, the use of chl-*a* fluorescence parameters as a complementary set of biofilm biomarkers to the more traditional diatom indices applied so far is recommended due to its capacity to provide different warning signals of early and late toxicity effects. Its application is of special interest in the context of the WFD, where the development of new structural and functional parameters from the BQEs is required [88].

Acknowledgments The authors would like to thank the authors of publications [20, 23, 29] to give us the permission to use some of their figures to illustrate PAM applications on biofilms. Also we thank Ulrich Schreiber to give us the permission to use a figure of his publication [89] to illustrate the typical chl-*a* fluorescence kinetics. This study was financed by the European project KEYBIOEFFECTS (MRTN-CT-2006-035695) and the Spanish project FLUVIALMULTISTRESS (CTM2009-14111-CO2-01).

References

1. Taiz L, Zeiger E (eds) (1999) Physiologie der Pflanzen. Spektrum Akad. Vlg., Heidelberg
2. Juneau P, Dewez D, Matsui S, Kim SG, Popovic R (2001) Evaluation of different algal species sensitivity to mercury and metolachlor by PAM-fluorometry. Chemosphere 45:589–598
3. Krause GH, Weis E (1991) Chlorophyll fluorescence and photosynthesis: the basics. Annu Rev Plant Physiol 42:313–349
4. Schreiber U, Bilger W, Neubaurer C (1994) Chlorophyll fluorescence as a non-invasive indicator for rapid assessment of in vivo photosynthesis. In: Schulze ED, Calswell MM (eds) Ecophysiology of photosynthesis. Springer, Berlin, pp 49–70
5. Kautsky H, Franck U (1943) Chlorophyll fluoreszenz und Kohlensäureassimilation. Biochemistry 315:139–232
6. Kautsky H, Appel W, Amann H (1960) Die Fluoreszenzkurve und die Photochemie der Pflanze. Biochemistry 332:277–292
7. Sabater S, Admiraal W (2005) Periphyton as Biological Indicators in Managed Aquatic Ecosystems. Periphyton: Ecology, Exploitation and Management. Cambridge, Massachusetts: Azim ME, Verdegem MCJ, AA van Dam, MCM Beveridge (eds.), pp. 159–177

8. Clements WH, Newman MC (2002) Community ecotoxicology. Wiley, West Sussex, p 336
9. Guasch H, Serra A, Corcoll N, Bonet B, Leira M (2010) Metal ecotoxicology in fluvial biofilms: potential influence of water scarcity. In: Sabater S, Barceló D (eds) Water scarcity in the Mediterranean. Perspecives under global change. Springer, Berlin, pp 41–54
10. Freeman C, Lock MA (1995) The biofilm polysaccharide matrix—a buffer against changing organic substrate supply. Limnol Oceanogr 40:273–278
11. Gray BR, Hill WR (1995) Nickel sorption by periphyton exposed to different light intensities. North Am Benthol Soc 14:29–305
12. Guasch H, Paulsson M, Sabater S (2002) Effect of copper on algal communities from oligotrophic calcareous streams. J Phycol 38:241–248
13. Ivorra N, Barranguet C, Jonker M, Kraak MHS, Admiraal W (2002) Metal-induced tolerance in the freshwater microbenthic diatom *Gomphonema parvulum*. Environ Pollut 16:147–157
14. Pinto J, Sigaud-Kutner TCS, Leitao MAS, Okamoto OK, Morse D, Colepicolo P (2003) Heavy metal-induced oxidative stress in algae. J Phycol 39:1008–1018
15. Guasch H, Admiraal W, Sabater S (2003) Contrasting effects of organic and inorganic toxicants on freshwater periphyton. Aquat Toxicol 64:165–175
16. Leboulanger C, Rimet F, Hème de Lacotte M, Bérard A (2001) Effects of atrazine and nicosulfuron on freshwater microalgae. Environ Int 26:131–135
17. Pesce S, Fajon C, Bardot C, Bonnemoy F, Portelli C, Bohatier J (2006) Effect of the phenylurea herbicide diuron on natural riverine microbial communities in an experimental study. Aquat Toxicol 78:303–314
18. Schmitt-Jansen M, Altenburger R (2005) Toxic effects of isoproturon on periphyton communities—a microcosm study. Estuar Coast Shelf Sci 62:539–545
19. Schmitt-Jansen M, Altenburger R (2005) Predicting and observing responses of algal communities to photosystem II-herbicide exposure using pollution-induced community tolerance and species-sensitivity distributions. Environ Toxicol Chem 24:304–312
20. Bonnineau C, Guasch H, Proia L, Ricart M, Geiszinger A, Romaní A, Sabater S (2010) Fluvial biofilms: a pertinent tool to assess β-blockers toxicity. Aquat Toxicol 96:225–233
21. Franz S, Altenburger R, Heilmeier H, Schmitt-Jansen M (2008) What contributes to the sensitivity of microalgae to triclosan? Aquat Toxicol 90:102–108
22. Lawrence JR, Swerhone GDW, Wassenaar LI, Neu TR (2005) Effects of selected pharmaceuticals on riverine biofilm communities. Can J Microbiol 51:655–669
23. Ricart M, Guasch H, Alberch M, Barceló D, Bonnineau C, Farré M, Ferrer J, Geiszinger A, Morin S, Proia L, Ricciardi F, Romaní AM, Sala L, Sureda D, Sabater S (2010) Triclosan persistence through wastewater treatment plants and its potential toxic effects on river biofilms. Aquat Toxicol 100:346–353
24. Sabater S, Guasch H, Ricart M, Romaní A, Vidal G, Klünder C, Schmitt-Jansen M (2007) Monitoring the effect of chemicals on biological communities. The biofilm as an interface. Anal Bioanal Chem 387:1425–1434
25. Barranguet C, Charantoni E, Pland M, Admiraal W (2000) Short-term response of monospecific and natural algal biofilms to copper exposure. Eur J Phycol 35:397–406
26. Navarro E, Guasch H, Sabater S (2002) Use of microbenthic algal communities in ecotoxicological tests for the assessment of water quality: the Ter river case study. J Appl Phycol 14:41–48
27. Guasch H, Navarro E, Serra A, Sabater S (2004) Phosphate limitation influences the sensitivity to copper in periphytic algae. Freshw Biol 49:463–473
28. Serra A, Corcoll N, Guasch H (2009) Copper accumulation and toxicity in fluvial periphyton. Chemosphere 5:633–641
29. Ricart M, Guasch H, Barceló D, Geiszinger A, López De Alda M, Romaní AM, Vidal G, Villagrasa M, Sabater S (2009) Effects of low concentrations of the phenylurea herbicide diuron on biofilm algae and bacteria. Chemosphere 76:1392–1401
30. Ricart M, Guasch H, Barceló D, Brix R, Conceição MH, Geiszinger A, López De Alda MJ, López-Doval JC, Muñoz I, Romaní AM, Villagrasa M, Sabater S (2010) Primary and complex stressors in polluted Mediterranean rivers: pesticide effects on biological communities. J Hydrol 383:52–61

31. Corcoll N, Bonet B, Leira M, Guasch H (2011) Chl-*a* fluorescence parameters as biomarkers of metal toxicity in fluvial periphyton: an experimental study. Hydrobiologia 673:119–136
32. Strasser BJ (1997) Donor side capacity of photosystem II probed by chlorophyll a fluorescence transients. Photosynth Res 52:147–155
33. Schmitt-Jansen M, Altenburger R (2007) The use of pulse-amplitude modulated (PAM) fluorescence-based methods to evaluate effects of herbicides in microalgal systems of different complexity. Toxicol Environ Chem 89:665–681
34. McClellan K, Altenburger R, Schmitt-Jansen M (2008) Pollution-induced community tolerance as a measure of species interaction in toxicity assessment. J Appl Ecol 45:1514–1522
35. Schreiber U, Quayle P, Schmidt S, Escher I, Mueller JF (2007) Methodology and evaluation of a highly sensitive algae toxicity test based on multiwall chlorophyll fluorescence imaging. Biosens Bioelectron 22:2554–2563
36. Schreiber U, Müller JF, Haugg A, Gademann R (2002) New type of dual-channel PAM chlorophyll fluorometer for highly sensitive water toxicity biotests. Photosynth Res 74:317–330
37. Coquery M, Morin A, Bécue A, Lepot B (2005) Priority substances of the European Water Framework Directive: analytical challenges in monitoring water quality. Trac-Trends Anal Chem 24:2
38. Farré M, Petrovica M, Gros M, Kosjekc T, Martinez H, Osvald P, Loos R, Le Menach K, Budzinski H, De Alencastrog F, Müller J, Knepper T, Finki G, Ternes TA, Zuccato E, Kormali P, Gans O, Rodil R, Quintana JB, Pastori F, Gentili A, Barceló D (2008) First interlaboratory exercise on non-steroidal anti-inflammatory drugs analysis in environmental samples. Talanta 76:580–590
39. Furse MT, Hering D, Brabec K, Buffagui A, Sandin L, Verdouschot PFM (eds) (2006) The Ecological status of European Rivers: evaluating and intercalibration of assessment methods. Hydrobiologia 566: 3–29
40. Samson G, Morisette JC, Popovic R (1988) Copper quenching of the variable fluorescence in *Dunaliella tertiolecta.* New evidence for a copper inhibition effect on PSII photoinhibitory. Photochem Photobiol 48:329–332
41. Singh DP, Khare P, Bisen PS (1989) Effects of Ni^{2+}, Hg^{2+} and Cu^{2+} on growth, oxygen evolution and photosynthetic electron transport in Cylindrospermum IU 942. Plant Physiol 134:406–412
42. Ralph PJ, Smith RA, Macinnis-Ng CMO, Seery CR (2007) Use of fluorescence-based ecotoxicological bioassays in monitoring toxicants and pollution in aquatic systems: review. Toxicol Environ Chem 89:589–607
43. Schreiber U (1998) Chlorophyll fluorescence: new instruments for special applications. In: Garab G (ed) Photosynthesis: mechanisms and effects, vol V. Kluwer, Dordrecht, pp 4253–4258
44. Govindjee R (1995) Sixty-three years since Kautsky: chlorophyll a fluorescence. Aust J Plant Physiol 22:131–160
45. Melis A, Homann PH (1975) Kinetic analysis of the fluorescence in 3-(3,4-dichlorophyenyl)-1,1 dimethylurea poisoned chloroplasts. Photochem Photobiol 21:431–437
46. Melis A, Schreiber U (1979) The kinetic relationship between the absorbance change, the reduction of Q (ΔA_{320}) and the variable fluorescence yield change in chloroplasts at room temperature. Acta Biochim Biophys 547:47–57
47. Lu CM, Chau CW, Zhang JH (2000) Acute toxicity of excess mercury on the photosynthetic performance of cyanobacterium, *S. platensis*—assessment by chlorophyll fluorescence analysis. Chemosphere 41:191–196
48. Havaux M, Strasser RJ, Greppin H (1991) A theoretical and experimental analysis of the qP and qN coefficients of chlorophyll fluorescence quenching and their relation to photochemical and nonphotochemical events. Photosynth Res 27:41–55
49. Schreiber U, Schliwa U, Bilger W (1986) Continuous recording of photochemical and non-photochemical chlorophyll fluorescence quenching with a new type of modulation fluorometer. Photosynth Res 10:51–62

50. Srivastava A, Greppin H, Strasser R (1995) The steady state chlorophyll *a* fluorescence exhibits in vivo an optimum as function of light intensity which reflects the physiological state of the plant. Plant Cell Physiol 36:839–848
51. Rysgaard S, Kuhl M, Glud RN, Hansen JW (2001) Biomass, production and horizontal patchiness of sea ice algae in a high-Arctic fjord (Young Sound, NE Greenland). Mar Ecol Prog Ser 223:15–26
52. Serôdio J, Silva JM, Catarino F (1997) Nondestructive tracing of migratory rhythms of intertidal benthic microalgae using in vivo chlorophyll *a* fluorescence. J Phycol 33:542–553
53. Schmitt-Jansen M, Altenburger R (2008) Community-level microalgal toxicity assessment by multiwavelength-excitation PAM fluorometry. Aquat Toxicol 86:49–58
54. Genty B, Briantais JM, Baker NR (1989) The relationship between the quantum yield of photosynthetic electron transport and quenching of chlorophyll fluorescence. Acta Biochim Biophys 99:87–92
55. García-Meza C, Barranguet C, Admiraal W (2005) Biofilm formation by algae as a mechanism for surviving on mine tailing. Environ Toxicol Chem 3:575–581
56. Barranguet C, Van den Ende FP, Rutgers M, Breure AM, Greijdanus M, Sinke JJ, Admiraal W (2003) Copper-induced modifications of the trophic relations in riverine algal-bacterial biofilms. Environ Toxicol Chem 22:1340–1349
57. Horton P, Ruban AV, Young AJ (1999) Regulation of the structure and function of the light harvesting complexes of photosytem II by the xanthophyll cycle. In: Frank HA, Young AJ, Cogdell RJ (eds) The photochemistry of carotenoids. Kluwer, Dordrecht, pp 271–291
58. Buschmann C (1995) Variation of the quenching of chlorophyll fluorescence under different intensities of the actinic light in wild-type plants of tabacco and in a Aurea mutant deficient of light-harvesting-complex. Plant Physiol 145:245–252
59. Pospísil P (1997) Mechanisms of non-photochemical chlorophyll fluorescence quenching in higher plants. Photosynthetica 34:343–355
60. Juneau P, Green BR, Harrison PJ (2005) Simulation of Pulse-Amplitude-Modulated (PAM) fluorescence: limitations of some PAM-parameters in studying environmental stress effects. Photosynthetica 43:75–83
61. Mallick N, Mohn FH (2003) Use of chlorophyll fluorescence in metalstress research: a case study with the green microalga Scenedesmus. Ecotoxicol Environ Saf 55:64–69
62. Kolbowski J, Schreiber U (1995) Computer-controlled phytoplankton analyzer based on a 4-wavelengths PAM chlorophyll fluorometer. In: Mathis P (ed) Photosynthesis: from light to biosphere, vol 5. Kluwer, Dordrecht, pp 825–828
63. Beutler M, Wiltshire KH, Meyer B, Moldaenke C, Lüring C, Meyerhöfer M, Hansen UP, Dau H (2002) A fluorometric method for the differentiation of algal populations in vivo and in situ. Photosynth Res 72:39–53
64. Conrad R, Buchel C, Wilhelm C, Arsalane W, Berkaloff C, Duval JC (1993) Changes in yield in-vivo fluorescence of chlorophyll a as a tool for selective herbicide monitoring. J Appl Phycol 5:505–516
65. Dorigo U, Leboulanger C (2001) A PAM fluorescence-based method for assessing the effects of photosystem II herbicides on freshwater periphyton. J Appl Phycol 13:509–515
66. Trapmann S, Etxebarria N, Schnabl H, Grobecker K (1998) Progress in herbicide determination with the thylakoid bioassay. Environ Sci Pollut Res Int 5:17–20
67. Dewez D, Marchand M, Eullaffroy P, Popovic R (2002) Evaluation of the effects of diuron and its derivatives on *Lemna gibba* using a fluorescence toxicity index. Environ Toxicol 17:493–501
68. Dewez D, Geoffroy L, Vernet G, Popovic R (2005) Determination of photosynthetic and enzymatic biomarkers sensitivity used to evaluate toxic effects of copper and fludioxonil in alga *Scenedesmus obliquus*. Aquat Toxicol 74:150–159
69. Laviale M, Prygiel J, Créach A (2010) Light modulated toxicity of isoproturon toward natural stream periphyton photosynthesis: a comparison between constant and dynamic light conditions. Aquat Toxicol 97:334–342

70. Tlili A, Bérard A, Roulier JL, Volat B, Montuelle B (2010) PO_4^{3-} dependence of the tolerance of autotrophic and heterotrophic biofilm communities to copper and diuron. Aquat Toxicol 98:165–177
71. López-Doval JC, Ricart M, Guasch H, Romaní AM, Sabater S, Muñoz I (2010) Does grazing pressure modify diuron toxicity in a biofilm community? Arch Environ Contam Toxicol 58:955–962
72. Blanck H, Wänkberg SÅ, Molander S (1988) Pollution-induced community tolerance—a new ecotoxicological tool. In: Cairs J Jr, Pratt JR (eds) Functional testing of aquatic biota for estimating hazards of chemicals, 988. ASTM STP, Philadelphia, PA, pp 219–230
73. Teisseire H, Guy V (2000) Copper-induced changes in antioxidant enzymes activities in fronds of duckweed (Leman minor). Plant Sci 153:65–72
74. Nalewajko C, Olaveson MM (1994) Differential responses of growth, photosynthesis, respiration, and phosphate uptake to copper in copper-tolerant and copper-intolerant strains of *Scenedesmus acutus* (Chlorophyceae). Can J Bot 73:1295–1303
75. Halling-Sørensen B, Nors Nielsen S, Lanzky PF, Ingerslev F, Holten Lützhoff HC, Jorgensen SE (1998) Occurrence, fate, and effects of pharmaceutical substances in the environment—A review. Chemosphere 36:357–393
76. Marwood CA, Solomon KR, Greenberg BM (2001) Chlorophyll fluorescence as a bioindicator of effects on growth in aquatic macrophytes from mixtures of polycyclic aromatic hydrocarbons. Environ Toxicol Chem 20:890–898
77. Metcalfe CD, Koenig BG, Bennie DT, Servos M, Ternes TA (2003) Occurrence of neutral and acidic drugs in the effluents of Canadian seawater treatment plants. Environ Toxicol Chem 22:2872–2880
78. Morin S, Pesce S, Tlili A, Coste A, Montuelle B (2010) Recovery potential of periphytic communities in ariver impacted by a vineyard watershed. Ecol Indic 10:419–426
79. Pesce S, Lissalde S, Lavieille D, Margoum C, Mazzella N, Roubeix V, Montuelle B (2010) Evaluation of single and joint toxic effects of diuron and its main metabolites on natural phototrophic biofilms using a pollution-induced community tolerance (PICT) approach. Aquat Toxicol 99:492–499
80. Pesce S, Margoum C, Montuelle B (2010) In situ relationship between spatio-temporal variations in diuron concentrations and phtotrophic biofilm tolerance in a contaminated river. Water Res 44:1941–1949
81. Rotter S, Sans-Piché F, Streck G, Altenburger R, Schmitt-Jansen M (2011) Active biomonitoring of contamination in aquatic systems—an in situ translocation experiment applying the PICT concept. Aquat Toxicol 101:228–236
82. Dorigo U, Bourrain X, Bérard A, Leboulanger C (2004) Seasonal changes in the sensitivity of river microalgae to atrazine and isoproturon along a contamination gradient. Sci Total Environ 318:101–114
83. Bonnineau C, Bonet B, Corcoll N, Guasch H (2011) Catalase in fluvial biofilms: a comparison between different extraction methods and example of application in a metal-polluted river. Ecotoxicology 20(1):293–303
84. Guasch H, Bonet B, Bonnineau C, Corcoll N, López-Doval JC, Muñoz I, Ricart M, Serra A, Clements W (2012) How to link field observations with causality? Field and experimental approaches linking chemical pollution with ecological alterations. In: Guasch H, Ginebreda A, Geiszinger A (eds) Emerging and priority pollutants in rivers, vol 19, The handbook of environmental chemistry. Springer, Heidelberg
85. Muñoz I, Sabater S, Barata C (2012) Evaluating ecological integrity in multi-stressed rivers: from the currently used biotic indices to newly developed approaches using biofilms and invertebrates. In: Guasch H, Ginebreda A, Geiszinger A (eds) Emerging and priority pollutants in rivers, vol 19, The handbook of environmental chemistry. Springer, Heidelberg
86. Hiraki M, van Rensen JJ, Vredenberg WJ, Wakabayashi K (2003) Characterization of the alterations of the chlorophyll a fluorescence induction curve after addition of Photosystem II inhibiting herbicides font. Photosynth Res 78:35–46

87. Guasch H, Muñoz I, Rosés N, Sabater S (1997) Changes in atrazine toxicity throughout succession of stream periphyton communities. J Appl Phycol 9:137–146
88. Johnson RK, Hering D, Furse MT, Verdouschot PFM (2006) Indicators of ecological change: comparison of the early response of four organism groups to stress gradients. Hydrobiologia 566:139–152
89. Schreiber U (2004) Pulse-amplitude-modulation (PAM) fluorometry and saturation pulse method: an overview. In: Papageorgiou GC, Govindjee R (eds) Chlorophyll fluorescence: a signature of photosynthesis. Springer, The Netherlands, pp 279–319
90. Müller P, Li XP, Niyogi KK (2001) Non-photochemical quenching. A response to excess light energy. Plant Physiol 125:1558–1566
91. Bilger W, Björkman O (1990) Role of the xanthophyll cycle in photoprotection elucidated by measurements of light-induced absorbance changes, fluorescence and photosynthesis in leaves of *Hedera canariensis*. Photosynth Res 25:173–185
92. Bolhàr-Nordenkampt HR, Öquist G (1993) Chlorophyll fluorescence as a tool in photosynthesis research. In: Hall DO, Scurlock JMO, Bolhàr-Nordenkampf HR, Leegood RC, Long SP (eds) Photosynthesis and production in a changing environment: a field and laboratory manual. Chapman & Hall, London, pp 193–206
93. Kriedemann PF, Graham RD, Wiskich JT (1985) Photosynthetic dysfunction and in vivo chlorophyll a fluorescence from manganese-deficient wheat leaves. Aust J Agric Res 36:157–169
94. Horton P, Ruban AV, Walters RG (1996) Regulation of light harvesting in green algae. Annu. Rev. Plant Physiol. Plant Mol. Biol. 47:655–684
95. Schreiber U (1998) Chlorophyll fluorescence: new instruments for special applications. In Grab, G. [Ed] Photosynthesis: Mechanisms and Effects. Vol V. Kluwer Academic Publishers, Dordrecht, pp 4258–8

Consistency in Diatom Response to Metal-Contaminated Environments

Soizic Morin, Arielle Cordonier, Isabelle Lavoie, Adeline Arini, Saul Blanco, Thi Thuy Duong, Elisabet Tornés, Berta Bonet, Natàlia Corcoll, Leslie Faggiano, Martin Laviale, Florence Pérès, Eloy Becares, Michel Coste, Agnès Feurtet-Mazel, Claude Fortin, Helena Guasch, and Sergi Sabater

Abstract Diatoms play a key role in the functioning of streams, and their sensitivity to many environmental factors has led to the development of numerous diatom-based indices used in water quality assessment. Although diatom-based monitoring of metal contamination is not currently included in water quality monitoring programs, the effects of metals on diatom communities have been studied in many polluted watersheds as well as in laboratory experiments, underlying their high potential for metal contamination assessment. Here, we review the response of

---

S. Morin (✉) • M. Coste
Irstea, UR REBX, 50 avenue de Verdun, 33612 Cestas, cedex, France
e-mail: soizic.morin@irstea.fr

A. Cordonier
Service cantonal de l'écologie de l'eau (SECOE), 23 avenue Sainte-Clothilde, Genève 1205, Switzerland

I. Lavoie • C. Fortin
Institut national de la recherche scientifique, Centre Eau Terre Environnement, 490 rue de la Couronne, Québec, QC G1K 9A9, Canada

A. Arini
Irstea, UR REBX, 50 avenue de Verdun, 33612 Cestas, cedex, France

Université de Bordeaux 1, CNRS, UMR 5805 EPOC, Place du Dr Peyneau, 33120 Arcachon, France

S. Blanco • E. Becares
Department of Biodiversity and Environmental Management, University of León, 24071 León, Spain

T.T. Duong
Institute of Environmental Technology, Vietnam Academy of Science and Technology, 18 Hoang Quoc Viet Road, Cau Giay, Hanoi, Viet Nam

E. Tornés • S. Sabater
Catalan Institute for Water Research (ICRA), Emili Grahit 101, 17003 Girona, Spain

B. Bonet • N. Corcoll • L. Faggiano • H. Guasch
Institute of Aquatic Ecology, University of Girona, Campus Montilivi, 17071 Girona, Spain

H. Guasch et al. (eds.), *Emerging and Priority Pollutants in Rivers*, Hdb Env Chem (2012) 19: 117–146, DOI 10.1007/978-3-642-25722-3_5,

diatoms to metal pollution from individual level (e.g. size, growth form, and morphological abnormalities) to community structure (replacement of sensitive species by tolerant ones). These potential effects are then tested using a large, multi-country database combining diatom and metal information. Metal contamination proved to be a strong driver of the community structure, and enabled for the identification of tolerant species like *Cocconeis placentula* var. *euglypta, Eolimna minima*, *Fragilaria gracilis*, *Nitzschia sociabilis*, *Pinnularia parvulissima*, and *Surirella angusta*. Among the traits tested, diatom cell size and the occurrence of diatom deformities were found to be good indicators of high metal contamination. This work provides a basis for further use of diatoms as indicators of metal pollution.

Keywords Deformities • Metals • Periphytic diatoms • Rivers • Species distribution • Species traits

Contents

M. Laviale
Laboratoire de Génétique et Evolution des Populations Végétales UMR CNRS 8016, Université des Sciences et Technologies de Lille, 59655 Villeneuve d'Ascq cedex, France

Departamento de Biologia and CESAM - Centro de Estudos do Ambiente e do Mar, Universidade de Aveiro, Campus de Santiago 3810–193 Aveiro, Portugal

F. Pérès
ASCONIT Consultants, Parc Scientifique Tony Garnier, 6-8 Espace Henry Vallée, 69366 Lyon Cedex 07, France

A. Feurtet-Mazel
Université de Bordeaux 1, CNRS, UMR 5805 EPOC, Place du Dr Peyneau, 33120 Arcachon, France

1 Introduction

Biomonitoring has been increasingly used to assess water quality due to the more time-integrative characteristic of the approach compared with punctual chemical measurements. Among the tools used in biomonitoring, diatoms are cosmopolitan aquatic organisms, and are a major component of benthic biofilms. Because diatoms are at the basis of the trophic chain, these microscopic algae respond quickly to environmental changes and are considered good indicators of environmental conditions [1, 2]. They are included in numerous water quality monitoring programs worldwide, and river diatom-based indices have been developed in numerous countries [3–8]. Diatom species distribution is driven by environmental factors acting at different scales, from local (general water quality) to larger scale determinants (biogeography). Therefore, biological monitoring is best achieved considering both local and larger scales. Diatom-based indices developed to assess ecosystems' health often include these different scales of variability. For example, geology has a strong influence on water chemistry, which in turn affects diatom community structure [9–11]. To overcome the natural variability associated with the geological characteristics of the region, diatom-based indices must include larger scale determinants to effectively provide information on the local environment. This explains the fact that diatom-based water quality monitoring is usually country- or even ecoregion-dependent (e.g. [8, 11, 12]). Diatom indices assess the biological status of streams, with reference to trophy, acidity, conductivity, etc., but generally do not take into account toxic pollution. Field studies dealing with metal contaminations in various regions and countries showed quite consistent responses of diatom communities such as higher abundances of small-sized species [13, 14], increasing proportions of metal-tolerant species or significantly higher occurrences of valve deformities.

The main research results dealing with diatoms exposed to metals are reviewed in the first part of this chapter. The impacts most frequently observed on periphytic algae are addressed at different organization levels (from the individual cell to the community structure). Diatom communities may respond similarly to metal pollution, regardless of the region investigated. Consequently, we built a database with diatom species composition and corresponding information on water metal content from six different countries (France, Switzerland, Spain, Vietnam, China and Canada) to investigate the relationships between diatom communities and metal contamination, without considering other determinants (nutrient bioavailability, geographical location, seasonality, stream order, etc). The main goals of this case study based on a multi-country dataset were (1) to investigate how metal exposure drives diatom community patterns in a comparable way among countries, (2) to assess the information brought by nontaxonomical indicators (diatom deformations, cell size and diatom growth forms) for the monitoring of metal pollution, and (3) to determine the indicative value of the most representative species occurring with significant abundances in the studied streams.

2 Effects of Metals on Freshwater Diatom Communities

Metal toxicity on diatoms is linked to different steps in the circulation of the toxicant (Fig. 1) across the membrane (especially uptake mechanisms) and inside the cell, inducing perturbations in the normal functioning of structural/functional intracellular components. Diatom communities exposed to metals have, therefore, variable capacities to tolerate the stress caused by the toxicant. Tolerance (or resistance) is developed at the individual scale (with different levels of sensitivity among species) and also at the community scale where the biofilm acts as a coherent and protective matrix.

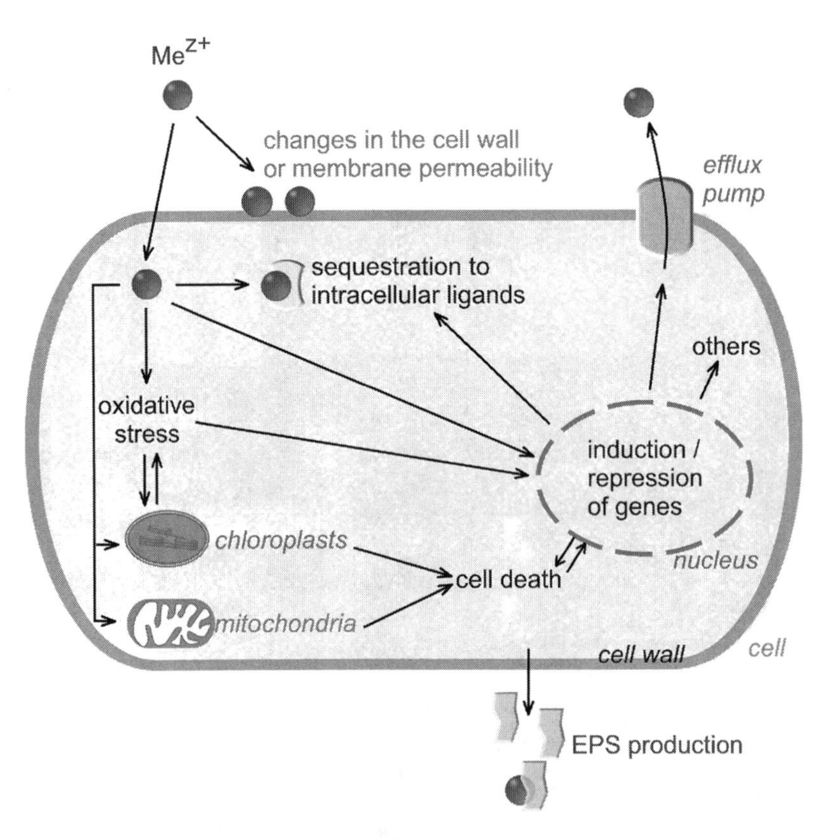

Fig. 1 Metal circulation in the cell and resulting potentially harmful effects at the cellular level. Metal influx can alter the membrane permeability; once in the cell, metals can induce an oxidative stress, affect the photosynthetic apparatus or mitochondria, and modify genetic expression, eventually leading to apoptosis. Several mechanisms are known to protect the cell against these toxic effects, such as metal binding by intracellular ligands, active expulsion, or EPS production for intracellular binding of the metals

2.1 Community Size Reduction

Exposure to metals leads to malfunction of cell metabolic processes (primary productivity, respiration, nutrient and oligoelement fluxes) (e.g. [15–17]), and reproductive characteristics (vegetative versus sexual reproduction) [13, 18], as well as increase in cell mortality [19]. Community size may be impaired through three complementary ways: (1) reduction of cell number, (2) selection for small-sized species, and (3) diminution of cell sizes within a given species.

Diatoms can accumulate high amounts of metals [20–24], which affects phosphorus metabolism [16], photosynthesis by production of reactive oxygen species [25] or by alteration of the functioning of the xanthophyll cycle [15, 26, 27], and homeostasis [28]. Thus, diatom growth can be delayed, or inhibited, leading to a reduction of diatom biomass [16, 29, 30]. In addition to lower survival and growth rates [31, 32], changes in emigration/immigration strategies [33] could also be responsible for the reduction in diatom cell densities and biomass [34].

Metal uptake depends on cell surface area exposed to the medium [35, 36], and can be reduced by physical protection offered by the exopolysaccharidic matrix [37] within the biofilm. This mechanical protection can be more effective for small-sized species, and thus this might be a mechanistic explanation for their positive selection under heavy metal pollution [18, 38, 39].

Reduction of cell size within taxa with metal exposure is probably linked to the mitotic division peculiar to diatoms, an important feature distinguishing these organisms from other algae. Hence, each division results in two daughter cells, one of which has the same size as the mother cell, and the other being smaller. As a consequence, average cell size at the population level is reduced with each successive round of mitosis [40]. Because the vegetative reproduction is the dominant mode of multiplication in diatoms [41], the decrease in size of many taxa observed in metal-contaminated environments [14, 42–44] could be a result of higher cell division rate inherent to organisms inhabiting in stressed ecosystems [45, 46].

Altogether, these combined effects of metal stress on diatom community would explain the significantly lower diatom biomass that is often observed in metal-contaminated environments.

2.2 Selection of Diatom Growth Forms

Metal exposure may modify the three-dimensional architecture of the diatom community by favouring some growth forms and constraining the development of others. Species colonization/growth strategies are driven by metal levels [47]: the communities that develop in such heavily impacted environments are dominated, even over long-term periods, by pioneer, substrate-adherent species that are more metal tolerant according to Medley and Clements [38]. It is the case, for example, of the cosmopolitan diatom *Achnanthidium minutissimum* frequently dominating in lotic environments exposed to toxic events, and considered as an indicator of metal pollution [48, 49]. Subsequent colonizers are generally stalked or filamentous, even motile species, and constitute the external layers of the biofilm under undisturbed conditions. Their development is less important in case of metal exposure, which results in the formation of thinner biofilms [50–52].

2.3 Diatom Teratologic Forms

The appearance of abnormal individuals is among the most striking effects of metals on diatom metabolism and has been widely reported in highly contaminated environments (see review in [53]). The deformities can affect the general shape of the frustule, and/or its ornamentations, which led Falasco et al. [54] to the description of seven types of abnormalities (Table 1). The most frequently observed are distortions of the cell outline (in particular in Araphid diatoms) [53–56], and changes in striation patterns. Deformities can be initiated at different stages throughout the diatom life cycle, and the processes leading to abnormal cell formation are yet unsolved. Current knowledge of diatom morphogenesis suggests direct and indirect effects consecutive to metal uptake on many cytoplasmic components involved in valve formation. The most documented negative effects of metal contamination on diatoms are nucleus alterations (e.g. [57]) and/or poisoning of the microtubular system involved in the transport of silica towards silica deposition vesicles [58].

Although no standard based on diatom teratologies has yet been established, many authors suggested using their occurrence as potential indicators of high metal pollution [48, 59–62].

2.4 Selection of Tolerant Species

Diatom species composition is driven by several environmental factors. Among the chemical parameters, the exposure to toxic agents such as metals can be a major determinant. Metal contamination selects for species able to tolerate metal-related stresses, whereas sensitive species tend to decrease in number or ultimately disappear. This is the conceptual basis of the Pollution-Induced Community Tolerance (PICT) concept developed by Blanck et al. [63], where the structure of a stressed community is rearranged in a manner that increases the overall community tolerance to the toxicant. Because the impacts specifically caused by metals are generally difficult to separate from other stressors, there is no agreement on the sensitivity or tolerance for some particular species. The influence on biofilm structure of other environmental parameters such as physical characteristics, nutrient availability or even biological interactions [64–68] may be one of the major reasons of these contradictory results.

A list of the presumptive sensitivity or tolerance of species based on those reported in the literature to occur, or disappear, in metal-contaminated

Table 1 Types of diatom deformities (from [54]). Scale bar: 10 μm

Teratology	Description	Example of normal vs. deformed individuals
Type 1	Deformed valve outline (loss of symmetry, pentagonal or trilobate shapes, abnormal outline)	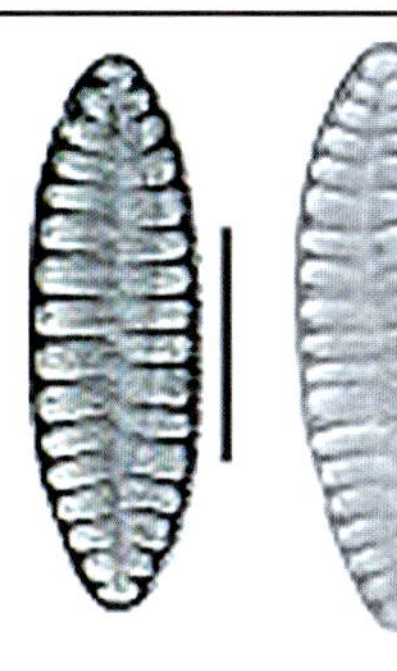 *Surirella angusta*, Charest river, Canada
Type 2	Changes in striation pattern, costae and septae	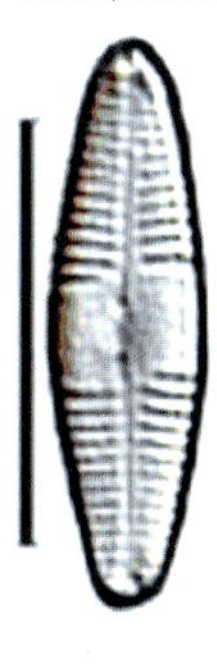 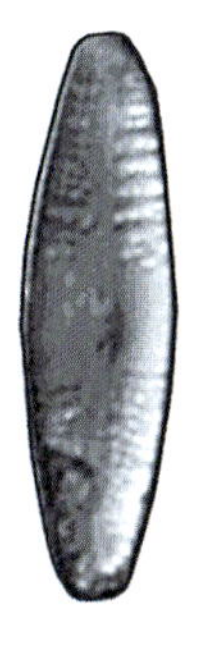 *Caloneis bacillum*, Nant d'Avril river, Switzerland
Type 3	Changes in shape, size and position of the longitudinal and central area (e.g. displaced, doubled, abnormally enlarged, absent)	  *Planothidium frequentissimum*, Riou Mort river, France
Type 4	Raphe modifications (split, sinuate or fragmented, changes in orientation, occasionally absent)	

(continued)

Table 1 (continued)

Teratology	Description	Example of normal vs. deformed individuals
		  *Gomphonema micropus* (SEM), Riou-Mort river, France
Type 5	Raphe canal system modifications (distorted, displaced, stretched out fibulae)	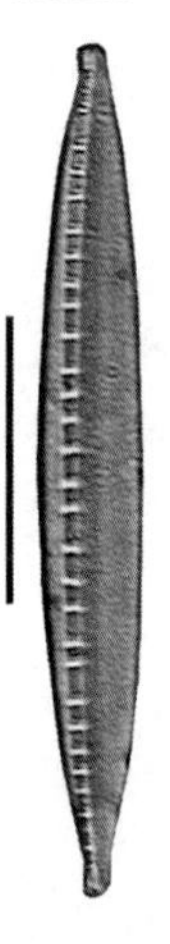  *Nitzschia dissipata*, Osor river, Spain
Type 6	Unusual arrangement of the cells forming colonies	
Type 7	Mixed type in which one valve shows more than one kind of teratology	 *Gomphonema truncatum*, Deûle river, France

Table 2 List of species that are described in the literature to disappear or be favoured in metal-contaminated environments

Species	Decrease/disappear with metals	Increase/still present with metals
Achnanthidium minutissimum (Kützing) Czarnecki	++	++++
Asterionella formosa Hassall	++	
Cocconeis placentula Ehrenberg var. *placentula*	++	+
Cyclotella meneghiniana Kützing	+	++
Diatoma vulgaris Bory	+	+
Encyonema minutum (Hilse *in* Rabhenhorst) D.G. Mann	+	++
Eolimna minima (Grunow) Lange-Bertalot		++++
Eunotia exigua (Brébisson ex Kützing) Rabenhorst		++
Fragilaria capucina Desmazières var. *capucina*	+	+++
Fragilaria capucina Desmazières var. *vaucheriae* (Kützing) Lange-Bertalot		++
Fragilaria crotonensis Kitton	+	++
Fragilaria rumpens (Kützing) G.W.F.Carlson	+	+
Gomphonema parvulum (Kützing) Kützing var. *parvulum*	+	++++
Mayamaea permitis (Hustedt) Bruder & Medlin	+	++
Melosira varians Agardh	+++	+
Navicula lanceolata (Agardh) Ehrenberg	+	+
Navicula tripunctata (O.F.Müller) Bory	+	++
Naviculadicta seminulum (Grunow) Lange Bertalot	+	+++
Nitzschia dissipata (Kützing) Grunow var. *dissipata*	++	++
Nitzschia linearis (Agardh) W.M.Smith var. *linearis*	+	++
Nitzschia palea (Kützing) W.Smith		++++
Pinnularia parvulissima Krammer		++
Planothidium frequentissimum (Lange-Bertalot) Lange-Bertalot	+	++
Planothidium lanceolatum (Brébisson ex Kützing) Lange-Bertalot		++
Staurosira construens Ehrenberg	++	+
Surirella *angusta* Kützing		+++
Surirella brebissonii Krammer & Lange-Bertalot var. *brebissonii*	+	++
Tabellaria flocculosa (Roth) Kützing	++	
Ulnaria ulna (Nitzsch) Compère	++	+++

Cited: +: more than once, ++: in more than five references, +++: in more than 10 references, ++++: in more than 20 references

environments [13, 14, 21, 29, 38, 39, 52, 56, 60, 61, 64–67, 69–125] is presented in Table 2. This table includes only the species that were cited in at least five papers; species that were found in both categories (either to tolerate or to be sensitive, depending on the references) are also listed. The diatoms *Eolimna minima,*

Gomphonema parvulum or *Nitzschia palea* are described as metal tolerant in a number of studies, whereas the sensitivity of species seems to be more difficult to determine, with only *Melosira varians* cited more than ten times to disappear in metal-contaminated environments. For some controversial species, such as *Ulnaria ulna*, Duong et al. [78] found different sensitivities that they ascribed to seasonal variability. The species *Achnanthidium minutissimum*, frequently dominant in lotic environments subjected to toxic events, is generally considered as indicator of metal pollution [14, 48, 49], but also indicates good general water quality (e.g. [3, 8]). Indeed, this species has been found to remain in highly metal-polluted conditions but to disappear with increasing trophy [61, 126, 127].

2.5 Tolerance Mechanisms

Diatoms present constitutive (phenotypic) and adaptative mechanisms to cope with elevated metal concentrations [128]. Defence and detoxification mechanisms mitigate perturbations in cell homeostasis caused by metal exposure. Regulation of metal fluxes through the cell may be driven by limitation of the influx, storage of the metal in the cytosol in insoluble form, neutralization of oxidative stress and active expulsion out of the cell (Fig. 1).

The limitation of the amount of metal entering the cell is linked to the decrease in free, i.e. bioavailable, ion concentration. Exposure enhances the production of polysaccharidic exudates (e.g. extracellular polymeric substances, EPS) able to bind metals outside of the cell [37, 106], in general proportionally to the concentration of metal exposed [129], thus leading to immobilization of the complexes outside the cell in a less bioavailable form. Recent studies indicate that frustulines, membrane-bound peptides linked to the diatom frustule resistance, may also play a role in metal binding [130]. Other regulation mechanisms have been described to occur at the cell surface. Some of the metals may be entrapped by iron or manganese hydroxides covering the cell wall [32]. Pokrovsky et al. [108] described the saturation of ligands (phosphoryl, sulfydryl) on diatom surfaces in highly polluted environments, leading to reduced adsorption capacities. Alterations of the membrane during metal internalization can lead to a decrease in membrane permeability [131, 132].

Internal mechanisms of storage contribute to efficient tolerance to metals. Metal induces the production of thiol-rich polypeptides known as phytochelatins [133, 134] or polyphosphate bodies [135], which are polymers that sequester intracellularly the excess of metal in a stable, detoxified form [27, 136, 137]. Resulting tolerance is variable among diatom species [138] and metals: sequestration capacities of the metal/protein complex depend, for example, on their valence characteristics [22, 137].

Cell defence against harmful effects of oxidative stress caused by metals relies on two main mechanisms. Increase in the production of proline [139, 140] and low-molecular-weight thiols (especially glutathione) [141–143] plays an antioxidant and detoxifying role. Metal exposure also induces activation of enzymes like superoxide-dismutase [142, 144] that convert superoxide anions into a less toxic form.

Excretion mechanisms of complexing compounds contribute to tolerance to toxicants [145]. Exposure to metals leads to increasing production of polysaccharides, which can bind metals externally after being exported in the extracellular environment [106]. Moreover, Lee et al. [146] described efflux of phytochelatin/cadmium complexes in *Thalassiosira weissflogii* exposed to high cadmium concentrations. Active expulsion by ATPase pumps as described in bacteria could also play a role in detoxification and survival of phytoplankton species [147].

The protective role of the matrix towards metals has been attributed to many features of the biofilm: metal-binding capacities of the polysaccharidic secretions [106, 148], local pH and hypoxia conditions in the internal layers of thick biofilms [149, 150], species interactions [151] and reduction of the exchanges between the inner cells and the environment [93, 152] partially linked to the presence of a superficial layer of dead cells [145].

As presented in this literature review, many studies generally performed at the watershed scale described the effects of metal contamination on cell size, growth forms, cell morphology, as well as diatom community structure. The second part of this chapter consists in assessing the relevance of these endpoints on a larger scale, using a multi-country dataset of diatom samples.

3 Case Study: A Multi-Country Database

3.1 Sites Studied

The diatom database consists of 202 samples of mature biofilms collected from hard substrates in rivers of circumneutral pH between 1999 and 2009. At each sampling unit, benthic diatoms were scraped from randomly collected substrates to form one composite sample, and preservatives (formaldehyde 4% or concentrated Lugol's iodine) were added to stop cell division and prevent organic matter decomposition. Most of them come from different parts of Europe: France (61 samples), Switzerland (15 samples) and Spain (71 samples); data from Eastern Canada (23 samples), Vietnam (18 samples) and China (14 samples) were also included. Some of the data have already been published for other purposes [12, 39, 61, 77, 90, 126, 153–155].

3.2 Diatom Analyses

Periphytic samples were cleaned of organic material before mounting permanent slides with Naphrax® (Brunel Microscopes Ltd, UK) for diatom identification based on observation of the frustule. Transects were scanned randomly under light microscopy at a magnification of ×1,000 until at least 400 valves were identified. Taxa were identified to the lowest possible taxonomical level, according to standard floras [156–162] and recent nomenclature updates that are listed in https://hydrobio-dce.cemagref.fr/en-cours-deau/cours-deau/Telecharger/indice_biologique_diatomee-ibd/. The diatom database was harmonized for taxonomy, leading to a final list of 640 taxa.

3.3 Determination of Metal Exposure

The metal data used come from surveys conducted simultaneously with diatom sampling. The water samples were collected in streams from various watersheds, where diatom communities were exposed to mixtures of dissolved metals (mainly Al, As, Cd, Cr, Cu, Fe, Hg, Ni, Pb, Se and Zn) in variable concentrations. To facilitate data interpretation, we used an estimate of metal concentration and toxicity developed by [163] and already used to investigate the responses of aquatic organisms to metals [163, 90, 91]. The CCU (cumulative criterion unit) is a score based on the sum of the ratios between metal concentrations measured in filtered waters and the corresponding criterion value (US EPA's National Recommended Water Quality Criteria, http://www.epa.gov/waterscience/criteria/wqctable/). Four categories of CCU were used following the thresholds determined by Guasch et al. [90]. CCU below 1.0 corresponded to background levels (B), low metal category (L) was characterized by CCUs between 1.0 and 2.0, and intermediate metal category (M) by CCUs between 2.0 and 10.0. For scores above 10.0, we added a high metal category (H) to Guasch´s classification.

Metals in the water samples ranged from undetectable concentrations (leading to CCU values of 0.0) to CCU scores higher than 1,000 (Spain, Rio Cea, June 2007). The distribution of the samples according to CCU categories in the different countries is given in Table 3, showing a balanced repartition of the samples in the different categories. The four CCU classes were all found in samples from France, Spain and Switzerland, but were unequally distributed among the other countries. Globally, CCU scores were due to various metals that were different between countries. Indeed, in some cases, we found two main metals contributing to CCU scores, such as in the samples from China (Cr > Pb), France (Cd > Zn), Switzerland (Cu > Zn) and Vietnam (Cd > Pb). On the contrary, much more metals were found in Canada (Zn > Cd > Al > Pb) and Spain (Se > Ni > Zn > Pb > Al, with one different dominant contributor in most of the samples). Metals were also unequally distributed between CCU categories. The highest values (H) were generally due to Cd, Zn, and to a lower extent, Pb, whereas in the L and M categories contributions were quite balanced and mostly involved Se, Pb, Cu and Cd.

Table 3 Data are average ± standard error

CCU category	B	L	M	H
Distribution of samples				
Number of samples	58	37	50	57
Minimum CCU value; metal mixture (μg/L)		1.02 Cu (3.2), Se (3.0), Zn (8.0)	2.04 Se(10.0), Zn (4.8)	7.04 Al (15.3), Cd (0.6), Cu (0.9), Fe (92.4), Ni (1.6), Zn (514.3)
Maximum CCU value; metal mixture (μg/L)		1.96 Cu (5.7), Se (6.0), Zn (15.7)	6.26 Pb (15.4), Zn (13.6)	1271.0 As (50.0), Cd (315.0), Hg (8.2)
Countries	CA, CH, CN, ES, FR	CH, CN, ES, FR	CA, CH, CN, ES, FR, VN	CA, CH, ES, FR
Distribution of cell sizes				
Mean biovolume per cell (μm^3)	774 ± 67^{a}	676 ± 79^{a}	$1,293 \pm 134^{b}$	524 ± 51^{a}
% of small-sized taxa (<100 μm^3)	25.7 ± 2.5^{a}	$31.4 \pm 3.3^{a,b}$	24.7 ± 3.5^{a}	43.1 ± 3.9^{b}
% of medium-sized taxa (100–1,000 μm^3)	59.1 ± 2.0^{a}	$50.9 \pm 3.5^{a,b}$	42.3 ± 2.6^{b}	42.2 ± 3.5^{b}
% of larger taxa (>1,000 μm^3)	15.2 ± 1.8^{a}	17.6 ± 2.7^{a}	33.0 ± 3.6^{b}	14.7 ± 2.2^{a}
Occurrences of valve deformities				
Abnormal diatom valves (‰)	3.6 ± 0.7^{a}	2.7 ± 0.7^{a}	4.3 ± 1.0^{a}	9.4 ± 1.7^{b}
Distribution of growth forms and postures				
Solitary not attached (%)	47.6 ± 3.4	36.0 ± 4.7	50.0 ± 4.3	45.1 ± 4.0
Solitary prostrate (%)	$17.6 \pm 2.6^{a,b}$	22.6 ± 3.6^{a}	9.5 ± 1.7^{b}	21.2 ± 3.2^{a}
Solitary erected (%)	19.5 ± 2.1	29.4 ± 3.7	16.8 ± 2.6	22.1 ± 3.0
Clump-forming (%)	0.8 ± 0.2	0.7 ± 0.2	0.7 ± 0.3	1.5 ± 0.5
Filamentous (%)	10.1 ± 1.7^{a}	$10.9 \pm 2.7^{a,b}$	21.8 ± 3.4^{b}	9.2 ± 1.7^{a}
Diversity indices				
Species richness	42.0 ± 2.8^{a}	44.9 ± 3.9^{a}	47.8 ± 3.5^{a}	29.1 ± 2.5^{b}
Shannon diversity index	2.46 ± 0.07^{a}	$2.39 \pm 0.13^{a,b}$	$2.29 \pm 0.11^{a,b}$	2.01 ± 0.11^{b}

Countries' abbreviations: CA Canada, CH Switzerland, CN China, ES Spain, FR France, VN Vietnam. Homogenous groups are noted with the same letters (a, b)

3.4 Non Taxonomical Indicators

Traits like diatom cell size, distribution of growth forms and postures (diatoms forming filaments, clumps or solitary forms including erected, prostrate and not attached cells) and proportion of valve abnormalities were investigated. Those descriptors present the major interest of being independent of the biogeographical variability of the natural communities. They strongly suggested in many cases to be indicative of perturbations such as toxicant exposure [13, 14, 38, 39, 60, 61], and were thus tested using the complete dataset based on the exhaustive list of 640 taxa. Specific biovolumes were calculated from the average dimensions provided in the floras for each taxon and using the formulae of Hillebrand et al. [164] established for the different geometrical shapes. The proportion of valve abnormalities was directly inferred from the taxonomical counts, by adding up the relative abundances of the individuals that had unusual shape and/or ornamentation of the frustule, and expressed in ‰. The distribution of growth forms and postures (in the case of solitary cells) was determined for each genera or occasionally species, according to the observations of Hoagland et al. [165], Hudon and Bourget [166], Hudon et al. [167], Katoh [168], Kelly et al. [169] and Tuji [170].

The patterns of those traits in the four CCU categories were compared using 1-way ANOVA (Statistica v5.1, StatSoft Inc., Tulsa, USA) after checking for normality of the data. Statistically significant probability level was set at $p < 0.01$. Spjotvoll/Stoline HSD tests for unequal sample sizes were performed for post hoc comparisons.

Valve abnormalities. Abnormal diatoms are generally observed in very low relative abundances, and authors agree that an average value of 10‰ is a significant threshold for metal-induced teratologies [14, 60, 61, 78, 171]. Indeed, occurrences of 3.5 ± 0.5‰ (i.e. values recorded in the B, L and M metal categories that were not discriminated by post hoc tests) can be considered as naturally occurring, or "background" levels. Previous laboratory experiments with Cd demonstrated that the percentage of valve abnormalities was not linearly correlated with metal concentration, but could be attributable, when above 10‰, to toxicity caused by concentrations above a given threshold [21, 100]. Thus we can suppose that teratologies occur in nature when a certain level of metal contamination is reached. When examining the distribution of abnormalities of this dataset along CCU scores (Fig. 2), statistical tests separated two sets of data: CCU values higher than 7.0 with average abnormalities frequency reaching values of about 10‰, and CCU values below 7.0 with average abnormalities frequency of ca. 3.5‰. This field-based evidence allowed us to refine the arbitrary threshold of the H category (CCU = 10.0) to a new threshold value of 7.0 that was used further on in the study.

When considering the type of diatom deformities, most of the cases concerned both global shape and ornamentation. In the Canadian samples, however, only the outline of the frustules was affected (Type 1 deformities as defined by [54]). The contribution of metals to the final CCU score was also different between the four countries where values higher than 7.0 were found. In most cases, deformities were

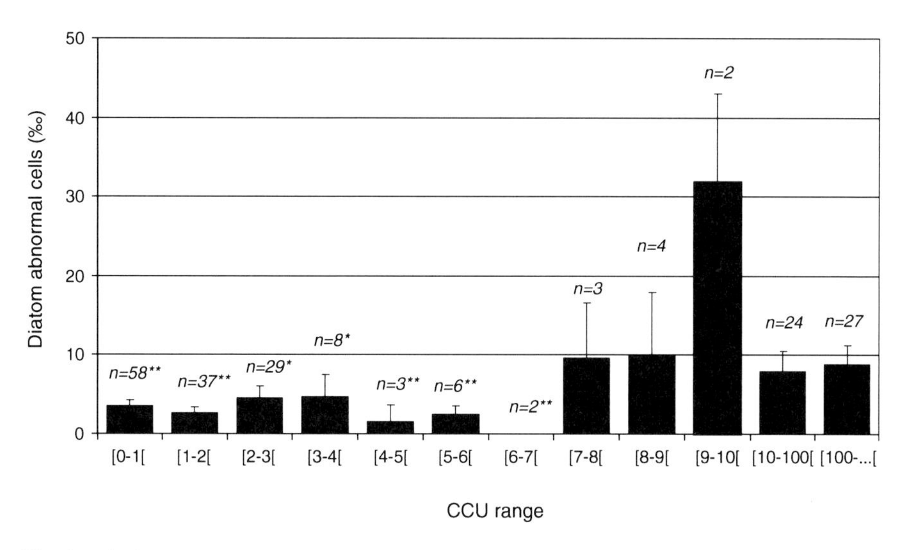

Fig. 2 Distribution of valves abnormalities within the CCU ranges. n = number of samples per CCU range. Statistically different from CCU range [9–10]: *: $p < 0.05$; **: $p < 0.01$

estimated to be caused by a "dominant" metal, as reflected by diversity indices based on metal contributions (metal diversity = 0.44 ± 0.06), whereas in Canada CCU scores were explained by a more balanced contribution of different metals (metal diversity = 1.31 ± 0.12; [172]). The calculation of CCUs is based upon the assumption that the adverse effects of metals are additive. However, the nature of the deformities observed indicates that there are differences in effects that could be linked to the balance between the metals contributing to the CCU scores, suggesting that alternative methods are needed to explain differences between the types of deformities. Guanzon et al. [22] evidenced competition between the metals that coexist in the medium, for the fixation on membrane binding sites. In their experimental exposures to binary and ternary mixtures of Cu, Zn and Cd, the diatom *Aulacoseira granulata* adsorbed and accumulated reduced quantities when compared to single-metal exposures. Since deformity formation is likely to be provoked by the metals absorbed, we can thus suppose that metal toxicities are not purely additive (in the particular case of abnormalities induction), with lower "teratogenic" power in the case of mixtures, or that some metals are more "teratogenic" than others and can also have different pathways. On the other hand, deformities have been widely described in long-term cultures [173, 174], and have been ascribed to somatic alterations linked to artificial conditions. The balance between metals was not taken into consideration, as metal concentrations in the culture medium were generally low. However, the consumption by the cultured cells of some of the oligoelements may modify, in the long term, the balance between essential and non-essential metals in the environment, which could be an alternative explanation of the occurrence of teratology in laboratory cultures.

Cell biovolumes. It has been demonstrated that small-sized species dominate in metal-contaminated environments (Sect. 2.1). Using this large database, we tried to link mean community biovolume with the gradient in CCUs, but there was no significant trend of cell size reduction with increasing metal pollution (Table 3). Diatom mean biovolume was, moreover, significantly higher in the M metal category than in the other ones, linked to higher abundances of larger taxa. Medium-sized taxa were found in higher abundances in the B and L metal categories, and a quite significant increase in small-sized taxa abundances was observed in the H categories. Indeed, there is not necessarily a decrease in average community cell size with increasing metal pollution, but higher amounts of small-sized taxa, which could in many cases not be sufficient to result in a significant decrease in mean biovolume.

Growth forms. Diatom growth forms' distribution was highly variable within CCU groups, somehow more than between categories. Samples were dominated by motile, non-attached species. In the high metal categories (M and H), these species tended to be more abundant (47.4 ± 2.9 % vs. 43.1 ± 2.8 % in B and L categories), although this trend was not statistically significant. Some studies evidenced that motile species would be less disfavoured than attached ones in metal-stressed environments [175]; however, we were unable to demonstrate this clearly using our database. Mitigation of the effects of metals by the environmental conditions [68, 70] should also be considered and we can suppose that, in a given watershed, this estimate is a good indicator of increasing metal pollution. However, changes in the community structure and thus in growth forms depend on the pool of species present (i.e. constrained by environmental drivers). The results from this study suggest that general environmental differences are likely to have stronger effects than species selection by metal contaminations. For this reason, the use of growth forms for biomonitoring metal pollution would not represent a reliable approach applicable in a large-scale context.

3.5 Global Patterns of Diatom Communities in Response to Metal Contamination

To investigate the common patterns in diatom communities between countries, the taxa that were only present in one country were removed from the analysis to exclude species that are either endemic or identified differently by operators/countries, and to reduce errors associated with the morphological approach to diatom identification, especially in the case of ambiguous species (e.g. [176]). Moreover, rare species (i.e., those that were observed in less than 5% of the samples and/or that had maximal relative abundances $< 1\%$) were not included in the analysis. This selection of data led to a final set of 152 taxa out of the initial list of 640. Prior to analysis, diatom counts were log-transformed, centred and scaled. A linear discriminant analysis (LDA) was then performed using the ade4 package

[177] implemented in the R statistical software [178]. The LDA was used to classify the dataset into exclusive groups corresponding to the four CCU categories as described above, using a M/H boundary of 7.0. The IndVal method of Dufrêne and Legendre [179] was used to identify the indicator value of each species to determine the most structuring ones for each CCU category.

Among the 640 taxa, 16 were found in the six countries covered in this study (see Fig. 3b), and 56 were observed in at least five of the countries, pointing out high "cosmopolitanism". An overall decrease in community complexity, i.e. declining species richness (ANOVA, $p < 0.001$) and diversity ($p = 0.0077$), was observed in the H metal category. Indeed, in the most contaminated cases, communities were dominated by one single species, representing on average 43.3% relative abundances and reaching in many cases values higher than 90%. The concomitant loss of sensitive species with development of more resistant species allowed for a clear discrimination by the LDA of three subsets of data, grouping B and L communities together, and separating them from M and H categories (Fig. 3). The species that were found in at least three countries and that had highly significant IndVals ($p < 0.05$) are given in Table 4. The tolerant indicator species identified were generally not those that showed most deformities, and many taxa that were not structuring the dataset also exhibited teratologies. Globally, the taxa characterizing M and H categories were in accordance with literature data (see Table 2), whereas contradictory results were found for some of the B and L categories. For example *Encyonema minutum, Mayamea permitis* and *Planothidium lanceolatum* are generally described as tolerant taxa, but were mostly found in the B and L categories. The extinction of sensitive species would be a strong signal to use for biomonitoring purposes; however, it seems that sensitivity to metals is more difficult to unequivocally determine than tolerance, maybe because of the importance of other environmental factors. Indeed, under non-contaminated conditions, competition for resource utilization selects for the species best adapted to their specific environment, whereas in metal-polluted conditions sensitive taxa tend to disappear leading to reduced competitive exclusion among species and the selective development of tolerant species, whatever their resource-competitive abilities. Diatom-based monitoring of metal pollutions would then be more relevant using the occurrences of metal-tolerant species than using a metric combining both sensitivities and tolerances of all species in the community.

The high cosmopolitanism observed indicates that metal-tolerant species derived from this study could be used to develop metal pollution diatom indices with a broad geographical application. Moreover, the specific information obtained in each non-taxonomical endpoint (e.g. relative abundances of small-sized species, of morphological abnormalities) could be used to improve sensitivity of such indicator for regional applications.

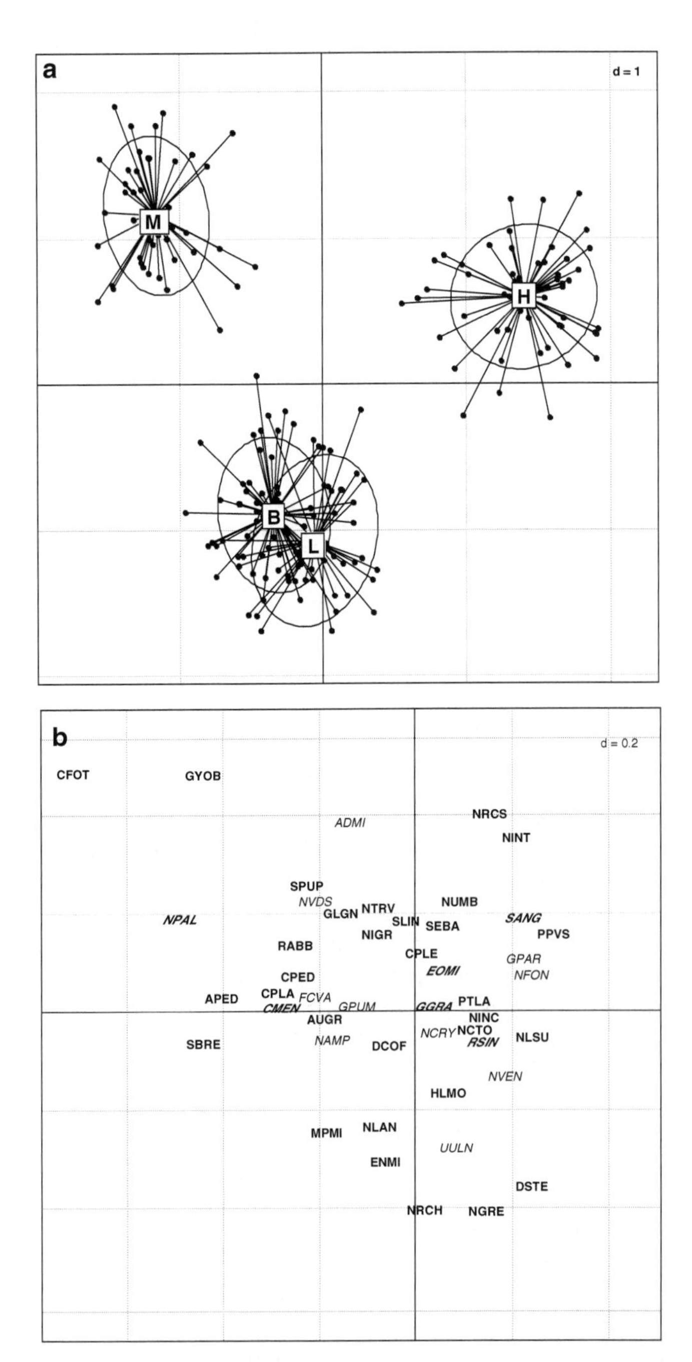

Fig. 3 Linear Discriminant Analysis of diatom community structure, constrained by CCU categories. (**a**) Projection of the samples, grouped by CCU category; (**b**) diatom species with highest indicative values (*in bold*) and the taxa common to the six countries (*in italics*). Species abbreviations: ADMI *Achnanthidium minutissimum*, FCVA *Fragilaria capucina* var. *vaucheriae*, NCRY *Navicula cryptocephala*, NVEN *N. veneta*, NVDS *Naviculadicta seminulum*, NAMP *Nitzschia amphibia*, NFON *N. fonticola*, UULN *Ulnaria ulna* and see Table 4

Table 4 Main structuring species of the different metal categories

Species	Abbreviation	CCU category	IndVal (%)	*p*-value
*Amphora pediculus**	APED	B-L	28.5	0.029
*Cocconeis placentula**	CPLA	B-L	29.5	0.003
Cocconeis pseudolineata	COPL	B-L	10.7	0.039
Encyonema minutum	ENMI	B-L	31.2	0.010
Fragilaria virescens	FVIR	B-L	9.8	0.018
Gomphonema gracile	GGRA	B-L	16.3	0.039
Gomphonema pumilum	GPUM	B-L	14.0	0.027
*Mayamaea permitis**	MPMI	B-L	28.7	0.005
Navicula gregaria	NGRE	B-L	40.5	0.001
Navicula lanceolata	NLAN	B-L	27.3	0.005
Navicula notha	NNOT	B-L	12.0	0.033
Navicula reichardtiana	NRCH	B-L	20.3	0.038
Nitzschia hantzschiana	NHAN	B-L	9.4	0.044
*Nitzschia inconspicua**	NINC	B-L	25.9	0.010
Parlibellus protracta	PPRO	B-L	13.6	0.019
*Planothidium lanceolatum**	PTLA	B-L	32.2	0.004
*Reimeria sinuata**	RSIN	B-L	23.4	0.037
*Rhoicosphenia abbreviata**	RABB	B-L	25.3	0.015
*Achnanthidium subatomus**	ADSU	M	15.0	0.016
Aulacoseira ambigua	AAMB	M	13.4	0.007
Aulacoseira granulata	AUGR	M	44.9	0.001
Bacillaria paxillifera	BPAX	M	11.7	0.017
Cocconeis pediculus	CPED	M	21.3	0.022
Cyclotella fottii	CFOT	M	32.4	0.001
*Cyclotella meneghiniana**	CMEN	M	26.5	0.026
Cyclostephanos invisitatus	CINV	M	12.1	0.034
Cymbella tumida	CTUM	M	12.9	0.024
Cymbella turgidula	CTGL	M	12.0	0.003
Diadesmis confervacea	DCOF	M	19.8	0.001
Discostella pseudostelligera	DPST	M	11.3	0.031
Discostella stelligera	DSTE	M	19.2	0.001
Gomphonema lagenula	GLGN	M	15.8	0.007
Gyrosigma obtusatum	GYOB	M	33.2	0.001
Halamphora montana	HLMO	M	31.1	0.001
*Lemnicola hungarica**	LHUN	M	12.3	0.002
Luticola mutica	LMUT	M	11.9	0.016
Navicula catalanogermanica	NCAT	M	13.5	0.018
Navicula cryptotenelloides	NCTO	M	19.0	0.008
*Navicula recens**	NRCS	M	31.8	0.001
Navicula trivialis	NTRV	M	27.2	0.001
Nitzschia filiformis	NFIL	M	9.5	0.049
Nitzschia gracilis	NIGR	M	15.2	0.004
Nitzschia intermedia	NINT	M	30.9	0.001
Nitzschia linearis var. *subtilis*	NLSU	M	16.5	0.001
*Nitzschia palea**	NPAL	M	36.3	0.007
*Nitzschia umbonata**	NUMB	M	28.5	0.001

(continued)

Table 4 (continued)

Species	Abbreviation	CCU category	IndVal (%)	*p*-value
Sellaphora bacillum	SEBA	M	17.6	0.001
Sellaphora pupula	SPUP	M	24.1	0.006
Surirella brebissonii	SBRE	M	28.7	0.010
Surirella linearis	SLIN	M	20.1	0.001
Surirella minuta	SUMI	M	12.5	0.003
Tabularia fasciculata	TFAS	M	13.2	0.003
Cocconeis placentula var. *euglypta*	CPLE	H	20.2	0.002
*Eolimna minima**	EOMI	H	35.3	0.006
*Fragilaria gracilis**	FGRA	H	12.3	0.032
Nitzschia sociabilis	NSOC	H	12.3	0.015
*Pinnularia parvulissima**	PPVS	H	18.2	0.003
*Surirella angusta**	SANG	H	26.6	0.039

*Species for which deformities were observed

4 Conclusions

Diatoms are ubiquitous and often predominant constituents of the primary producers in streams, and are sensitive to many environmental changes including metal concentrations. Even if many other environmental factors have a predominant influence on diatom community structure, the broad-scale patterns observed proved that diatom-based approaches are adequate for the monitoring of metal pollution, bringing ecological relevance based on specific sensitivities/tolerances at the community level.

The effects of metals can be observed at different levels, from the individual (deformations of the frustule) to the structure of the community. From a literature review and the analysis of indicator values determined from our large database, we provide lists of species that are likely to disappear, or to develop, with increasing metal contamination, thus providing a basis for the development of monitoring methods.

The CCU approach used in this study offers a satisfactory alternative for assessing the relationships between diatom communities and complex mixtures of metals in the field. A further step would be the development of indices taking into account metal diversity and potential toxicity to improve metal assessment in the field.

Finally, we can observe that, on the contrary to what was expected, the responses of diatoms were markedly different between rivers with intermediate and high metal pollution. In particular the increasing percentage of valve deformities proposed by many authors to be indicative of metal pollutions is, in fact, observed above all in cases of high metal contamination (H category). Our database allowed for the determination of a naturally occurring, or "background", abundance of deformed cells (3.5 ± 0.5‰) in environmental samples. A significant increase of abnormal cells was used to re-define a new threshold value for the H category from CCU $=$ 10 to 7.

In the natural environment, conditions corresponding to the H category are expected to happen less and less frequently, especially with the development of sustainable practices of industries and of site remediation. However, worldwide, many rehabilitation programs are being implemented in historical mining sites, and there is public demand for the evaluation of restoration success. Multiple abiotic and biotic criteria can be used to qualify/quantify the changes in "stream health" during and after rehabilitation programs. Diatom-based indicators would thus be an appropriate tool for assessing the success of rehabilitation actions and justify, through a recovery of the aquatic biota, the rehabilitation programs undergone and corresponding investments.

Acknowledgements The data used in this paper are based on the results of studies partially founded by the following programs: ANR ReSyst 08-CES-2009, EC2CO-CYTRIX (2008–2009), CNRS ACI ECCO-ECODYN (2003–006), Cemagref "PestExpo", ESPOIR (French Ministry of Foreign Affairs), FASEP n°694, MODELKEY 511237-2 GOCE and KEYBIOEFFECTS MRTN-CT-2006-035695. The authors also acknowledge the Agence de l'eau Artois Picardie, the Agència Catalana de l'Aigua, the Duero Basin Authority (Confederación Hidrográfica del Duero, CHD), the Natural Sciences and Engineering Research Council of Canada (NSERC), the Direction Générale de l'Eau (Département de l'Intérieur et de la Mobilité, Geneva, Switzerland), the Pearl River Water Resources Commission (PRWRC), Eric Baye (Asconit Consultants) and Paul B. Hamilton (Canadian Museum of Nature). Thanks to Marius Bottin and Elisa Falasco for useful comments and suggestions.

References

1. Hill BH, Herlihy AT, Kaufmann PR, Stevenson RJ, McCormick FH, Johnson CB (2000) Use of periphyton assemblage data as an index of biotic integrity. J N Am Benthol Soc 19(1):50–67
2. Potapova MG, Charles DF (2002) Benthic diatoms in USA rivers: distributions along spatial and environmental gradients. J Biogeogr 29(2):167–187
3. Coste M, Boutry S, Tison-Rosebery J, Delmas F (2009) Improvements of the Biological Diatom Index (BDI): description and efficiency of the new version (BDI-2006). Ecol Indic 9(4):621–650
4. Dell'Uomo A (2004) L'indice diatomico di eutrofizzazione/polluzione (EPI-D) nel monitoraggio delle acque correnti. Linee guida, APAT, ARPAT, CTN_AIM, Roma, Firenze, p 101
5. Hürlimann J, Niederhauser P (2007) Méthodes d'analyse et d'appréciation des cours d'eau. Diatomées Niveau R (région). État de l'environnement n° 0740. Office fédéral de l'environnement, Berne, p 132
6. Kelly MG, Whitton BA (1995) The Trophic Diatom Index: a new index for monitoring eutrophication in rivers. J Appl Phycol 7:433–444
7. Lange-Bertalot H (1979) Pollution tolerance of diatoms as a criterion for water quality estimation. Nova Hedwigia 64:285–304
8. Lavoie I, Campeau S, Grenier M, Dillon PJ (2006) A diatom-based index for the biological assessment of eastern Canadian rivers: an application of correspondence analysis (CA). Can J Fish Aquat Sci 63(8):1793–1811
9. Cordonier A, Gallina N, Nirel PM (2010) Essay on the characterization of environmental factors structuring communities of epilithic diatoms in the major rivers of the canton of Geneva, Switzerland. Vie Milieu 60(3):223–232

10. Rimet F, Goma J, Cambra J, Bertuzzi E, Cantonati M, Cappelletti C, Ciutti F, Cordonier A, Coste M, Delmas F, Tison J, Tudesque L, Vidal H, Ector L (2007) Benthic diatoms in western European streams with altitudes above 800 M: characterisation of the main assemblages and correspondence with ecoregions. Diatom Res 22:147–188
11. Tison J, Park YS, Coste M, Wasson JG, Ector L, Rimet F, Delmas F (2005) Typology of diatom communities and the influence of hydro-ecoregions: a study on the French hydrosystem scale. Water Res 39(14):3177–3188
12. Tornés E, Cambra J, Gomà J, Leira M, Ortiz R, Sabater S (2007) Indicator taxa of benthic diatom communities: a case study in Mediterranean streams. Annal Limnol 43(1):1–11
13. Cattaneo A, Asioli A, Comoli P, Manca M (1998) Organisms' response in a chronically polluted lake supports hypothesized link between stress and size. Limnol Oceanogr 43(8):1938–1943
14. Cattaneo A, Couillard Y, Wunsam S, Courcelles M (2004) Diatom taxonomic and morphological changes as indicators of metal pollution and recovery in Lac Dufault (Québec, Canada). J Paleolimnol 32:163–175
15. Bertrand M, Schoefs B, Siffel P, Rohacek K, Molnar I (2001) Cadmium inhibits epoxidation of diatoxanthin to diadinoxanthin in the xanthophyll cycle of the marine diatom *Phaeodactylum Tricornutum*. FEBS Lett 508(1):153–156
16. Guanzon NG, Nakahara H, Yoshida Y (1994) Inhibitory effects of heavy-metals on growth and photosynthesis of 3 freshwater microalgae. Fish Sci 60(4):379–384
17. Husaini Y, Rai LC (1991) Studies on nitrogen and phosphorus-metabolism and the photosynthetic electron-transport system of *Nostoc linckia* under cadmium stress. J Plant Physiol 138(4):429–435
18. Joux-Arab L, Berthet B, Robert JM (2000) Do toxicity and accumulation of copper change during size reduction in the marine pennate diatom *Haslea ostrearia*? Mar Biol 136(2): 323–330
19. Torres E, Cid A, Herrero C, Abalde J (1998) Removal of cadmium ions by the marine diatom *Phaeodactylum tricornutum* Bohlin accumulation and long-term kinetics of uptake. Bioresour Technol 63(3):213–220
20. Chang SI, Reinfelder JR (2000) Bioaccumulation, subcellular distribution and trophic transfer of copper in a coastal marine diatom. Environ Sci Technol 34(23):4931–4935
21. Duong TT, Morin S, Coste M, Herlory O, Feurtet-Mazel A, Boudou A (2010) Experimental toxicity and bioaccumulation of cadmium in freshwater periphytic diatoms in relation with biofilm maturity. Sci Total Environ 408(3):552–562
22. Guanzon NG, Nakahara H, Nishimura K (1995) Accumulation of copper, zinc, cadmium, and their combinations by 3 freshwater microalgae. Fish Sci 61(1):149–156
23. Sunda WG, Huntsman SA (1998) Control of Cd concentrations in a coastal diatom by interactions among free ionic Cd, Zn, and Mn in seawater. Environ Sci Technol 32(19): 2961–2968
24. Wang W-X, Dei RC (2001) Metal uptake in a coastal diatom influenced by major nutrients (N, P, and Si). Water Res 35(1):315–321
25. Knauert S, Knauer K (2008) The role of reactive oxygen species in copper toxicity to two freshwater green algae. J Phycol 44(2):311–319
26. Hill WR, Bednarek AT, Larsen IL (2000) Cadmium sorption and toxicity in autotrophic biofilms. Can J Fish Aquat Sci 57(3):530–537
27. Soldo D, Behra R (2000) Long-term effects of copper on the structure of freshwater periphyton communities and their tolerance to copper, zinc, nickel and silver. Aquat Toxicol 47(3–4):181–189
28. Cardozo KHM, De Oliveira MAL, Tavares MFM, Colepicolo P, Pinto E (2002) Daily oscillation of fatty acids and malondialdehyde in the dinoflagellate *Lingulodinium polyedrum*. Biol Rhythm Res 33(4):371–381
29. Gold C, Feurtet-Mazel A, Coste M, Boudou A (2003) Impacts of Cd and Zn on the development of periphytic diatom communities in artificial streams located along a river pollution gradient. Arch Environ Contam Toxicol 44:189–197

30. Payne CD, Price NM (1999) Effects of cadmium toxicity on growth and elemental composition of marine phytoplankton. J Phycol 35(2):293–302
31. Pérès F (1996) Etude des effets de quatre contaminants: - herbicide (Isoproturon), dérivés du mercure (mercure inorganique, méthylmercure), cadmium – sur les communautés au sein de microcosmes d'eau douce. PhD thesis, Univ. Paul Sabatier, Toulouse, p 176
32. Perrein-Ettajani H, Amiard JC, Haure J, Renaud C (1999) Effects of metals (Ag, Cd, Cu) on the biochemical composition and compartmentalization of these metals in two microalgae *Skeletonema costatum* and *Tetraselmis suecica*. Can J Fish Aquat Sci 56(10): 1757–1765
33. Peterson CG (1996) Mechanisms of lotic microalgal colonization following space-clearing disturbances acting at different spatial scales. Oikos 77(3):417–435
34. Paulsson M, Nystrom B, Blanck H (2000) Long-term toxicity of zinc to bacteria and algae in periphyton communities from the river Göta Älv, based on a microcosm study. Aquat Toxicol 47(3–4):243–257
35. Campbell PGC, Errecalde O, Fortin C, Hiriart-Baer VR, Vigneault B (2002) Metal bioavailability to phytoplankton – applicability of the biotic ligand model. Comp Biochem Physiol C Toxicol Pharmacol 133(1–2):189–206
36. Khoshmanesh A, Lawson F, Prince IG (1997) Cell surface area as a major parameter in the uptake of cadmium by unicellular green microalgae. Chem Eng J 65(1):13–19
37. Vasconcelos MTSD, Leal MFC (2001) Adsorption and uptake of Cu by Emiliania huxleyi in natural seawater. Environ Sci Technol 35(3):508–515
38. Medley CN, Clements WH (1998) Responses of diatom communities to heavy metals in streams: the influence of longitudinal variation. Ecol Appl 8(3):631–644
39. Morin S, Vivas-Nogues M, Duong TT, Boudou A, Coste M, Delmas F (2007) Dynamics of benthic diatom colonization in a cadmium/zinc-polluted river (Riou-Mort, France). Fundam Appl Limnol 168(2):179–187
40. Drebes G (1977) Sexuality. In: Werner D (ed) The biology of diatoms (Botanical monographs). Blackwell, Oxford, pp 250–283
41. Chepurnov VA, Mann DG, von Dassow P, Vanormelingen P, Gillard J, Inzé D, Sabbe K, Vyverman W (2008) In search of new tractable diatoms for experimental biology. Bioessays 30(7):692–702
42. Cattaneo A, Galanti G, Gentinetta S, Romo S (1998) Epiphytic algae and macroinvertebrates on submerged and floating-leaved macrophytes in an Italian lake. Freshwat Biol 39(4):725–740
43. Gensemer RW (1990) Role of aluminium and growth rate on changes in cell size and silica content of silica-limited populations of *Asterionella ralfsii* var. *americana* (Bacillariophyceae). J Phycol 26(2):250–258
44. Morin S, Coste M (2006) Metal-induced shifts in the morphology of diatoms from the Riou Mort and Riou Viou streams (South West France). In: Acs E, Kiss KT, Padisák J, Szabó K (eds) Use of algae for monitoring rivers VI. Hungarian Algological Society, Göd, Hungary, Balatonfüred, pp 91–106
45. Gensemer RW, Smith REH, Duthie HC (1995) Interactions of pH and Aluminium on cell length reduction in *Asterionella ralfsii* var. *americana* Körner. In: Marino D, Montresor M (eds) Proceedings of the 13th International Diatom Symposium, 1–7 Sep 1994, Koeltz Scientific Books Königstein, Acquafredda di Maratea, Italy, pp 39–46
46. Potapova M, Snoeijs P (1997) The natural life cycle in wild populations of *Diatoma moniliformis* (Bacillariophyceae) and its disruption in an aberrant environment. J Phycol 33(6):924–937
47. Stevenson RJ, Peterson CG, Kirschtel DB, King CC, Tuchman NC (1991) Density-dependent growth, ecological strategies and effects of nutrients and shading on benthic diatom succession in streams. J Phycol 27(1):59–69
48. Stevenson RJ, Bahls L (1999) Periphyton protocols. In: Barbour MT, Gerritsen J, Snyder BD, Stribling JB (eds) Rapid bioassessment protocols for use in streams and wadeable rivers:

periphyton, benthic macroinvertebrates and fish, 2nd edn. U.S. Environmental Protection Agency; Office of Water, Washington, DC, pp 1–22
49. Takamura N, Hatakeyama S, Sugaya Y (1990) Seasonal changes in species composition and production of periphyton in an urban river running through an abandoned copper mining region. Jpn J Limnol 51(4):225–235
50. Gold C (2002) Etude des effets de la pollution métallique (Cd/Zn) sur la structure des communautés de diatomées périphytiques des cours d'eau. Approches expérimentales *in situ* et en laboratoire. PhD thesis – Univ. Bordeaux I Ecole Doct. Sciences du vivant, Géosciences et Sciences de l'Environnement, p 175
51. Ivorra N (2000) Metal induced succession in benthic diatom consortia. PhD thesis, University of Amsterdam, Faculty of Science, Department of Aquatic Ecology and Ecotoxicology, p 157.
52. Morin S, Duong TT, Boutry S, and Coste M (2008) Modulation de la toxicité des métaux vis-à-vis du développement des biofilms de cours d'eau (bassin versant de Decazeville, France). Cryptog Algol 29(3):201–216.
53. Falasco E, Bona F, Badino G, Hoffmann L, Ector L (2009) Diatom teratological forms and environmental alterations: a review. Hydrobiologia 623(1):1–35
54. Falasco E, Bona F, Ginepro M, Hlúbiková D, Hoffmann L, Ector L (2009) Morphological abnormalities of diatom silica walls in relation to heavy metal contamination and artificial growth conditions. Water SA 35(5):595–606
55. Adshead-Simonsen PC, Murray GE, Kushner DJ (1981) Morphological changes in the diatom Tabellaria flocculosa induced by very low concentrations of cadmium. Bull Environ Contam Toxicol 26:745–748
56. McFarland BH, Hill BH, Willingham WT (1997) Abnormal *Fragilaria* spp. (Bacillariophyceae) in streams impacted by mine drainage. J Freshwat Ecol 12(1):141–149
57. Debenest T, Silvestre J, Coste M, Delmas F, Pinelli E (2008) Herbicide effects on freshwater benthic diatoms: induction of nucleus alterations and silica cell wall abnormalities. Aquat Toxicol 88(1):88–94
58. Parkinson J, Brechet Y, Gordon R (1999) Centric diatom morphogenesis: a model based on a DLA algorithm investigating the potential role of microtubules. Biochim Biophys Acta 1452(1):89–102
59. Cordonier A (2006) Formes tératologiques de diatomées benthiques dans le Nant d'Avril, Genève, 2005-2006. Rapport d'analyses. Etat de Genève, Département de l'intérieur, de l'agriculture et de l'environnement, Service de l'écologie de l'eau, p 2
60. Dickman MD (1998) Benthic marine diatom deformities associated with contaminated sediments in Hong Kong. Environ Int 24(7):749–759
61. Morin S, Duong TT, Dabrin A, Coynel A, Herlory O, Baudrimont M, Delmas F, Durrieu G, Schäfer J, Winterton P, Blanc G, Coste M (2008) Long term survey of heavy metal pollution, biofilm contamination and diatom community structure in the Riou-Mort watershed, South West France. Environ Pollut 151(3):532–542
62. SECOE (2004) Etude du Nant d'Avril et ses affluents, état 2003 et évolution depuis 1997. Rapport d'état des cours d'eau. Etat de Genève, Département de l'intérieur, de l'agriculture et de l'environnement, p 30
63. Blanck H, Wängberg SA, Molander S (1988) Pollution-induced community tolerance – a new ecotoxicological tool. In: Cairns J Jr, Pratt JR (eds) Functional testing of aquatic biota for estimating hazards of chemicals. ASTM, Philadelphia, pp 219–230
64. Gold C, Feurtet-Mazel A, Coste M, Boudou A (2002) Field transfer of periphytic diatom communities to assess short-term structural effects of metals (Cd, Zn) in rivers. Water Res 36(14):3654–3664
65. Guasch H, Navarro E, Serra A, Sabater S (2004) Phosphate limitation influences the sensitivity to copper in periphytic algae. Freshwat Biol 49(4):463–473

66. Interlandi SJ (2002) Nutrient-toxicant interactions in natural and constructed phytoplankton communities: results of experiments in semi-continuous and batch culture. Aquat Toxicol 61(1–2):35–51
67. Ivorra N, Hettelaar J, Kraak MHS, Sabater S, Admiraal W (2002) Responses of biofilms to combined nutrient and metal exposure. Environ Toxicol Chem 21(3):626–632
68. Lozano RB, Pratt JR (1994) Interaction of toxicants and communities – the role of nutrients. Environ Toxicol Chem 13(3):361–368
69. Admiraal W, Ivorra N, Jonker M, Bremer S, Barranguet C, Guasch H (1999) Distribution of diatom species in a metal polluted Belgian-Dutch River: an experimental analysis. In: Prygiel J, Whitton BA, Bukowska J (eds) Use of algae for monitoring rivers III. Agence de l'Eau Artois-Picardie, Douai, pp 240–244
70. Barranguet C, Plans M, van der Grinten E, Sinke JJ, Admiraal W (2002) Development of photosynthetic biofilms affected by dissolved and sorbed copper in a eutrophic river. Environ Toxicol Chem 21(9):1955–1965
71. Besch WK, Ricard M, Cantin R (1970) Utilisation des diatomées benthiques comme indicateur de pollutions minères dans le bassin de la Miramichi N.W. Fisheries Research Board of Canada, p 72
72. Blanck H, Admiraal W, Cleven RFMJ, Guasch H, van den Hoop M, Ivorra N, Nystrom B, Paulsson M, Petterson RP, Sabater S, Tubbing GMJ (2003) Variability in zinc tolerance, measured as incorporation of radio-labeled carbon dioxide and thymidine, in periphyton communities sampled from 15 European river stretches. Arch Environ Contam Toxicol 44(1):17–29
73. Chanson F, Cordonier A, Nirel P (2005) Essai de mise au point d'un indice diatomique pour évaluer la pollution métallique des cours d'eau du Genevois (Genève, Suisse). In: 24ème Colloque de l'ADLaF, Bordeaux, p 37
74. Conway HL, Williams SC (1979) Sorption of cadmium and its effects on growth and the utilization of inorganic carbon and phosphorus of two freshwater diatoms. J Fish Res Board Can 36(5):579–586
75. Cunningham L, Stark JS, Snape I, McMinn A, Riddle MJ (2003) Effects of metal and petroleum hydrocarbon contamination on benthic diatom communities near Casey Station, Antarctica: an experimental approach. J Phycol 39(3):490–503
76. De Jonge M (2007) Respons van aquatische organismen op metaalverontreiniging in natuurlijke waterlopen. Universitaire Instelling Antwerpen, Faculteit Wetenschappen, Department Biologie, p 124
77. Duong TT, Feurtet-Mazel A, Coste M, Dang DK, Boudou A (2007) Dynamics of diatom colonization process in some rivers influenced by urban pollution (Hanoi, Vietnam). Ecol Indic 7(4):839–851
78. Duong TT, Morin S, Herlory O, Feurtet-Mazel A, Coste M, Boudou A (2008) Seasonal effects of cadmium accumulation in periphytic diatom communities of freshwater biofilms. Aquat Toxicol 90(1):19–28
79. Ferreira da Silva E, Almeida SFP, Nunes ML, Luís AT, Borg F, Hedlund M, de Sá CM, Patinha C, Teixeira P (2009) Heavy metal pollution downstream the abandoned Coval da Mó mine (Portugal) and associated effects on epilithic diatom communities. Sci Total Environ 407(21):5620–5636
80. Feurtet-Mazel A, Gold C, Coste M, Boudou A (2003) Study of periphytic diatom communities exposed to metallic contamination through complementary field and laboratory experiments. J Phys IV 107:467–470
81. Fisher NS, Jones GJ, Nelson DM (1981) Effects of copper and zinc on growth, morphology, and metabolism of *Asterionella japonica* (Cleve). J Exp Mar Biol Ecol 51:37–56
82. Gélabert A, Pokrovsky O, Reguant C, Schott J, Boudou A (2006) A surface complexation model for cadmium and lead adsorption onto diatom surface. J Geochem Explor 88: 110–113

83. Genter RB, Cherry DS, Smith EP, Jr JC (1987) Algal periphyton population and community changes from zinc stress in stream mesocosms. Hydrobiologia 153(3):261–275
84. Genter RB, Amyot DJ (1994) Freshwater benthic algal population and community changes due to acidity and aluminum-acid mixtures in artificial streams. Environ Toxicol Chem 13 (3):369–380
85. Genter RB (1995) Benthic algal populations respond to aluminium, acid, and aluminium-acid mixtures in artificial streams. Hydrobiologia 306(1):7–19
86. Genter RB, Lehman RM (2000) Metal toxicity inferred from algal population density, heterotrophic substrate use, and fatty acid profile in a small stream. Environ Toxicol Chem 19(4):869–878
87. Gold C (1998) Etude expérimentale des effets d'un contaminant métallique – le cadmium – sur les communautés de diatomées périphytiques, au sein de microcosmes plurispécifiques d'eau douce. Univ. Bordeaux I LEESA, p 23
88. Gold C, Feurtet-Mazel A, Coste M, Boudou A (2003) Effects of cadmium stress on periphytic diatom communities in indoor artificial streams. Freshwat Biol 48:316–328
89. Gómez N, Licursi M (2003) Abnormal forms in *Pinnularia gibba* (Bacillariophyceae) in a polluted lowland stream from Argentina. Nova Hedwigia 77(3–4):389–398
90. Guasch H, Leira M, Montuelle B, Geiszinger A, Roulier J-L, Tornés E, Serra A (2009) Use of multivariate analyses to investigate the contribution of metal pollution to diatom species composition: search for the most appropriate cases and explanatory variables. Hydrobiolgia 627(1):143–158
91. Hirst H, Jüttner I, Ormerod SJ (2002) Comparing the responses of diatoms and macro-invertebrates to metals in upland streams of Wales and Cornwall. Freshwat Biol 47(9): 1752–1765
92. Ivorra N, Hettelaar J, Tubbing GMJ, Kraak MHS, Sabater S, Admiraal W (1999) Translocation of microbenthic algal assemblages used for *in situ* analysis of metal pollution in rivers. Arch Environ Contam Toxicol 37(1):19–28
93. Ivorra N, Bremer S, Guasch H, Kraak MHS, Admiraal W (2000) Differences in the sensitivity of benthic microalgae to Zn and Cd regarding biofilm development and exposure history. Environ Toxicol Chem 19(5):1332–1339
94. Kocev D, Naumoski A, Mitreski K, Krstic S, Dzeroski S (2010) Learning habitat models for the diatom community in Lake Prespa. J Ecol Model 221:330–337
95. Laviale M (2008) Effet des polluants sur les communautés périphytiques naturelles : Apport des mesures de fluorescence chlorophyllienne en lumière modulée (PAM). Université des Sciences et Technologies de Lille – Lille 1, p 198
96. Lehmann V, Tubbing GMJ, Admiraal W (1999) Induced metal tolerance in microbenthic communities from three lowland rivers with different metal loads. Arch Environ Contam Toxicol 36(4):384–391
97. Lindstrøm E-A, Rørslett B (1991) The effects of heavy metal pollution on periphyton in a Norwegian soft-water river. Verh Internat Verein Limnol 24:2215–2219
98. Monteiro MT, Oliveira R, Vale C (1995) Metal stress on the plankton communities of Sado River (Portugal). Water Res 29(2):695–701
99. Morin S, Coste M, Delmas F (2008) From field studies to laboratory experiments for assessing the influence of metal contamination on relative specific growth rates of periphytic diatoms. In: Brown SE, Welton WC (eds) Heavy metal pollution. Nova Science, New York, pp 137–155
100. Morin S, Duong TT, Herlory O, Feurtet-Mazel A, Coste M (2008) Cadmium toxicity and bioaccumulation in freshwater biofilms. Arch Environ Contam Toxicol 54(2):173–186
101. Nakanishi Y, Sumita M, Yumita K, Yamada T, Honjo T (2004) Heavy-metal pollution and its state in algae in Kakehashi River and Godani River at the foot of Ogoya mine, Ishikawa prefecture. Anal Sci 20(1):73–78
102. Navarro E, Guasch H, Sabater S (2002) Use of microbenthic algal communities in ecotoxicological tests for the assessment of water quality: the Ter river case study. J Appl Phycol 14(1):41–48

103. Nunes ML, Ferreira Da Silva E, De Almeida SFP (2003) Assessment of water quality in the Caima and Mau River basins (Portugal) using geochemical and biological indices. Water Air Soil Pollut 149(1–4):227–250
104. Pérès F, Coste M, Ricard M, Boudou A, Ribeyre F (1995) Effets des métaux lourds (Cd, Hg) sur les communautés de diatomées périphytiques développées sur substrats artificiels en microcosmes. Vie Milieu 45(3/4):210–230
105. Pérès F, Coste M, Ribeyre F, Ricard M, Boudou A (1997) Effects of methylmercury and inorganic mercury on periphytic diatom communities in freshwater indoor microcosms. J Appl Phycol 9(3):215–227
106. Pistocchi R, Guerrini F, Balboni V, Boni L (1997) Copper toxicity and carbohydrate production in the microalgae *Cylindrotheca fusiformis* and *Gymnodinium* sp. Eur J Phycol 32(2):125–132
107. Pistocchi R, Mormile MA, Guerrini F, Isani G, Boni L (2000) Increased production of extra- and intracellular metal-ligands in phytoplankton exposed to copper and cadmium. J Appl Phycol 12(3–5):469–477
108. Pokrovsky OS, Feurtet-Mazel A, Martinez RE, Morin S, Baudrimont M, Duong T, Coste M (2010) Experimental study of cadmium interaction with periphytic biofilms. Appl Geochem 25(3):418–427
109. Pomian-Srzednicki I (2006) Relations entre la composition des communautés de diatomées et les concentrations des polluants métalliques dans les cours d'eau genevois, SECOE, Editor: Genève, p 29
110. Ruggiu D, Luglie A, Cattaneo A, Panzani P (1998) Paleoecological evidence for diatom response to metal pollution in Lake Orta (N. Italy). J Paleolimnol 20(4):333–345
111. Sabater S (2000) Diatom communities as indicators of environmental stress in the Guadiamar River, S-W. Spain, following a major mine tailings spill. J Appl Phycol 12(2):113–124
112. Sanders JG, Riedel GF (1998) Metal accumulation and impacts in phytoplankton. In: Langston W, Bebianno M (eds) Metal metabolism in aquatic environments. Chapman and Hall, London, pp 59–76
113. Say PJ (1978) Le Riou-Mort, affluent du Lot pollué par les métaux. I. Etude préliminaire de la chimie et des algues benthiques. Annls Limnol 14(1–2):113–131
114. Serra A, Corcoll N, Guasch H (2009) Copper accumulation and toxicity in fluvial periphyton: the influence of exposure history. Chemosphere 74(5):633–641
115. Serra A, Guasch H, Admiraal W, Van der Geest H, Van Beusekom SAM (2010) Influence of phosphorus on copper sensitivity of fluvial periphyton: the role of chemical, physiological and community-related factors. Ecotoxicology 19(4):770–780
116. Shehata SA, Lasheen MR, Kobbia IA, Ali GH (1999) Toxic effect of certain metals mixture on some physiological and morphological characteristics of freshwater algae. Water Air Soil Pollut 110(1–2):119–135
117. Szabó K, Kiss KT, Taba G, Ács É (2005) Epiphytic diatoms of the Tisza River, Kisköre Reservoir and some oxbows of the Tisza River after the cyanide and heavy metal pollution in 2000. Acta Bot Croat 64(1):1–46
118. Takamura N, Kasai F, Watanabe MM (1989) Effects of Cu, Cd and Zn on photosynthesis of freshwater benthic algae. J Appl Phycol 1(1):39–52
119. Tapia PM (2008) Diatoms as bioindicators of pollution in the Mantaro River, Central Andes, Peru. Int J Environ Health 2(1):82–91
120. Tien CJ (2004) Some aspects of water quality in a polluted lowland river in relation to the intracellular chemical levels in planktonic and epilithic diatoms. Water Res 38(7):1779–1790
121. Tien CJ, Sigee DC, White KN (2005) Copper adsorption kinetics of cultured algal cells and freshwater phytoplankton with emphasis on cell surface characteristics. J Appl Phycol 17(5): 379–389
122. van Dam H, Mertens A (1990) A comparison of recent epilithic diatom assemblages from the industrially acidified and copper polluted lake Orta (Northern Italy) with old literature data. Diatom Res 5(1):1–13

123. Verb RG, Vis ML (2005) Periphyton assemblages as bioindicators of mine-drainage in unglaciated western allegheny plateau lotic systems. Water Air Soil Pollut 161(1–4):227–265
124. Whitton BA (1975) River ecology. B.S. Publications, Oxford, 725
125. Whitton BA (2003) Use of plants for monitoring heavy metals in freshwaters. In: Ambasht RS, Ambasht NK (eds) Modern trends in applied aquatic ecology. Kluwer, New-York, pp 43–63
126. Morin S, Pesce S, Tlili A, Coste M, Montuelle B (2010) Recovery potential of periphytic communities in a river impacted by a vineyard watershed. Ecol Indic 10(2):419–426
127. Rimet F, Ector L, Cauchie H-M, Hoffmann L (2009) Changes in diatom-dominated biofilms during simulated improvements in water quality: implications for diatom-based monitoring in rivers. Eur J Phycol 44(4):567–577
128. Meharg AA (1994) Integrated tolerance mechanisms – constitutive and adaptive plant responses to elevated metal concentrations in the environment. Plant Cell Environ 17(9): 989–993
129. Serra A, Guasch H (2009) Effects of chronic copper exposure on fluvial systems: linking structural and physiological changes of fluvial biofilms with the in-stream copper retention. Sci Total Environ 407(19):5274–5282
130. da Costa Santos JA (2010) Cadmium effects in Nitzschia palea frustule proteins (Efeitos do cádmio nas proteínas da frústula de Nitzschia palea). Universidade de Aveiro, Departamento de Biologia, p 35
131. De Filippis LF, Pallaghy CK (1994) Heavy metals: sources and biological effects. In: Rai LC, Gaur JP, Soeder CJ (eds) Algae and water pollution. E. Schweizerbart'sche Verlagsbuchhandlung, Stuttgart, pp 31–77
132. Gaur JP, Rai LC (2001) Heavy metal tolerance in algae. In: Rai LC, Gaur JP (eds) Algal adaptation to environmental stresses: physiological, biochemical and molecular mechanisms. Springer, Berlin, pp 363–388
133. Ahner BA, Morel FMM (1995) Phytochelatin production in marine algae. 2. Induction by various metals. Limnol Oceanogr 40(4):658–665
134. Le Faucheur S, Behra R, Sigg L (2005) Thiol and metal contents in periphyton exposed to elevated copper and zinc concentrations: A field and microcosm study. Environ Sci Technol 39(20):8099–8107
135. Wong SL, Wainwright JF, Pimenta J (1995) Quantification of total and metal toxicity in wastewater using algal bioassays. Aquat Toxicol 31(1):57–75
136. Gonzalezdavila M (1995) The role of phytoplankton cells on the control of heavy-metal concentration in seawater. Mar Chem 48(3–4):215–236
137. Scarano G, Morelli E (2002) Characterization of cadmium- and lead- phytochelatin complexes formed in a marine microalga in response to metal exposure. Biometals 15(2): 145–151
138. Rijstenbil JW, Sandee A, Vandrie J, Wijnholds JA (1994) Interaction of toxic trace-metals and mechanisms of detoxification in the planktonic diatoms *Ditylum brightwellii* and *Thalassiosira pseudonana*. FEMS Microbiol Rev 14(4):387–396
139. Wu JT, Chang SC, Chen KS (1995) Enhancement of intracellular proline level in cells of *Anacystis nidulans* (Cyanobacteria) exposed to deleterious concentrations of copper. J Phycol 31(3):376–379
140. Wu JT, Hsieh MT, Kow LC (1998) Role of proline accumulation in response to toxic copper in *Chlorella* sp. (Chlorophyceae) cells. J Phycol 34(1):113–117
141. Ahner BA, Wei LP, Oleson JR, Ogura N (2002) Glutathione and other low molecular weight thiols in marine phytoplankton under metal stress. Mar Ecol Prog Ser 232:93–103
142. Rijstenbil JW, Derksen JWM, Gerringa LJA, Poortvliet TCW, Sandee A, Mvd B, Jv D, Wijnholds JA (1994) Oxidative stress induced by copper: defense and damage in the marine planktonic diatom *Ditylum brightwellii*, grown in continuous cultures with high and low zinc levels. Mar Biol 119:583–590

143. Rijstenbil JW, Gerringa LJA (2002) Interactions of algal ligands, metal complexation and availability, and cell responses of the diatom *Ditylum brightwellii* with a gradual increase in copper. Aquat Toxicol 56(2):115–131
144. Pinto E, Sigaud-Kutner TCS, Leitao MAS, Okamoto OK, Morse D, Colepicolo P (2003) Heavy metal-induced oxidative stress in algae. J Phycol 39(6):1008–1018
145. Teitzel GM, Parsek MR (2003) Heavy metal resistance of biofilm and planktonic *Pseudomonas aeruginosa*. Appl Environ Microbiol 69(4):2313–2320
146. Lee JG, Ahner BA, Morel FMM (1996) Export of cadmium and phytochelatin by the marine diatom *Thalassiosira weissflogii*. Environ Sci Technol 30(6):1814–1821
147. Rosen BP (1996) Bacterial resistance to heavy metals and metalloids. J Biol Inorg Chem 1(4): 273–277
148. Decho AW (2000) Microbial biofilms in intertidal systems: an overview. Continent Shelf Res 20(10–11):1257–1273
149. Revsbech NP, Nielsen LP, Christensen PB, Sørensen J (1988) Combined oxygen and nitrous oxide microsensors for denitrification studies. Appl Environ Microbiol 54(9):2245–2249
150. Teissier S, Torre M (2002) Simultaneous assessment of nitrification and denitrification on freshwater epilithic biofilms by acetylene block method. Water Res 36(15):3803–3811
151. Sabater S, Guasch H, Ricart M, Romaní A, Vidal G, Klünder C, Schmitt-Jansen M (2007) Monitoring the effect of chemicals on biological communities. The biofilm as an interface. Anal BioAnal Chem 387(4):1425–1434
152. Barranguet C, van den Ende FP, Rutgers M, Breure AM, Greijdanus M, Sinke JJ, Admiraal W (2003) Copper-induced modifications of the trophic relations in riverine algal-bacterial biofilms. Environ Toxicol Chem 22(6):1340–1349
153. Asconit Consultants (2007) Methodological study for biological monitoring of surface water quality in the Pearl River Basin. In: Report to the Pearl River Water Resources Commission (Guangzhou, China), p 159
154. Blanco S, Bécares E (2010) Are biotic indices sensitive to river toxicants? A comparison of metrics based on diatoms and macro-invertebrates. Chemosphere 79(1):18–25
155. Pérès F (1999) Mise en évidence des effets toxiques des métaux lourds sur les diatomées par l'étude des formes tératogènes. Rapport d'étude, Agence de l'Eau Artois Picardie, p 24
156. Coste M (1999) Atlas des diatomées pour la mise en œuvre de l'Indice Biologique Diatomées (IBD). Agences de l'Eau – Cemagref QEBX Bordeaux, p 130
157. Krammer K, Lange-Bertalot H (1986–1991) Bacillariophyceae 1. Teil: Naviculaceae. 876 p.; 2. Teil: Bacillariaceae, Epithemiaceae, Surirellaceae, 596 p.; 3. Teil: Centrales, Fragilariaceae, Eunotiaceae, 576 p.; 4. Teil: Achnanthaceae. Kritische Ergänzungen zu *Navicula* (Lineolatae) und *Gomphonema*. 437 p. In: Ettl H, Gerloff J, Heynig H, Mollenhauer D (eds) Süßwasserflora von Mitteleuropa, vol Band 2/1-4. G. Fischer, Stuttgart
158. Krammer K (2002) Cymbella. In: Lange-Bertalot H (ed) Diatoms of Europe, vol 3. A.R.G. Gantner Verlag, Ruggell, p 584
159. Krammer K (2003) Cymbopleura, Delicata, Navicymbula, Gomphocymbula, Gomphocymbellopsis, Afrocymbella. In: Lange-Bertalot H (ed) Diatoms of Europe: Diatoms of the European inland waters and comparable habitats, vol 4. A.R.G.Gantner Verlag K.G, Ruggell, p 530
160. Lange-Bertalot H, Metzeltin D (1996) Indicators of oligotrophy. 800 taxa representative of three ecologically distinct lakes types. In: Lange-Bertalot H (ed) Carbonated buffered – Oligodystrophic – Weakly buffered soft water. Iconographia Diatomologica – Annotated diatom micrographs, vol 2. Koeltz Scientific Books, Königstein, p 390
161. Lange-Bertalot H (2001) *Navicula sensu stricto* 10 genera separated from *Navicula sensu lato, Frustulia*. In: Lange-Bertalot H (ed) Diatoms of Europe: diatoms of the European inland waters and comparable habitats. A.R.G. Gantner Verlag K.G, Ruggell, p 526
162. Lavoie I, Hamilton PB, Campeau S, Grenier M, Dillon PJ (2008) Guide d'identification des diatomées des rivières de l'est du Canada, ed. Presses de l'Université du Québec (PUQ)
163. Clements WH, Carlisle DM, Lazorchak JM, Johnson PC (2000) Heavy metals structure benthic communities in Colorado mountain streams. Ecol Appl 10(2):626–638

164. Hillebrand H, Dürselen CD, Kirschtel D, Pollingher U, Zohary T (1999) Biovolume calculation for pelagic and benthic microalgae. J Phycol 35(2):403–424
165. Hoagland KD, Roemer SC, Rosowski JR (1982) Colonization and community structure of two periphyton assemblages, with emphasis on the diatoms (Bacillariophyceae). Am J Bot 69(2):188–213
166. Hudon C, Bourget E (1983) The effect of light on the vertical structure of epibenthic diatom communities. Bot Mar 26:317–330
167. Hudon C, Duthie HC, Paul B (1987) Physiological modifications related to density increase in periphytic assemblages. J Phycol 23(3):393–399
168. Katoh K (1992) Correlation between cell density and dominant growth form of epilithic diatom assemblages. Diatom Res 7:77–86
169. Kelly MG, Bennion H, Cox EJ, Goldsmith B, Jamieson J, Juggins S, Mann DG, Telford RJ (2005) Common freshwater diatoms of Britain and Ireland: an interactive key. Environment Agency, Bristol
170. Tuji A (2000) Observation of developmental processes in loosely attached diatom (Bacillariophyceae) communities. Phycol Res 48(2):75–84
171. Gómez N, Sierra MV, Cortelezzi A, Rodrigues Capítulo A (2008) Effects of discharges from the textile industry on the biotic integrity of benthic assemblages. Ecotoxicol Environ Saf 69(3):472–479
172. Shannon CE, Weaver W (1949) The mathematical theory of communication. University of Illinois Press, Urbana, IL, p 117
173. Estes A, Dute RR (1994) Valve abnormalities in diatom clones maintained in long-term culture. Diatom Res 9(2):249–258
174. Granetti B (1978) Struttura di alcune valve teratologiche di Navicula gallica (W. Smith) Van Heurck. Giornale botanico italiano 112:1–12
175. Townsend CR, Dolédec S, Scarsbrook MR (1997) Species traits in relation to temporal and spatial heterogeneity in streams: a test of habitat templet theory. Freshwat Biol 37(2): 367–387
176. Potapova M, Hamilton PB (2007) Morphological and ecological variation within the *Achnanthidium minutissimum* (Bacillariophyceae) species complex. J Phycol 43(3):561–575
177. Thioulouse J, Chessel D, Dolédec S, Olivier JM (1997) ADE-4: a multivariate analysis and graphical display software. Stat Comput 7(1):75–83
178. Ihaka R, Gentleman R (1996) R: A language for data analysis and graphics. J Comput Graph Stat 5:299–314
179. Dufrêne M, Legendre P (1997) Species assemblages and indicator species: the need for a flexible asymmetrical approach. Ecol Monogr 67(3):345–366

Advances in the Multibiomarker Approach for Risk Assessment in Aquatic Ecosystems

Chloé Bonnineau, Anja Moeller, Carlos Barata, Berta Bonet, Lorenzo Proia, Frédéric Sans-Piché, Mechthild Schmitt-Jansen, Helena Guasch, and Helmut Segner

Abstract Nowadays, the term biomarker has become widespread in environmental sciences and ecology as shown by the increasing amount of research articles published on this topic in these research areas (2,042 articles published between 2000 and 2010 over a total of 2,352 published since 1985, WOK, ISI Web of Knowledge, http://www.isiknowledge.com). Despite this increasing enthusiasm for biomarkers, they are still poorly used in regulatory monitoring or ecological risk assessment. Indeed, the interest and pertinence of using biomarkers in such approaches have been questionned. While biomarkers of various toxicants for many species can be found in the literature, the limited extrapolation of biomarker results is particularly criticized and may prevent routine application. In this chapter, we discuss how to link biomarker responses to chemical exposure but also how to correlate them to effects at higher levels of biological organization. In the first part, the word biomarker is defined in the context of risk assessment. Then, the historical and current use of biomarkers in river biofilms, macroinvertebrates, and fish is shortly presented focusing on the specificity of biomarkers' utilization in each of these biological entities. Next, laboratory and field studies are used to exemplify

C. Bonnineau (✉) • B. Bonet • L. Proia • H. Guasch
Department of Environmental Sciences, Institute of Aquatic Ecology, University of Girona, Campus de Montilivi, 17071 Girona, Spain
e-mail: chloebonnineau@hotmail.com

A. Moeller • H. Segner
Centre for Fish and Wildlife Health, University of Bern, Laenggasstrasse 122, 3012 Bern, Switzerland

C. Barata
Department Environmental Chemistry, IDAEA-CSIC, Jordi Girona 18, 08034 Barcelona, Spain

F. Sans-Piché • M. Schmitt-Jansen
Department Bioanalytical Ecotoxicology, Helmholtz-Centre for Environmental Research – UFZ, Permoserstraße 15, 04318 Leipzig, Germany

H. Guasch et al. (eds.), *Emerging and Priority Pollutants in Rivers*, Hdb Env Chem (2012) 19: 147–180, DOI 10.1007/978-3-642-25722-3_6,

relationships between stressors and biomarkers but also between biomarkers at different levels of biological organization. Finally, current limitations of biomarkers are discussed and some propositions are made to overcome these limitations and to apply a multibiomarker approach to environmental risk assessment.

Keywords Biofilm • Fish • Macroinvertebrates • Molecular biomarkers • Rivers

Contents

Abbreviations

APX	Ascorbate peroxidase
CAT	Catalase
CYP1A	Cytochrome P4501A enzyme
DOC	Dissolved organic carbon
EC_{50}	Effect concentration 50
EDCs	Endocrine disruptor compounds
EPS	Extracellular polymeric substances
EROD	Ethoxyresorufin-O-deethylase
FGA	Functional gene array
LC_{50}	Lethal concentration 50
MoA	Mode of action

MXR	Multidrug transporters
PAHs	Polycyclic aromatic hydrocarbons
PCBs	Coplanar polychlorobiphenyls
PICT	Pollution-induced community tolerance
PLS	Partial least square
PNEC	Predicted no effect concentrations
RDA	Redundancy direct analysis

1 What Is a Biomarker?

1.1 A Word Widely Used Across Different Scientific Fields

The term biomarker became popular through its use in the field of human medicine. In this context, a biomarker is a biological indicator of the body state especially useful as it allows noninvasive diagnostic to be performed. Natural or synthetic substances may indicate a cancer, the beginning of a disease or an infection, or help to locate a tumor (Fig. 1, [1]); for instance, the PSA (prostate-specific antigen) is a biomarker of cancer in humans [2]. A good knowledge of human biology and of interactions occurring inside human body makes biomarkers especially efficient in human medicine. By extension, the term biomarker has become popular in other fields such as geology, ecology, and ecotoxicology. For instance in geology, molecular fossils found in oils extracts are considered as helpful biomarkers to determine the hydrocarbon potential of a petroleum source rock [3]. In microbial ecology, different biological molecules such as lipids (membrane lipids, sterols, etc.) or amino acids are considered as useful biomarkers. Since they are specific of groups or species, they provide information on the identity and function of microbial communities (Fig. 1, [4]). However, in ecology, the term biomarker has been

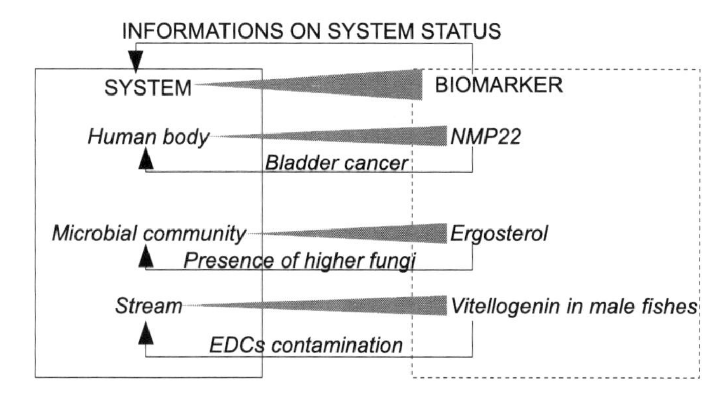

Fig. 1 Biomarkers belong to a system. Their responses characterize changes in the system status. Examples of biomarkers used in medicine (*NMP22* nuclear membrane proteine 22, [153]), in microbial ecology [4], or in ecotoxicology (*EDCs* endocrine disruptor compounds, [7])

used a lot in association with the study of contamination effects. Indeed in the areas of "environmental sciences" and "ecology", 2,352 scientific articles (research articles, reviews, and letters) containing the keyword "biomarker" were published since 1985 (WOK, ISI Web of Knowledge, http://www.isiknowledge.com) and more than half of these papers (1,259) also contained the keywords "contamination," "pollution," "toxic," "pesticides," or "metals." In ecotoxicology, biomarkers usually refer to biochemical, physiological, or histological indicators of exposure or toxicant effects [5, 6]. For instance, in several fish species, exposure to endocrine disruptors led specifically to an increase in the lipoprotein vitellogenin in male fish; thus, vitellogenin is considered as a biomarker for endocrine disruption in fish (Fig. 1, [7]). Whole-organism responses (i.e., survival, growth, reproduction, or behavior) may also be considered as biomarkers for a population or an ecosystem [8–10].

A common definition of biomarker can be deduced from their use in the aforementioned fields of application. *A biomarker is a biological element belonging to a wider system whose observation is expected to give information on the wider system based on the a priori knowledge of the links and interactions existing between the biomarker and this system* (Fig. 1). Indeed a biomarker informs not on itself but on the system it is belonging to. For instance, the biomarker vitellogenin is not relevant as itself but because it informs on the fact that the vitellogenin producing "system" (e.g., a fish) is exposed to an estrogenic compound. Therefore, the type of biomarker can vary from molecular ones (e.g., enzymes) to whole-organism ones (e.g., bacteria mortality) depending on the size of the system studied and the importance of the links throughout biological complexity.

Three classes of biomarkers are usually discriminated based on the fact that biomarkers can provide information on the exposure to, the effects of, or the susceptibility to a perturbation. The biomarkers of exposure may indicate that the biological system is exposed to stressor and, in certain cases, these biomarkers may also inform on the intensity of the stressor. An example would be the induction of vitellogenin in male fish, indicating that the organism is exposed to estrogen-active compounds, and provided that benchmarking data on vitellogenin induction by estrogens in the fish species under investigation are available, the vitellogenin induction observed in the field can be translated—with caution—into "estrogen equivalents" (e.g., [11, 12]). The exposure to an environmental stressor may be associated with functional or structural impairment or damage of the exposed biological system (organism, population, and community), and a biomarker of effect would indicate the adverse outcome. For instance, induction of cytochrome P4501A enzyme (CYP1A) is used as a biomarker of exposure to arylhydrocarbon receptor binding chemicals (e.g., polycyclic aromatic hydrocarbons, PAHs; coplanar polychlorobiphenyls, PCBs; polychlorinated dibenzofurans and dibenzodioxins), but induction of CYP1A is not yet indicative of an adverse effect [13]. However, CYP1A-catalyzed metabolism can lead to reactive chemical metabolites eventually causing cancer, and thus, neoplastic changes in the tissues of exposed organisms would be a biomarker of effect. Finally, the biomarkers of susceptibility indicate the inherent or acquired predisposition of an individual or a community to respond to a stressor exposure [14]. For instance, in humans genetic markers are established which indicate the ability of an individual to generate cancerogenic metabolites from a given toxicant. Nevertheless, biomarkers of susceptibility are rarely used in ecotoxicology.

In this chapter, we focus on biomarker (from molecular to whole-organism response) related to exposure to or effects of toxicants. Ecological indexes that integrate the responses of various biomarkers (e.g., IBI, IPS) are not considered as biomarkers in this chapter. Their use in risk assessment of aquatic ecosystems is developed in other chapters of this book [15, 16]. To avoid confusion, the term biomarker should be distinguished from those of response, endpoint, or indicator. The biological response is a general term corresponding to any variation of a biological system exposed to a toxicant, but not each response is already a biomarker. Different criteria can be defined to discriminate a biological response from a biomarker such as the specificity, the dependence with concentration (linear or not), and the reproducibility. The word "endpoint" usually refers to something that is measured as a result of an experiment. This word can be especially misleading since in laboratory experiments an endpoint is any defined measure (it can also be a molecular endpoint) while in risk assessment an endpoint defines an ecological function and/or structure to be preserved by risk management actions and corresponds therefore to the goal of risk assessment [17–19]. It is worth noting that the adverse outcome due to the presence of a toxicant (i.e., its toxicity) depends, in risk assessment, on the endpoints selected. Therefore, biomarkers can truly indicate toxicity when their response can be linked to the endpoints selected a priori. Finally, the term indicator usually refers to an organism or a population which is strictly associated with some specific environmental conditions and whose characteristics are indicative of these conditions [18]. Indicator species are then biological elements informing by their abundance, presence/absence, on the status of a wider system; consequently, they can be considered as biomarkers of the ecosystem.

1.2 Biomarkers in Site-Directed Risk Assessment of Aquatic Ecosystems

In Europe, contamination of aquatic ecosystems by toxic chemicals has changed in the last decades, on the one hand, due to a reduction of the toxicant concentration in water (for instance, because of improved wastewater treatment technologies), and on the other hand, due to an increase in the diversity of toxic compounds [20]. The improvement of detection and identification techniques of chemicals allows very small concentrations (e.g., pg L^{-1}) to be detected and a wide array of molecules to be identified as presented in previous chapters [21, 22]. However, the chemical analyses always detect only a fraction of all the compounds present in the aquatic ecosystem; thus, these analyses may greatly underestimate risks. In addition, even though state-of-the art chemical analyses allow minute quantities of chemicals in the environment to be detected it remains difficult to determine their impact on the environment [23, 24]. The standard approach used in assessing the risk of chemicals to aquatic biota is based on the determination of effect concentrations (e.g., LC_{50} values) of chemicals in single-species laboratory tests. Different extrapolation approaches (in the simplest case a convention-based extrapolation factor) are then used to derive "safe" environmental levels (predicted no effect concentrations, PNEC) of the chemicals. By comparing predicted or analytically measured concentrations of the chemical in the aquatic environment with the derived PNEC values, the risk of this compound to be harmful to aquatic biota is estimated [25].

The approach described above has been criticized for its lack of ecological relevance. Various studies showed that effect values established for selected "model" species in the lab under standardized and optimized conditions are not always meaningful for the many species that are exposed under real environmental conditions [26]. A first concern is the use of a restricted number of species in laboratory test, which means that the lab data might not be representative for the species in the aquatic ecosystem. A second limitation is linked with the use of single-species tests while under natural conditions species are always found in communities; i.e., the approach does not take into consideration possible effects arising from toxic impact on species interactions [27]. A third limitation is that the laboratory experiment considers exposure to a single chemical, while in the field, organisms are usually exposed to mixtures of chemicals, which can modify the toxicity of the individual compound. Finally, ecosystems experience multiple stress situations which are not taken into account in lab experiments [28]. The interactions between various biological (e.g., multi-species interactions), physical (e.g., light intensity, flow velocity), and chemical (e.g., mixture of toxicants) stressors are not considered in current risk assessment approaches, although they may substantially modulate the toxic effects of chemicals on the ecosystem. As a consequence, the impact of contamination in the field may be different from the effects predicted on the basis of laboratory tests [29].

To partially overcome these limitations and better approach the real complexity of contamination scenarios in aquatic ecosystems, biomarkers were proposed as diagnostic tools for biological systems. They have the promise to more closely link responses observed in organisms or communities at a given study site to the presence of chemicals at this site. Indeed, a biomarker can be linked either to the perturbation (which is known to induce effect) or to the effect (which is the result of the perturbation). The simultaneous measure of various biomarkers may then help in detecting both chemical exposure and toxic effects in multiple stress situations. Moreover, molecular biomarkers are often presented as early warning systems [14] since changes at molecular level are expected to be detected before changes at population or community levels.

While numerous biomarkers of exposure and effects have been described in the literature for various compounds and are used in research activities, including monitoring (see for example, [14, 30–33]), they are not routinely used in regulatory risk assessment. Since we believe that biomarkers can provide valuable information in regulatory risk assessment, in this chapter, we discuss the interests and limitations of biomarkers in detecting chemical exposure and assessing toxic effects at different levels of biological organization. To illustrate the situations in which biomarkers are more likely to be useful, we present various laboratory and field experiments involving river biofilms, macroinvertebrates, and fish. These three biological systems represent important components of the aquatic ecosystem and represent three levels of biological complexity, i.e., community for biofilms, population for invertebrates, and individual for fishes. Since a lot of reviews have been published on fish biomarkers, most of the examples presented here concern biofilm and macroinvertebrates biomarkers.

2 Biomarkers in River Biofilms, Macroinvertebrates, and Fish

The historical development of the biomarker approach had a strong link with medicine and vertebrate biology (NRC 1987 [34]); thus, most of the field studies and applications in the aquatic environment have been focused on fish. However, biomarker measurements are equally feasible on river biofilm communities and aquatic invertebrates.

2.1 River Biofilms

River biofilms (also known as periphyton) have a key role in the aquatic ecosystem, in particular in mid-size order streams in which they are considered as the main primary producers [35]. Freshwater biofilms are complex and structured benthic communities. The numerous species of these communities form a 3D structure as they live closely together in a matrix composed of extracellular polymeric substances (EPS matrix). Biofilms can be found in various microhabitats as they are able to attach to different solid substrates (cobbles, wood, sand, etc.). In this community, two functional components can be distinguished. Green algae, diatoms, and cyanobacteria form the autotrophic component of biofilms, while bacteria, fungi, and protozoa compose the heterotrophic one [36]. Due to their omnipresence, their important role in primary production, nutrient fluxes, and trophic cascades as well as their sensitivity to organic and inorganic pollutants river biofilms have been recognized as pertinent indicators of integrated ecosystem health [37–39].

Because of its complexity and its structure, a river biofilm can be considered as a microecosystem with different levels of biological organization. Therefore, multiple biomarkers have been developed to assess both structure and function of the different components of biofilms. In addition, most of the functional biomarkers developed focused on functions directly linked with processes essential for the whole aquatic ecosystem such as primary production. Indeed, the measure of various photosynthetic parameters (photosynthetic efficiency, capacity, and quenching processes) is specific of the autotrophic component of biofilm and is directly linked with its key role of primary producer [40]. The activities of the extracellular enzymes peptidase and glucosidase are also biomarkers in biofilms, specific of the heterotrophic component, and directly linked with the role of biofilms within organic matter decomposition ([26, 41]). Other biofilm biomarkers such as antioxidant enzyme activities and amino acid composition integrate the response of the whole biofilm [42, 43]. Since biofilm communities are assemblages of different components, a multibiomarker approach, including component-specific and species-specific biomarkers, is recommended to determine both direct and indirect effects of toxicants. This approach may not only point out some direct and indirect underlying mechanisms of toxicity, but also specific impairment in biofilm functioning. For example, Ricart et al. [26] highlighted the indirect effect of the herbicide diuron, inhibitor of photosynthesis, on bacteria within biofilm communities.

2.2 *Macroinvertebrates*

Aquatic invertebrates are commonly used in biological monitoring programs, but their use in biomarker studies has been limited to few species mostly bivalves. For example, the number of scientific publications (WOK, ISI Web of Knowledge, http://www.isiknowledge.com) published annually, since 1980, containing the keyword "biomarker," "environment," and "aquatic or marine or water" is about 3,561. From these ones, 1,600 were about fish and only 1,041 publications included the term "invertebrate or *Daphnia* or insect or crustacean or amphipod or *Chironomus* or mussel or clam or oyster". It is worth noting that less than 400 were conducted in species different than bivalves. This means that, despite the large variability of phylogenetically and hence physiologically distinct groups of aquatic invertebrates, only a few of them have been considered in biomarker studies. Aquatic invertebrates offer distinct advantages for biomonitoring, including (a) their ubiquitous occurrence; (b) their huge species richness, which offers a spectrum of environmental responses; (c) their basic sedentary nature, which facilitates spatial analysis of pollution effects; (d) the long life cycles of some species, which can be used to trace pollution effects over longer periods; (e) their compatibility with inexpensive sampling equipment; (f) the well-described taxonomy for genera and families; (g) the sensitivities of many common species, which have been established for different types of pollution; and (h) the suitability of many species for experimental studies of pollution effects [44]. In field studies, aquatic invertebrates have further advantages. As invertebrate populations are often numerous, samples can readily be taken for analyses without a significant impact on the population dynamics. Also, the application of biomarkers in invertebrate species allows the linkage between biomarkers responses and adverse effects on populations and communities. Currently, increasing knowledge of the biochemistry and physiology of invertebrates permits a reasonable interpretation of biomarker responses.

2.3 *Fish*

Fish are probably most frequently used for biomarker measurements in aquatic monitoring. As fish are vertebrates, the use of biomarkers in this animal group has been promoted by the in-depth knowledge on biomarkers being available in mammalian toxicology. Fish biomarkers have been successfully developed and adopted to be used in monitoring by national and international monitoring programs (e.g., International Council for the Exploration of the Sea ICES, OSPAR Convention for the Protection of the Marine Environment of the North East Atlantic). Frequently applied biomarkers in fish include vitellogenin induction in male fish as a marker for exposure to estrogens, the induction of hepatic CYP1A1 or EROD (ethoxyresorufin-O-deethylase) activity, or the observation of neoplasia in organisms exposed to carcinogenic compounds; an in-depth review on fish biomarkers has been published recently [14]. The purposes fish biomarkers analyses' are used for include (a) unraveling exposure of organisms to chemicals (e.g., acetyl cholinesterase inhibition as marker of organophosphate exposure, induction of EROD activity as marker of exposure to dioxin-like chemicals), (b) surveying spatial and temporal changes in aquatic contamination levels (e.g., by analyzing body burdens of contaminants in migratory, long-lived fish species), and (c) providing warning of potential adverse ecological consequences of aquatic pollution (e.g., malformation of reproductive organs may lead to reduced reproduction). A well-known example of aquatic pollution where biomarker studies in fish played a major role in detecting and defining the case is endocrine disruption [45]: it was the observation that male fish in many rivers exhibited elevated levels of vitellogenin, a female-characteristic protein, that brought environmental contamination by endocrine-active compounds to awareness.

Despite the intensive use of biomarkers with fish, the currently available set of markers is still rather limited, being indicative for comparatively few stressors or MoA. However, technologies such as genomics and proteomics may generate a broader suite of diagnostic tools.

2.4 *Target-Oriented Choice of Biomarkers*

Due to the specific characteristics of each aquatic compartment, biomarkers of river biofilms, macroinvertebrates, and fish would obviously give different information. Assessing these three compartments is essential to determine the aquatic ecosystem status; however, biomarkers of each compartment should be carefully selected to represent its specificity. Biomarkers of river biofilms and macroinvertebrates integrate response on shorter spatial and temporal scale than fish. Biomarkers of macroinvertebrates and fish are more species specific while biomarkers of river biofilms integrate community response and hence detect indirect effects. Biomarkers of macroinvertebrates and fish can account for specific effects of toxicants on organism's life cycle while biofilm biomarkers can account for toxic effects on community succession and/or diversity.

3 Biomarkers in Laboratory Studies

As laboratory studies allow an important experimental control and a higher replication [27], they have been widely used to establish cause–effect relationships in particular between toxicant exposure and biological effects. This type of ecotoxicological experiment includes the majority of laboratory work on biomarkers and can be especially useful to elucidate underlying mechanisms of action of toxicants. This mechanism-oriented research allows then the selection among different biomarkers of the most specific and sensitive to the toxicant tested. Therefore, these specific biomarkers developed in laboratory studies can then be used to help identifying candidate causes for ecological impairment in the field.

3.1 Biomarkers to Assess Mode of Action

Investigating the effects of toxicants at different levels of biological organization allows MoA to be determined but also direct and indirect toxic effects to be elucidated. Here, biomarkers can make an important contribution as one element in a weight-of-evidence approach to establish relationships between chemical exposure and biological effects.

Many fish biomarkers reflect MoA of toxic chemicals; e.g., EROD induction indicates that the fish is exposed to arylhydrocarbon receptor-activating chemicals. In invertebrate species, there are a few interesting case studies that have identified key MoA of particular pollutants. Concerning the macroinvertebrates, Livingstone and Viarengo can be considered pioneer researchers with important original and review articles on xenobiotic metabolism and oxidative stress in mussels [46, 47]. Recently, a review of Porte et al. [48] provided good examples of research work conducted with radiometric and biochemical techniques to identify different steroidogenic paths in mollusks and crustacean species and their modulation by pollutants that disrupt those paths and hence alter sex hormones. The previous review showed that now by using radiometric techniques it is possible to follow precisely hormonal metabolic paths and to evaluate accurately enzymatic activities involved in those paths. In the past, the use of unspecific enzymatic substrates did not allow to characterize precisely the rate of synthesis and catabolism of sex hormones. These types of studies are particularly useful since in the last decades most studies conducted in endocrine disruption effects were focused on agonist and antagonistic effects on sexual receptors and hence underestimate many endocrine disruptors that affect the metabolism of sexual hormones. Many review articles on invertebrates' biomarkers have also evaluated and discussed previous work conducted on detoxification and sexual hormonal metabolic paths (i.e., phase I, II, and III metabolizing and steroidogenic paths), oxidative stress, cholinesterases, metallothionein and other stress proteins, and endocrine disruption markers (i.e., vitellogenin-like proteins, sexual hormones, and their receptors; [34, 48–58]). These reviews have pointed out the lack of knowledge on invertebrate physiology and hence the difficulty to interpret suborganism biomarker alteration and so to understand the biological effects. Indeed only few biomarkers can be directly linked with key organism functions such as the acetylcholinesterase, protein peroxidation, or vitellogenin, among others. For instance, in many arthropod species the inhibition of acetylcholinesterases by organophosphorous pesticide directly caused the death of the individual. Indeed, this inhibition causes the accumulation of the neurotransmitter acetylcholine in the synaptic neuronal gab which causes an over and permanent stimulation of neuronal cells that ultimately kills the organism [59]. Markers of cellular damage such as increased levels of lipid and protein peroxidation, DNA damage, or micronuclei can also be linked to tissue damage (necrosis) and death or organ malfunctions [56, 58, 60]; depletion levels of energy reserves such as carbohydrates, glycogen, lipids, and proteins can be related with reductions in growth and reproduction [61, 62]; and vitellogenin levels are related to egg production in oviparous invertebrate and vertebrate species [33]. In these types of studies, to be able to clearly understand the biological effects it is very important to establish physiological and toxicological relationships by using different exposure doses and periods and if possible by including model agonists or antagonists of the studied system. For example, Barata et al. [63] and Damásio et al. [64] characterized the physiological role of B esterases (Cholinesterase and carboxylesterases) in *Daphnia magna* combining acute responses of organophosphorous pesticides co-administered with specific inhibitors of certain esterases, phase I metabolic inhibitors, and enzymatic assays. Recently in sea urchin (*Strongylocentrotus purpuratus*) and zebra mussel (*Dreissena polymorpha*) embryos, Bosnjak et al. [65] and Faria et al. [66] using toxic substrates, model inhibitors and agonists, and transporter efflux assays were able to establish a physiological link between the inhibition of transmembrane multidrug transporters (MXR) and toxicity. Transmembrane MXR transporters have an active role effluxing out pollutants or their metabolites from cells and hence are considered as an important defensive system against pollution. Their inhibition or chemosensitization by certain emergent compounds, thus, is expected to increase the accumulation of toxic pollutants and hence the toxicity of mixtures [67]. Therefore in fish and invertebrates, biomarkers of exposure highly specific to a toxicant can be developed by investigating the mechanism of action of a toxicant while physiology studies help in understanding the biological effects of toxicants on biota.

In river biofilms, the complexity of the community may prevent the precise determination of toxicant MoA. Nevertheless, the use of biomarkers in river biofilm communities has elucidated indirect effects on nontarget organisms as shown in several studies [68–71]. The selection of an appropriate set of biomarkers is a key step directly related with the known or expected mode of action of the compound to be tested and the time of exposure (acute or chronic effects).

Biofilm biomarkers can be used to detect both direct and indirect effects. Ricart et al. [26] used the photosynthetic efficiency as a biomarker of direct effects of the photosynthesis inhibitor diuron on river biofilm, while bacterial membrane integrity (live/dead bacteria ratio) was used as a biomarker of direct effects of the broad-spectrum bactericide triclosan in another experiment performed on river biofilms [70]. Previous knowledge about the interactions between target and nontarget organisms in complex communities is an important prerequisite to select biomarkers able to highlight indirect effects. For example, positive interactions between algae and bacteria in fluvial biofilms have been widely described [41, 72–75]. In fact, biomarkers related to heterotrophs highlighted indirect effects of diuron on bacteria [26, 71, 76, 77] as well as biomarkers related to autotrophs revealed indirect effects of the bactericide triclosan on algae [70, 78, 79]. Since river biofilm communities have a relatively short time of colonization and succession, acute and chronic effects of toxicants at community level can also be assessed in laboratory under controlled conditions. Comparison between acute and chronic effects can then be used to discriminate early warning biomarkers, useful in case of a spillage, from biomarkers of chronic contamination. Different experiments underlined the importance of selecting biomarkers adapted to the timescale of the experiment to be able to detect both direct and indirect effects throughout time. In the conceptual framework established by Guasch et al. [80], after the study of different scenarios of copper exposure, physiological responses are expected after acute exposure while changes in community composition may result from chronic exposure. However, transient biomarker responses to pollutant exposure also occur. In a microcosm study, Bonet et al. (personnal communication) exposed biofilms from a natural freshwater system (Llémena River, NE Spain; [81]) to realistic zinc concentrations (400 $\mu g\ L^{-1}$) during 5 weeks. They observed temporal differences in the activation of two antioxidant enzymes: ascorbate peroxidase (APX) and catalase (CAT) in response to Zn exposure. While APX activity increased after few days of exposure as a response to a transitory physiological effect, CAT activity increased significantly after chronic exposure. The activation of CAT and APX activity at different times might be matched with the shift in communities due to the tolerance acquisition in biofilms exposed to Zn. This study shows how biomarkers can provide information about transitory and persistent effects of toxicants. Finally, recovery experiments allow to identify those biomarkers that can integrate toxicant effects over a longer time than exposure. For example, Proia et al. [82] investigated short- and long-term effects of repeated pulses, during 48 h, of the herbicide diuron, or of the bactericide triclosan on river biofilms. The acute effects observed just after exposure were directly related to the MoA of these toxicants as photosynthesis was inhibited by diuron exposure and bacterial mortality increased after triclosan exposure. However, indirect effects of both toxicants were more persistent. Indeed, the 48 h exposure to diuron pulses led to a decrease in diatom viability that persisted

2 weeks, suggesting some indirect effects not related with photosynthesis inhibition. One week after the 48 h exposure to pulses of triclosan, diatom viability significantly decreased before coming back to control level 2 weeks after the end of exposure. This delayed increase of diatoms' mortality could be an indirect response consecutive of the bacterial mortality occurring immediately after exposure. The positive interactions between diatoms and bacteria within biofilms communities [83] reinforce the possibility of indirect effects of triclosan on diatoms via direct effects on bacteria (Proia et al. [82]). Though the use of these natural consortia provides a high ecological relevance to ecotoxicological studies, the general lack of knowledge of the mechanisms of action of many chemicals associated with the unclear relationships existing between organisms of complex communities may complicate the interpretation of the biomarkers' response at the community level.

Recently, the use of toxicogenomic, proteomic, or metabolomic approaches has aided in identifying novel mechanisms of action of known and emerging pollutants in fish [84–93] and aquatic invertebrates [94–97]. Due to their sensitivity and their comprehensive analysis of the genes, transcripts, proteins, or metabolites reacting to exposures, omics technologies have the potential to help unraveling precise mechanisms of action of several pollutants. Nevertheless, despite the great effort invested in omics in the last decades, only few studies performed on macroinvertebrate taxa and fish have identified successfully hidden or unknown mechanisms of action of pollutants. This lack of success is related to the fact that the full genome has only been sequenced in few species that are not always ecologically relevant. In addition within the macroinvertebrate species sequenced (i.e., *Caenorhabditis elegans*, [98]; *Daphnia pulex* and shortly *Daphnia magna*, [99]; sea urchin *Strongylocentrotus purpuratus*, [100]) only a few of these genes are annotated (at most 50% in *D. pulex*). This means that often only a few of the putative genes up- and downregulated in transcriptomic studies can be identified and hence related with metabolic paths. The problem is even worse in proteomic studies since posttranscriptional modifications (e.g., phosphorylation, tridimensional structure) often increase the complexity in proteins compared to genes. An additional problem inherent of working with small species is the use of the whole organism. Transcriptomic and proteomic profiles are tissue specific and thus by using a mixture of tissues it is difficult to identify specific genes or proteins that might be altered by a particular pollutant acting in a target tissue. In relation to that, today most successful toxicogenomic studies have been focused in pollutants causing major effects. For example, Connon et al. [94], Poynton et al. [97], and Heckmann et al. [95] worked with custom microarrays in *Daphnia magna* and identified the major metabolic paths affected by cadmium and ibuprofen in *Daphnia magna*. A similar procedure but using proteomic profiles was used in *Gammarus pulex* exposed to PCBs [96]. Future research should be focused in tissue-specific gene or protein expression profiles using immunohistochemistry and in situ hybridization methods. To do that however more genomes need to be characterized and annotated and RNA/DNA probes and antibodies should be more available to researchers. The application of -omics technologies in river biofilms is even more challenging due to the complexity of the community. Indeed, river biofilms are composed of hundreds of species with unsequenced genome. An approach to overcome these limitations and understand -omics data from river biofilm communities is discussed in the last part of this chapter (Sect. 5). Omics technology may also be particularly useful to select specific biomarkers of exposure among a wide array of candidates. This step would obviously require an anchoring of omics result to understandable effects.

To sum up, biomarkers provide information whether an organism or a community has been actually exposed to chemicals with a specific MoA. Therefore, a multibiomarker approach targeting different river compartments but also various levels of biological organization (i.e., tissue, individual, population, and community) is essential to assess the effects of potential toxic compounds and eventually discriminate between direct and indirect effects on the ecosystem. Thus, biomarkers help to target detailed chemical and biological analysis of water, sediments, and biota. They may also help to prioritize discharges of concern with regard to municipal and industrial effluents.

3.2 Biomarkers to Assess Toxic Effects Meaningful at Ecosystem Level

Laboratory studies as described previously allow identifying specific biomarkers of exposure and of effects that can be used as diagnostic tools to help identifying toxic pressures in the ecosystem. Biomarkers of effect are also expected to inform on the consequences of toxic exposure either because they directly measure some essential processes of the ecosystem (e.g., primary production) or because a causal relationships has been established between biomarkers' alterations and effects at higher levels of biological organization (individual, population, and ecosystem).

Several community biomarkers indeed can directly inform on the resources and the processes of the aquatic ecosystems, especially in river biofilms, and thus bear functional implications. Many studies on river biofilms have highlighted the tight link between their structure and function and the processes of aquatic ecosystems and explored the relative importance of the different biofilm components in these processes [41, 101, 102]. Therefore, the biomarkers developed in biofilms concerned already the community level and are linked to higher trophic levels, through food chain, or to aquatic ecosystem, through their implication in important processes such as primary production, organic matter decomposition, and nutrient recycling. For instance, biomass, chlorophyll *a* concentration, or C, N, P composition of river biofilms are some biomarkers of its quality as food supply for macroinvertebrates and fish [103, 104]. Stelzer and Lamberti [104] showed that for low quantity of biofilms, growth of *Elimia livescens* snails fed on river biofilms grown under high P was higher (40–60%) than for snails fed on biofilms grown under low P. Variations in photosynthesis and respiration of river biofilms have direct implications on primary production of aquatic ecosystem [35]. Therefore, different techniques have been developed to measure photosynthetic activity (dissolved oxygen variations, incorporation of radioactive carbon, nonradioactive ^{13}C-labeled bicarbonate, or pulse amplitude modulated fluorescence (PAM); [37, 40, 105]) and the heterotrophic activity (incorporation of radiolabeled nucleic acid, substrate-induced respiration; [105, 106]). The measure of extracellular enzymes activities such as peptidase, glucosidase, and phosphatase of river biofilms can also be directly linked with the carbon and nutrient fluxes of rivers. For instance, Romaní et al. [41] evidenced the positive relationships between extracellular enzymatic activities (glucosidase, phosphatase, and lipase) of river biofilms and the content in dissolved organic carbon (DOC) and biodegradable DOC of flowing waters. Finally, the measure of the diversity of both the autotroph and heterotroph components is an indicator of the response capacity of the community as a loss in diversity is expected to reduce community resistance and resilience to further stressors [27, 107].

The situation is more controversial if it comes to the predictive value of suborganismic biomarkers with respect to individual or population effects. It has been postulated that suborganismic biomarkers can function as early warning indicators that respond before measurable effects on individuals and populations occur. While this may be true in certain cases, the available literature does not support a generally higher sensitivity or earlier responsiveness of biomarkers over organism-level endpoints (see the excellent review by [108]). Furthermore, the ability of biomarkers to predict ecologically relevant effects, i.e., adverse changes of organism or population parameters, has to be considered with care. The reasons for this have been outlined in detail in the study of Forbes et al. [5]. Even if in a field study, an empirical correlation between a biomarker and an organism/population response has been observed, this does not imply that this biomarker is a general predictor of the ecological change. To make biomarkers useful as predictors of ecologically relevant effects, it would need integrated mechanistic models that establish a functional and quantitative link between the suborganismic biomarker response and the adverse outcome. One example of such a relationship is the study of Miller et al. [109] who established—for a laboratory setting—a relationship between the induction of vitellogenin and the reproductive output of fathead minnows. In this case, the relationship was built on a mechanistic relationship between the biomarker effect (binding to the estrogen receptor and subsequent activation of the vitellogenin synthesis), the organism-level effect (reduced reproductive output due to estrogen receptor-mediated neuroendocrine disturbance), and a population model that translated altered individual rates into altered population growth: the population outcome [110, 111]. However, even with such a model available, under field conditions the relationship can be confounded by a variety of factors [5]. While molecular, cellular, and physiological responses are directly involved in the toxic mechanisms induced by the environmental contaminants or are at least closely associated with the initial chemico-biological interactions, the causal relationship between the biological change and the toxicant action is getting increasingly confounded at higher levels of biological organization, due to the action of compensatory processes and of new, level-specific properties [112]. In turn, this implicates that the ecological relevance of biomarkers is limited. Demonstration of an effect at the molecular, cellular, or physiological levels does by no means imply that this effect will propagate into organism or population effects. Thus, the greater potential of biomarkers is in using them as "signposts" rather than as predictors, that is, to help targeting detailed chemical analysis, prioritizing study sites, or to guide further effects' assessment [113]. As a consequence, biomarkers should not be used as stand-alone tools but should be embedded in an integrated monitoring strategy combining the biomarkers with analytical (bioanalytics and chemical analytics), experimental, and ecological tools [114–116].

3.3 Bridging the Gap Between Laboratory Studies and Field Studies

Biomarkers may help in identifying putative toxic environmental stressors. This, however, would include confirmatory lab exposures to demonstrate that the responses obtained in the field can be induced in the laboratory by the candidate stressor or stressor combination. Often such studies fail due to three problems:

1. Lab and field populations are different and hence are acclimated or/and adapted to distinct environmental factors [117];
2. In the field many different environmental factors affect biomarker responses whereas in the lab only few of them are present;
3. Lab and field exposures are different in terms of exposure duration, exposure variation (constant versus fluctuating exposure), or toxicant bioavailability.

The first problem can be solved by using transplanted/translocated organisms, that is, by exposing lab or reference populations in the field using community colonized on artificial substrates for river biofilms or caged organisms for macroinvertebrates. While the use of natural biofilm colonized over artificial substrata allows the heterogeneity occurring on natural substrates to be reduced [118], the use of natural substrata (cobbles or stones) is more realistic but also more complex. After a period of colonization (4–5 weeks) in a reference site or in the laboratory, the substrata are translocated in the exposed sites [119, 120]. To study recovery of biofilm from pollution, river biofilms adapted to a polluted site can also be translocated to a reference site (Rotter et al. 2010 [122]) This translocation approach is also often used with bivalve species and quite recently has been applied to benthic macroinvertebrates and model lab species. Barata et al. [121] combined field-transplanted and laboratory-exposed assays with the crustacean *D. magna* and showed that the organophosphorate fenitrothion inhibited the enzyme cholinesterase in field-exposed individuals. The second problem can be addressed by investigating the role of different environmental factors on chemicals' toxicity and biomarkers' response, although it will remain inherently difficult to "mimic" the complexity of the various environmental influences. For instance, light intensity was found to modulate herbicides' toxicity on river biofilms [123, 124]. Guasch and Sabater [123] showed that biofilms from open sites were more sensitive to atrazine than biofilms from shaded rivers. Influence of light intensity can also be investigated in laboratory under controlled conditions. Indeed, biofilms grown in microcosms under constant suboptimal, saturating, and high-light intensities showed characteristics common to shade/light adaptation which constrain their capacity to answer to further stressors. In particular, river biofilms colonized in artificial streams and adapted to high-light intensity were more tolerant to glyphosate exposure ($EC_{50} = 35.6$ mg L^{-1} for photosynthetic efficiency) than those of shade-adapted ($EC_{50} = 11.7$ mg L^{-1} for photosynthetic efficiency) [125]. Not only the light intensity but also the daily variation in light can affect toxicity; for instance, Laviale et al. [124] showed that a dynamic light regime increased the sensibility of river biofilms to the herbicide isoproturon. The influence of other environmental factors such as flow velocity, temperature, and nutrient concentrations has also been studied to better understand multiple stress situations [126–129]. In the field using transplanted and field-collected invertebrate species, Damásio et al. [130–132] showed that water temperature, suspended solids, and salinity affected many of the biomarkers used to detect pollutant effects on aquatic communities. Biomarkers of tissue damage such as DNA strand breaks and lipid peroxidation levels were dramatically affected by salinity, whereas more specific ones such as the activities of antioxidant and cholinesterase enzymes were more affected by particular pollutants such as organophosphorous pesticides and persistent organic pollutants [132].

Also the third problem represents major challenges in the field–lab comparison. Indeed, a variety of factors can confound the relation between analytically determined chemical concentrations at a given study site and the biomarker response of the fish living thereof. These confounders include—to name a few—toxicokinetic processes such as chemical bioavailability, uptake and metabolism, combination effects from chemical mixtures, different time-effect profiles, or modulation of the biomarker response by physiological processes (compensatory responses, physiological changes related to reproduction, pathogen infection or nutrition, etc.). Another factor that can cause disparate findings of chemical analytics and biomarker measurements in fish is different scales of time integration in the two approaches. While chemical analytics detect the actual concentration of the chemical being present in the water or sediment at the moment when the sample is taken, the biomarker response in the fish integrates the long-term exposure of fish to the contaminants in the environment. Finally, the physiology of the lab and the field fish may differ, and this in turn affects the biomarker response. For instance, in the field, many biomarkers show seasonal variability (e.g., [133]), while in the lab, this seasonality may not be observed.

4 Biomarkers as Complementary Tools for Ecological Water Quality Assessment

One of the particular and unique aspects of the European Union's Water Framework Directive (WFD, Directive 2000/60/EC) is the use of ecologically based instruments to assess and predict the ecological impacts of environmental pressures on water quality. In general, ecological status assessment involves sampling the aquatic community, and comparing against a reference prediction for that water body type. This approach is currently used in many countries (see review of [44, 134]). Various biological metrics exist to quantify changes in community composition and these are often combined in multimetric indices to improve the chances of detecting adverse changes. Among them, those focusing on assemblages of benthic macroinvertebrates or on diatom community composition are widely used [38, 44, 134]. Although structural community metrics can detect the degradation of surface waters, they detect neither specific effects of water pollutants nor moderate changes in chemical water quality [39, 130]. Therefore, there is a need to complement the biological metrics actually used with other biological measures that may serve as descriptors of cause–effect or may inform about further degradation (or improvement) of the water. In this context, biomarkers can complement the information given by classical endpoints to identify hot spots of pollution by detecting toxic effects in the field through different levels of biological organization and identifying causes of impairment. To do so, two valuable methods can be identified: the active biomonitoring [135] using transplanted communities and the integrated use of chemical analyses and biochemical and cellular responses.

4.1 Active Monitoring: Translocation Experiments and In Situ Bioassays

A key advantage of translocation experiments or of in situ bioassays over whole effluent toxicity tests and biological surveys of communities is their greater relevance to the natural situation, especially with respect to the contamination scenario. Additionally, in situ bioassays are able to detect effects in transplanted individuals more rapidly (hours to days) than the time taken to observe changes in community structure (months to years) measured during macroinvertebrate or river biofilm sampling.

Several studies have shown the usefulness of river biofilm translocation to detect both acute and chronic effects of different types of perturbations: acidification [136], metal contamination [137–139], and pesticide and herbicide contamination [119, 120, 140]. Short-term translocations allow determining the acute effects of perturbations while during long-term translocations the time of adaptation can be measured and acute effects can be discriminated from chronic ones. This type of experiment can also be useful to assess the recovery potential of river biofilms by translocating contaminated river biofilms to reference site. Following this approach, Dorigo et al. [119] translocated river biofilms from an herbicide-polluted site of the river Morcille (France) to an upstream reference site. After 9 weeks of acclimatization, the structure of the translocated eukaryotic, bacterial, and diatom communities was more similar to the downstream communities than to the reference one. In addition, the tolerance to the herbicides diuron and copper was higher in translocated biofilms than in reference site, indicating that chronic contamination by herbicides can affect durably river biofilm communities. Indeed, the chronic exposure of a community to a critical level of a chemical is expected to exert a selective pressure on the community by selecting more resistant individuals to the chemical following the PICT concept (pollution-induced community tolerance) as demonstrated by various authors [106, 141–143]. The acquisition of tolerance may be supported by both structural and functional adaptation [144]. Bioassays can be performed to detect and identify this tolerance acquisition. To do so adapted communities (collected from the field) are exposed to a range of toxicant concentrations and toxicity thresholds are compared between reference and polluted sites. Higher toxicity thresholds indicate an adaptation of the community to this toxicant. By measuring the tolerance of river biofilms from polluted sites to various classes of toxicants, it may then be possible to point out which class of toxicant the community is adapted to and consequently to determine the nature of the chronic contamination affecting river biofilms at this site. Rotter et al. [140] followed this approach in a field study on the river Elbe basin (Germany). They could discriminate river biofilms of a site polluted by the herbicide prometryn from reference biofilms based on their resistance to prometryn. The use of PICT bioassays could then contribute to investigative monitoring to identify the causes of ecological impairment.

A set of in situ and cost-effective bioassays with caged, single species of macroinvertebrates based on feeding and growth responses have permitted detecting lethal and sublethal responses that are biologically linked with key ecological processes such as detritus processing and algal grazing rates, and with specific toxicological mechanisms [145]. Here we also provide two examples of the use of in situ bioassays to identify major pesticides that may cause detrimental effects in crustacean and bivalve species affected by agricultural pollution. Investigations were carried out using freshwater clams (*Corbicula fluminea*) and *Daphnia magna* individuals transplanted in the main drainage channels that collect the effluents coming from agriculture fields in the Ebro Delta (NE Spain) during the main growing season of rice (from May to August) [121, 131]. The Ebro Delta, located at the end of the largest Spanish river "The Ebro river", holds 21,600 Ha of rice fields, producing 113,500 Tons of rice per year. Barata et al. [121], using transplanted *Daphnia magna* individuals, found a good correlation between the levels of the organophorous insecticide fenitrothion in water, toxicity responses, and the inhibition of acethylcholinesterase. Damásio et al. [131] in caged clams reported that eight out of the nine biomarkers' analyses showed significant differences among sites within and/or across months. Antioxidant and esterase enzyme responses were in most cases inhibited in clams transplanted in drainage channels heavily polluted with pesticides. Conversely, lipid peroxidation levels increased steadily from May in upstream stations to August in drainage channels. Damásio et al. [131] used a Multivariate Partial Least Square Projections to Latent Structures regression analyses (PLS), which aims to establish relationships between the environmental variables and biological responses, to identify chemical contaminants causing the most adverse effects in the studied species. According to PLS analyses, the environmental parameters that affected most biological responses and had higher effects (high Vip scores) were endosulfan, DDTs, Hexachlobenzene, Alkylphenols, Cu and Hg concentration levels in organisms, and the concentrations of the acidic herbicides, propanil, phenylureas, and temperature in water. PLS results also revealed the presence of additive or antagonistic effects among environmental parameters. For example, glutathione-S-transferase activities were negatively correlated to endosulfan, acidic herbicides, and temperature and were positively correlated with Hg and PAHs. Catalase activities only showed negative relationships. Glutathione peroxidase activities despite being affected by different factors shared in common their negative relationship with levels of acidic herbicides and endosulfan. Cholinesterase activities showed a different pattern, being negatively correlated to temperature, Cu, and Hg and positively affected by the rest of parameters. Levels of lipid peroxidation were positively correlated to the concentrations of the major pesticides including phenylureas, propanil, and endosulfan with negative contributions of Hg and Cu. In summary, PLS results performed on biomarker responses of transplanted clams and contaminant levels indicated that endosulfan, propanil, and phenylureas were the chemical contaminants causing the most adverse effects in the studied species.

4.2 *Multivariate Analyses of Chemical and Biomarkers Response in Passive Monitoring*

The integrated use of chemical analyses and biochemical and cellular responses to pollutants can be very useful for detecting exposure to anthropogenic contaminants in freshwater systems and to guide studies on cause–effect relationships. Moreover, since in real field situations aquatic organisms are exposed to multiple chemical contaminants involving different toxicity mechanisms, each contributing to a final overall adverse effect, the use of a large set of biochemical responses may allow us to identify which are the potential hazardous contaminants in the field [131, 146, 147].

Field monitoring studies on river biofilms focused on structural parameters such as diatom community composition rather than on more functional parameters. The use of diatom community composition as an indicator of water quality is discussed in a following chapter [16] and here, we focus on the use of other biomarkers in a multivariate approach to determine impacts of pollutant and indicate potential main sources of contamination. These approaches are rare but still promising. One of these studies was conducted in the highly polluted river Llobregat (Spain) by Ricart et al. [148]. Along with the analyses of water physicochemistry and of pesticides concentrations, they combined the measure of different biomarkers of the autotroph (chlorophyll *a* concentration, photosynthetic parameters, and EPS content) and heterotroph (extracellular enzymatic activities of glucosidase, peptidase, and phosphatase) components of river biofilms with the determination of diatom community composition. A partial redundancy direct analysis (RDA) showed that pesticides' concentration explained a major part of variance of chlorophyll *a* response (91.57%) and of photosynthetic efficiency (77.75%) and capacity (60.94%), while variance in extracellular enzymatic activities was mostly explained by physicochemical parameters. Among the pesticides detected, organophosphates and phenylureas accounted for the major part of the explained variance. Based on these results, the authors pointed out *the need of further laboratory tests to confirm causality* [148]. In this example, the correlation between an impact parameter (reduction of primary production) and the presence of some chemicals (pesticides and more precisely organophosphates and phenylureas) could be identified thanks to an integrated use of chemical analyses, biomarkers, and appropriate statistics.

Several studies also showed that the use of individual and physiological responses of keystone macroinvertebrates may provide additional information about water quality characteristics [130, 131, 149]. Recently, a large set of biochemical markers were developed and used in field-collected caddisflies of the tolerant species *Hydropsyche exocellata* and in transplanted *Daphnia magna* to monitor pollution and ecological quality of water in Mediterranean Rivers [130, 149–151]. The studies were conducted in two rivers, the Llobregat and Besós river basins (NE Spain), whose natural resources have been greatly affected by human activities such as agriculture, urbanization, salinization, and an intensive water use for human consumption (supplying water to many urban areas including Barcelona city). The results obtained indicated that biomarkers responded more specifically to water pollutants than individuals or species responses. In one of these

Table 1 Biomarkers of river biofilms, macroinvertebrates, and fish recommended to be used in routine and investigative monitoring

Routine monitoring	River biofilms	Photosynthetic efficiency
	Macroinvertebrates	Biotransformation enzymes (GST)
		Oxidative stress (CAT), cholinesterases (ChE)
		Damage (lipid peroxidation)
	Fish	Cytochrome P4501A
		Metallothioneins
		Vitellogenin
Investigative monitoring	River biofilms	PICT bioassays based on photosynthetic efficiency
		PICT bioassays based on extracellular enzymatic activities (peptidase, glucosidase, and phosphatase)
	Macroinvertebrates	Biotransformation enzymes (GST, MT)
		Oxidative stress (SOD, CAT, GPX, GR, GSH), cholinesterase
		Damage (lipid peroxidation, DNA strand breaks, carboryl)

studies, Damásio et al. [130] used a Multivariate Partial Least Square Projections to Latent Structures regression analyses (PLS). From up to 20 environmental variables considered, six of them: habitat degradation, suspended solids, and nitrogenous and conductivity-related parameters affected macroinvertebrate assemblages. On the other hand, levels of organophosphorous compounds and polycyclic aromatic hydrocarbons were high enough to trigger cholinesterase activities and feeding rates of transplanted *D. magna,* respectively. More recently, Puertolas et al. [149] showed that biomarkers of transplanted *Daphnia* and of field-collected caddisflies but not species diversity responded to the application of glyphosate. Effects included oxidative stress-related responses such as increased antioxidant enzyme activities related with the metabolism of glutathione and increased levels of lipid peroxidation. These results emphasize the importance of combining chemical, ecological, and specific biological responses to identify ecological effects of pesticides in the field.

5 Challenges and Perspectives

5.1 How to Use the Already Developed Biomarkers?

The data presented in this chapter showed that biomarkers can be used successfully to point to ecological risks of toxic pollutants in river biofilms, freshwater macroinvertebrate species, and fish in the field. In Table 1, we present a list of biomarkers which have been sufficiently investigated to be used as diagnostic tools in routine or investigative monitoring. To succeed in doing so the six following aspects have to be considered:

1. Select keystone species in macroinvertebrates and fish studies.
2. Choose a large battery of markers representing different metabolic detoxification-effect paths and different essential processes in rivers.
3. Select appropriate reference sites and if required use translocation experiments.
4. Combine biomarkers with other measurements of ecological quality (in situ bioassays, biological indexes, and biodiversity).
5. Complement biomarker work with a thorough physicochemical analysis of the studied sites and organisms.
6. Be careful in your data interpretation; *correlations* between chemicals, biomarkers, and ecological parameters *are not cause–effect relations.*

Although less important it is also highly recommended to use appropriate multivariate methods to relate environmental factors with biological responses. Provided that many pollutants are usually highly correlated (i.e., high co-correlations among the concentrations of the different pesticides applied simultaneously) the use of Principal Component and Partial Least Square regression (PLS) methods is highly recommended.

5.2 *Limitations of Biomarkers*

So far, biomarkers were not successfully implemented with ERA procedure because they did not provide evident links between exposure situations and the resulting ecological effect. Indeed, most of biomarkers studies still focused on a cause–effect relationship between one biomarker and one chemical. It is essential to be able to link biomarkers' changes with effects at higher levels of organization to derive ecological understanding. That is why we recommend biomarkers to be used as diagnostic tools in an integrated monitoring strategy combining the biomarkers with analytical (bioanalytics and chemical analytics), experimental, and ecological tools [114–116].

5.3 *What Is Still Missing?*

The studies presenting the successful application of biomarkers in an ecological risk assessment are rare and different points are still missing to make it possible.

Likewise biological indexes, standardized protocols for sampling, biomarker determination, and interpretation across species should be established before biomarkers could be used in ecological risk assessment. Moreover, biomarkers development should address different points including:

1. The evaluation of biomarkers specificity: which kind of toxicants can be linked to biomarker response?
2. The determination of biomarkers' sensitivity to environmental variations: how biomarker response is influenced by flow velocity, temperature, light intensity, salinity, oxygen levels, and ammonia, among others?
3. The exploration of the biomarkers' variations through different levels of biological organization: which are the consequences of biomarker variations at higher levels of biological organization?

Different physiological biomarkers showed nonlinear response; therefore, special effort should be put on modeling the response of such biomarkers and ease their interpretation. Indeed well-fit and understandable models of biomarkers' response combined with extensive knowledge on biomarkers' effects through increasing levels of biological organization may help to predict future scenarios.

Omics technologies are aiming at comprehensively studying responses of biological systems to a wide variety of factors. Thus, in principle these technologies offer a new alternative to discover biomarkers as shown by successful studies on macroinvertebrates. But the implementation of omics within ecological risk assessment was recognized as one of the biggest challenges for ecotoxicologist (Ankley 2006 [84]). Indeed, while omics technologies can be currently used in macroinvertebrates or fish studies, their use is still limited to the genetic information available (genome sequences and annotation). The application of transcriptomic to communities such as river biofilms represents major challenges due to the simultaneous presence and activity of hundreds of different species not genetically characterized [152]. Functional Gene Arrays (FGAs) based on consensus sequences from genes of key processes may partially overcome these limitations and allow gene expression of communities to be monitored [153]. Consensus sequences used to design FGAs' probes result from the alignment of various genes from known species and are also expected to be found in nonsequenced organisms [154]. One of the most complete FGAs developed is the GeoChip 2.0 that covers more than 10,000 microbial genes in more than 150 functional groups involved mainly in biogeochemical cycling processes [155]. Preliminary experiment showed that such an array can be successfully developed for the autotroph component of the river biofilm communities based on genetic information available for diatoms and chorophytes [125]. This FGA could be then used in ecotoxicological tests to determine the effects of a chemical on different important processes related to primary production, nutrient utilization, or antioxidant protection in biofilms. Quantification of the changes occurring in the genes detected can further be done by RT-qPCR (Dorak 2006 [158]). However, since the extrapolation of genes' variations to changes at higher levels of organization is likely to be influenced by many regulation mechanisms, this screening is a prognostic tool to indicate which processes are at risk and should be investigated at higher levels of biological organization (e.g. protein concentration, enzymatic activity, and photosynthesis/respiration).

Indeed, all omics technologies may be considered as prognostic tools to guide further detection of adverse outcome. Recently, the adverse outcome pathways were proposed as frame in which classical ecotoxicological knowledge and omics result could be organized to link exposure to a toxicant to subsequent adverse effect [111]. This is a first step toward the good integration of omics and the related biomarkers within the ecological risk assessment procedures.

Acknowledgments The research was funded by the Spanish Ministry of Science and Education (CTM2009-14111-CO2-01) and the EC Sixth Framework Program (KEYBIOEFFECTS MRTN-CT-2006-035695).

References

1. Bensalah K, Montorsi F, Shariat SF (2007) Challenges of cancer biomarker profiling. Eur Urol 52:1601–1609
2. van Gils MPMQ, Stenman UH, Schalken JA, Schröder FH, Luider TM, Lilja H, Bjartell A, Hamdy FC, Pettersson KSI, Bischoff R, Takalo H, Nilsson O, Mulders PFA, Bangma CH (2005) Innovations in serum and urine markers in prostate cancer: current european research in the P-mark project. Eur Urol 48:1031–1041
3. Peters KE, Walters CC, Moldowan JM (2005) The biomarker guide. Cambridge University Press, Cambridge, UK
4. Boschker HTS, Middelburg JJ (2002) Stable isotopes and biomarkers in microbial ecology. FEMS Microbiol Ecol 40:85–95

5. Forbes VE, Palmqvist A, Bach L (2006) The use and misuse of biomarkers in ecotoxicology. Environ Toxicol Chem 25:272–280
6. Huggett RJ, Kimerle RA, Mehrle Jr PM (1992) Biomarkers: biochemical, physiological and histological makers of anthropogenic stress. In: Harbor A (ed) Proceedings of the 8th Pellston Workshop. Keystone, Colorado
7. Hansen PD, Dizer H, Hock B, Marx A, Sherry J, McMaster M, Blaise C (1998) Vitellogenin – a biomarker for endocrine disruptors. TrAC-Trend Anal Chem 17:448–451
8. Depledge MH, Aagaard A, Györkös P (1995) Assessment of trace metal toxicity using molecular, physiological and behavioural biomarkers. Mar Pollut Bull 31:19–27
9. Peakall DB (1994) The role of biomarkers in environmental assessment (1) Introduction. Ecotoxicology 3:157–160
10. Peakall DB (1994) Biomarkers: the way forward in environmental assessment. Toxicol Ecotoxicol News 1:55–60
11. Aerni HR, Kobler B, Rutishauser BV, Wettstein FE, Fischer R, Giger W, Suter MJF, Eggen R (2004) Combined biological and chemical assessment of estrogenic activities in wastewater treatment plant effluents. Anal Bioanal Chem 378:688–696
12. Tyler CR, Spary C, Gibson R, Shears J, Santos E, Sumpter JP (2005) Accounting for differences in the vitellogenic responses of rainbow trout and roach exposed to oestrogenic effluents from wastewater treatment works. Environ Sci Technol 39:2599–2607
13. Bucheli TD, Fent K (1995) Induction of cytochrome P450 as a biomarker for environmental contamination in aquatic ecosystems. Crit Rev Environ Sci Tech 25:201
14. Van der Oost R, Beyer J, Vermeulen NPE (2003) Fish bioaccumulation and biomarkers in environmental risk assessment: a review. Environ Toxicol Pharmacol 13:57–149
15. Morin S, Cordonier A, Lavoie I, Arini A, Blanco S, Duong TT, Tornés E, Bonet B, Corcoll N, Faggiano L, Laviale M, Pérès F, Becares E, Coste M, Feurtet-Mazel A, Fortin C, Guasch H, Sabater S (2012) Consistency in diatom response to metal-contaminated environments. In: Guasch H, Ginebreda A, Geiszinger A (eds) Emerging and priority pollutants in rivers, The handbook of environmental chemistry, vol 19. Springer, Heidelberg
16. Muñoz I, Sabater S, Barata C (2012) Evaluating ecological integrity in multi-stressed rivers: from the currently used biotic indices to newly developed approaches using biofilms and invertebrates. In: Guasch H, Ginebreda A, Geiszinger A (eds) Emerging and priority pollutants in rivers, The handbook of environmental chemistry, vol 19. Springer, Heidelberg
17. Boyd JW (2007) The endpoint problem. Resources-washington, DC, 165:26–28
18. Suter GW (2001) Applicability of indicator monitoring to ecological risk assessment. Ecol Indic 1:101–112
19. Kontogianni A, Luck GW, Skourtos M (2010) Valuing ecosystem services on the basis of service-providing units: a potential approach to address the "endpoint problem" and improve stated preference methods. Ecol Econ 69:1479–1487
20. Schwarzenbach RP, Escher BI, Fenner K, Hofstetter TB, Johnson CA, von Gunten U, Wehrli B (2006) The challenge of micropollutants in aquatic systems. Science 313: 1072–1077
21. Jelić A, Gros M, Petrović M, Ginebreda A, Barceló D (2012) Occurrence and elimination of pharmaceuticals during conventional wastewater treatment. In: Guasch H, Ginebreda A, Geiszinger A (eds) Emerging and priority pollutants in rivers. The handbook of environmental chemistry, vol 19. Springer, Heidelberg
22. Akkanen J, Slootweg T, Mäenpää K, Leppänen MT, Agbo S, Gallampois C, Kukkonen JVK (2012) Bioavailability of organic contaminants in freshwater environments. In: Guasch H, Ginebreda A, Geiszinger A (eds) Emerging and priority pollutants in rivers, The handbook of environmental chemistry, vol 19. Springer, Heidelberg
23. Barceló D (2008) Advanced MS analysis of metabolites and degradation products. TrAC-Trend Anal Chem 27:805–806

24. Geiszinger A, Bonnineau C, Faggiano L, Guasch H, López-Doval JC, Proia L, Ricart M, Ricciardi F, Romaní AM, Rotter S, Muñoz I, Schmitt-Jansen M, Sabater S (2009) The relevance of the community approach linking chemical and biological analyses in pollution assessment. TrAC-Trend Anal Chem 28:619–626
25. Calow P, Forbes VE (2003) Does ecotoxicology inform ecological risk assessment? Environ Sci Technol 37:146A–151A
26. Ricart M, Barceló D, Geiszinger A, Guasch H, Alda ML, Romaní AM, Vidal G, Villagrasa M, Sabater S (2009) Effects of low concentrations of the phenylurea herbicide diuron on biofilm algae and bacteria. Chemosphere 76:1392–1401
27. Clements WH, Newman MC (2002) Community ecotoxicology. Wiley, West Sussex, UK
28. Matthaei CD, Piggott JJ, Townsend CR (2010) Multiple stressors in agricultural streams: interactions among sediment addition, nutrient enrichment and water abstraction. J Appl Ecol 47:639–649
29. Laetz CA, Baldwin DH, Collier TK, Hebert V, Stark JD, Scholz NL (2009) The synergistic toxicity of pesticide mixtures: implications for risk assessment and the conservation of endangered pacific Salmon. Environ Health Perspect 117:348–353
30. Bayley M, Nielsen JR, Baatrup E (1999) Guppy sexual behavior as an effect biomarker of estrogen mimics. Ecotoxicol Environ Saf 43:68–73
31. Chan KM (1995) Metallothionein: potential biomarker for monitoring heavy metal pollution in fish around Hong Kong. Mar Pollut Bull 31:411–415
32. Forget J, Beliaeff B, Bocquené G (2003) Acetylcholinesterase activity in copepods (*Tigriopus brevicornis*) from the Vilaine River estuary, France, as a biomarker of neurotoxic contaminants. Aquat Toxicol 62:195–204
33. Matozzo V, Gagné F, Marin MG, Ricciardi F, Blaise C (2008) Vitellogenin as a biomarker of exposure to estrogenic compounds in aquatic invertebrates: a review. Environ Int 34:531–545
34. NRC: Committee on Biological Markers of the National Research Council (1987) Biological markers in environmental health research. Environ Health Perspect 74:3–9
35. Vannote RL, Minshall GW, Cummins KW, Sedell JR, Cushing CE (1980) The river continuum concept. Can J Fish Aquat Sci 37:130–137
36. Romaní AM (2010) Freshwater biofilms. In: Dürr S, Thomason JC (eds) Biofouling. Wiley, NJ
37. Guasch H, Admiraal W, Sabater S (2003) Contrasting effects of organic and inorganic toxicants on freshwater periphyton. Aquat Toxicol 64:165–175
38. Sabater S, Admiraal W (2005) Periphyton as biological indicators in managed aquatic ecosystems. In: Azim ME, Verdegem MCJ, van Dam AA, Beveridge MCM (eds) Periphyton: ecology, exploitation and management. CABI Publising, Wallingford
39. Sabater S, Guasch H, Ricart M, Romaní AM, Vidal G, Klünder C, Schmitt-Jansen M (2007) Monitoring the effect of chemicals on biological communities. The biofilm as an interface. Anal Bioanal Chem 387:1425–1434
40. Schmitt-Jansen M, Altenburger R (2008) Community-level microalgal toxicity assessment by multiwavelength-excitation PAM fluorometry. Aquat Toxicol 86:49–58
41. Romaní AM, Guasch H, Muñoz I, Ruana J, Vilalta E, Schwartz T, Emtiazi F, Sabater S (2004) Biofilm structure and function and possible implications for riverine DOC dynamics. Microb Ecol 47:316–328
42. Bonnineau C, Bonet B, Corcoll N, Guasch H (2011) Catalase in fluvial biofilms: a comparison between different extraction methods and example of application in a metal-polluted river. Ecotoxicology 20:293–303
43. Ylla I, Sanpera-Calbet I, Muñoz I, Romaní AM, Sabater S (2011) Organic matter characteristics in a Mediterranean stream through amino acid composition: changes driven by intermittency. Aquat Sci 73:523–535
44. Bonada N, Prat N, Resh VH, Statzner B (2006) Developments in aquatic insect biomonitoring: a comparative analysis of recent approaches. Annu Rev Entomol 51:495–523

45. Sumpter JP, Johnson AC (2008) 10th Anniversary Perspective: reflections on endocrine disruption in the aquatic environment: from known knowns to unknown unknowns (and many things in between). J Environ Monit 10:1476
46. Livingstone DR, Garcia-Martinez P, Michel X, Narbone JF, O'Hara S, Ribera D, Winston GW (1990) Oxyradical prodiction as a pollution-mediated mechanisms of toxicity in the common mussel, *Mytilus edulis L*, and other molluscs. Funct Ecol 4:415–424
47. Viarengo A, Pertica M, Canesi L, Accomando R, Mancinelli G, Orunesu M (1989) Lipid peroxidation and level of antioxidant compounds (GSH, Vitamin E) in the digestive glands of mussels of three different age groups exposed to aneaerobic and aerobic conditions. Mar Environ Res 28:291–295
48. Porte C, Janer G, Lorusso LC, Ortiz-Zarragoitia M, Cajaraville MP, Fossi MC, Canesi L (2006) Endocrine disruptors in marine organisms: approaches and perspectives. Comp Biochem Physiol C 143:303–315
49. Amiard JC, Amiard-Triquet C, Barka S, Pellerin J, Rainbow PS (2006) Metallothioneins in aquatic invertebrates: their role in metal detoxification and their use as biomarkers. Aquat Toxicol 76:160–202
50. Depledge MH, Fossi MC (1994) The role of biomarkers in environmental assessment (2). Invertebrates. Ecotoxicology 3:161–172
51. Domingues I, Agra AR, Monaghan K, Soares AMVM, Nogueira AJA (2010) Cholinesterase and glutathione-S-transferase activities in freshwater invertebrates as biomarkers to assess pesticide contamination. Environ Toxicol Chem 29:5–18
52. Hutchinson TH (2007) Small is useful in endocrine disrupter assessment – four key recommendations for aquatic invertebrate research. Ecotoxicology 16:231–238
53. Hyne RV, Maher WA (2003) Invertebrate biomarkers: links to toxicosis that predict population decline. Ecotoxicol Environ Saf 54:366–374
54. Fulton MH, Key PB (2001) Acetylcholinesterase inhibition in estuarine fish and invertebrates as an indicator of organophosphorus insecticide exposure and effects. Environ Toxicol Chem 20:37–45
55. Livingstone DR (1993) Biotechnology and pollution monitoring: use of molecular biomarkers in the aquatic environment. J Chem Technol Biotechnol 57:195–211
56. Mitchelmore CL, Chipman JK (1998) DNA strand breakage in aquatic organisms and the potential value of the comet assay in environmental monitoring. Mutat Res-Fund Mol M 399:135–147
57. Solé M, Livingstone DR (2005) Components of the cytochrome P450-dependent monooxygenase system and 'NADPH-independent benzo[a]pyrene hydroxylase' activity in a wide range of marine invertebrate species. Comp Biochem Phys C 141:20–31
58. Valavanidis A, Vlahogianni T, Dassenakis M, Scoullos M (2006) Molecular biomarkers of oxidative stress in aquatic organisms in relation to toxic environmental pollutants. Ecotox Environ Saf 64:178–189
59. Eto M (1974) Organophosphorous pesticides: organic and biological chemistry. CRC, Ohio
60. Monserrat JM, Martinez PE, Geracitano LA, Amado LL, Martins CMG, Pinho GLL, Chaves IS, Ferreira-Cravo M, Ventura-Lima J, Bianchini A (2007) Pollution biomarkers in estuarine animals: critical review and new perspectives. Comp Biochem Physiol C 146:221–234
61. De Coen WM, Janssen CR (2003) The missing biomarker link: relationships between effects on the cellular energy allocation biomarker of toxicant-stressed *Daphnia magna* and corresponding population characteristics. Environ Toxicol Chem 22:1632–1641
62. Smolders R, Bervoets L, De Coen W, Blust R (2004) Cellular energy allocation in zebra mussels exposed along a pollution gradient: linking cellular effects to higher levels of biological organization. Environ Pollut 129:99–112
63. Barata C, Solayan A, Porte C (2004) Role of B-esterases in assessing toxicity of organophosphorus (chlorpyrifos, malathion) and carbamate (carbofuran) pesticides to *Daphnia magna*. Aquat Toxicol 66:125–139

64. Damásio J, Guilhermino L, Soares AMVM, Riva MC, Barata C (2007) Biochemical mechanisms of resistance in *Daphnia magna* exposed to the insecticide fenitrothion. Chemosphere 70:74–82
65. Bosnjak I, Uhlinger KR, Heim W, Smital T, Franekic-Colic J, Coale K, Epel D, Hamdoun A (2009) Multidrug efflux transporters limit accumulation of inorganic, but not organic, mercury in sea urchin embryos. Environ Sci Technol 43:8374–8380
66. Faria M, Navarro A, Luckenbach T, Piña B, Barata C (2010) Characterization of the multixenobiotic resistance (MXR) mechanism in embryos and larvae of the zebra mussel (*Dreissena polymorpha*) and studies on its role in tolerance to single and mixture combinations of toxicants. Aquat Toxicol 101:78–87
67. Luckenbach T, Epel D (2005) Nitromusk and polycyclic musk compounds as long-term inhibitors of cellular xenobiotic defense systems mediated by multidrug transporters. Environ Health Perspect 113:17–24
68. Bonnineau C, Guasch H, Proia L, Ricart M, Geiszinger A, Romaní AM, Sabater S (2010) Fluvial biofilms: a pertinent tool to assess β-blockers toxicity. Aquat Toxicol 96:225–233
69. Lawrence JR, Swerhone GDW, Wassenaar LI, Neu TR (2005) Effects of selected pharmaceuticals on riverine biofilm communities. Can J Microbiol 51:655–669
70. Ricart M, Guasch H, Alberch M, Barceló D, Bonnineau C, Geiszinger A, Farré M, Ferrer J, Ricciardi F, Romaní AM, Morin S, Proia L, Sala L, Sureda D, Sabater S (2010) Triclosan persistence through wastewater treatment plants and its potential toxic effects on river biofilms. Aquat Toxicol 100:346–353
71. Tlili A, Dorigo U, Montuelle B, Margoum C, Carluer N, Gouy V, Bouchez A, Bérard A (2008) Responses of chronically contaminated biofilms to short pulses of diuron. An experimental study simulating flooding events in a small river. Aquat Toxicol 87:252–263
72. Espeland EM, Wetzel RG (2001) Complexation, stabilization, and UV photolysis of extracellular and surface-bound glucosidase and alkaline phosphatase: implications for biofilm microbiota. Microb Ecol 42:572–585
73. Francoeur SN, Wetzel RG (2003) Regulation of periphytic leucine-aminopeptidase activity. Aquat Microb Ecol 31:249–258
74. Neely RK (1994) Evidence for positive interactions between epiphytic algae and heterotrophic decomposers during the decomposition of *Typha latifolia*. Arch Hydrobiol 129:443–457
75. Rier ST, Kuehn KA, Francoeur SN (2007) Algal regulation of extracellular enzyme activity in stream microbial communities associated with inert substrata and detritus. J N Am Benthol Soc 26:439–449
76. Lopez-Doval JC, Ricart M, Guasch H et al (2010) Does grazing pressure modify diuron toxicity in a biofilm community? Arch Environ Contam Toxicol 58:955–962
77. Pesce S, Fajon C, Bardot C, Bonnemoy F, Portelli C, Bohatier J (2006) Effects of the phenylurea herbicide diuron on natural riverine microbial communities in an experimental study. Aquat Toxicol 78:303–314
78. Franz S, Altenburger R, Heilmeier H, Schmitt-Jansen M (2008) What contributes to the sensitivity of microalgae to triclosan? Aquat Toxicol 90:102–108
79. Morin S, Proia L, Ricart M, Bonnineau C, Geiszinger A, Ricciardi F, Guasch H, Romaní AM, Sabater S (2010) Effects of a bactericide on the structure and survival of benthic diatom communities. Vie Milieu 60:109–116
80. Guasch H, Atli G, Bonet B, Corcoll N, Leira M, Serra A (2010) Discharge and the response of biofilms to metal exposure in Mediterranean rivers. Hydrobiologia 657:143–157
81. Serra A, Corcoll N, Guasch H (2009) Copper accumulation and toxicity in fluvial periphyton: the influence of exposure history. Chemosphere 74:633–641
82. Proia L, Morin S, Peipoch M, Romaní AM, Sabater S (2011) Resistance and recovery of river biofilms receiving short pulses of Triclosan and Diuron. Sci total environ 409:3129–3137
83. Murray RE, Cooksey KE, Priscu JC (1986) Stimulation of bacterial DNA synthesis by algal exudates in attached algal-bacterial consortia. Appl Environ Microbiol 52:1177–1182

84. Ankley GT, Daston GP, Degitz SJ, Denslow ND, Hoke RA, Kennedy SW, Miracle AL, Perkins EJ, Snape J, Tillitt DE, Tyler CR, Versteeg D (2006) Toxicogenomics in regulatory ecotoxicology. Environ Sci Technol 40:4055–4065
85. De Wit M, Keil D, van der Ven K, Vandamme S, Witters E, De Coen W (2010) An integrated transcriptomic and proteomic approach characterizing estrogenic and metabolic effects of 17 alpha-ethinylestradiol in zebrafish (*Danio rerio*). Gen Comp Endocrinol 167:190–201
86. Iguchi T, Watanabe H, Katsu Y (2006) Application of ecotoxicogenomics for studying endocrine disruption in vertebrates and invertebrates. Environ Health Perspect 114:101–105
87. Karlsen OA, Bjorneklett S, Berg K, Brattas M, Bohne-Kjersem A, Grosvik BE, Goksoyr A (2011) Integrative environmental genomics of cod (*Gadus morhua*): the proteomics approach. J Toxicol Environ Health A 74:494–507
88. Martyniuk CJ, Kroll KJ, Doperalski NJ, Barber DS, Denslow ND (2010) Genomic and proteomic responses to environmentally relevant exposures to dieldrin: indicators of neurodegeneration? Toxicol Sci 117:190–199
89. Park KS, Kim H, Gu MB (2005) Eco-toxicogenomics research with fish. Mol Cell Toxicol 1:17–25
90. Steinberg CEW, Sturzenbaum SR, Menzel R (2008) Genes and environment – striking the fine balance between sophisticated biomonitoring and true functional environmental genomics. Sci Total Environ 400:142–161
91. Wang RL, Bencic D, Biales A, Lattier D, Kostich M, Villeneuve D, Ankley GT, Lazorchak J, Toth G (2008) DNA microarray-based ecotoxicological biomarker discovery in a small fish model species. Environ Toxicol Chem 27:664–675
92. Wang RL, Biales A, Bencic D, Lattier D, Kostich M, Villeneuve D, Ankley GT, Lazorchak J, Toth G (2008) DNA microarray application in ecotoxicology: experimental design, microarray scanning, and factors affecting transcriptional profiles in a small fish species. Environ Toxicol Chem 27:652–663
93. Yoon S, Han SS, Rana SVS (2008) Molecular markers of heavy metal toxicity – a new paradigm for health risk assessment. J Environ Biol 29:1–14
94. Connon R, Hooper HL, Sibly RM, Lim FL, Heckmann LH, Moore DJ, Watanabe H, Soetaert A, Cook K, Maund SJ, Hutchinson TH, Moggs J, De Coen W, Iguchi T, Callaghan A (2008) Linking molecular and population stress responses in *Daphnia magna* exposed to cadmium. Environ Sci Technol 42:2181–2188
95. Heckmann LH, Callaghan A, Hooper HL, Connon R, Hutchinson TH, Maund SJ, Sibly RM (2007) Chronic toxicity of ibuprofen to *Daphnia magna*: effects on life history traits and population dynamics. Toxicol Lett 172:137–145
96. Leroy D, Haubruge E, De Pauw E, Thomé JP, Francis F (2009) Development of ecotoxicoproteomics on the freshwater amphipod *Gammarus pulex*: Identification of PCB biomarkers in glycolysis and glutamate pathways. Ecotoxicol Environ Saf 73:343–352
97. Poynton HC, Varshavsky JR, Chang B, Cavigiolio G, Chan S, Holman PS, Loguinov AV, Bauer DJ, Komachi K, Theil EC, Perkins EJ, Hughes O, Vulpe CD (2007) *Daphnia magna* ecotoxicogenomics provides mechanistic insights into metal toxicity. Environ Sci Technol 41:1044–1050
98. Menzel R, Swain SC, Hoess S, Claus E, Menzel S, Steinberg CEW, Reifferscheid G, Stürzenbaum SR (2009) Gene expression profiling to characterize sediment toxicity – a pilot study using *Caenorhabditis elegans* whole genome microarrays. BMC Genomics 10:160
99. Colbourne JK, Singan VR, Gilbert DG (2005) wFleaBase: the *Daphnia* genome database. BMC Bioinformatics 6:45
100. Goldstone JV, Hamdoun A, Cole BJ, Howard-Ashby M, Nebert DW, Scally M, Dean M, Epel D, Hahn ME, Stegeman JJ (2006) The chemical defensome: environmental sensing and response genes in the *Strongylocentrotus purpuratus* genome. Dev Biol 300:366–384
101. Hill WR, Roberts BJ, Francoeur SN, Fanta SE (2011) Resource synergy in stream periphyton communities. J Ecol 99:454–463

102. Stevenson RJ, Bothwell ML, Lowe RL (1996) Algal ecology: freshwater benthic ecosystems. Academic, San Diego
103. Sommer U (2001) Reversal of density dependence of juvenile *Littorina littorea* (Gastropoda) growth in response to periphyton nutrient status. J Sea Res 45:95–103
104. Stelzer RS, Lamberti GA (2002) Ecological stoichiometry in running waters: periphyton chemical composition and snail growth. Ecology 83:1039–1051
105. Admiraal W, Blanck H, Buckert-de Jong M, Guasch H, Ivorra N, Lehmann V, Nyström BAH, Paulsson M, Sabater S (1999) Short-term toxicity of zinc to microbenthic algae and bacteria in a metal polluted stream. Water Res 33:1989–1996
106. Tlili A, Bérard A, Roulier JL, Volat B, Montuelle B (2010) PO_4^{3-} dependence of the tolerance of autotrophic and heterotrophic biofilm communities to copper and diuron. Aquat Toxicol 98:165–177
107. Vinebrooke RD, Cottingham KL, Norberg J, Scheffer M, Dodson SI, Maberly SC, Sommer U (2004) Impacts of multiple stressors on biodiversity and ecosystem functioning: the role of species co-tolerance. Oikos 104:451–457
108. Wu RSS, Siu WHL, Shin PKS (2005) Induction, adapation and recovery of biological responses: implications for environmental monitoring. Mar Pollut Bull 51:623–634
109. Miller DH, Jensen KM, Vilelneuve DL, Kahl MD, Makynen EA, Durhan EJ, Ankley GT (2007) Linkage of biochemical responses to population-level effects: a case study with vitellogenin in the fathead minnow. Environ Toxicol Chem 26:521–527
110. Ankley GT, Bencic DC, Breen MS, Collette TW, Conolly RB, Denslow ND, Edwards SW, Ekman DR, Garcia-Reyero N, Jensen KM, Lazorchak JM, Martinovic D, Miller DH, Perkins EJ, Orlando EF, Villeneuve DL, Wang RL, Watanabe KH (2009) Endocrine disrupting chemicals in fish: developing exposure indicators and predictive models of effects based on mechanism of action. Aquat Toxicol 92:168–178
111. Ankley GT, Bennett RS, Erickson RJ, Hoff DJ, Hornung MW, Johnson RD, Mount DR, Nichols JW, Russom CL, Schmieder PK, Serrrano JA, Tietge JE, Villeneuve DL (2010) Adverse outcome pathways: a conceptual framework to support ecotoxicology research and risk assessment. Environ Toxicol Chem 29:730–741
112. Segner H (2007) Ecotoxicology – how to assess the impact of toxicants in a multi-factorial environment? In: Mothersill C, Mosse I, Seymour C (eds) Multiple stressors: a challenge for the future. Springer, Dordrecht, The Netherlands
113. Hutchinson TH, Ankley GT, Segner H, Tyler CR (2006) Screening and testing for endocrine disruption in fish – Biomarkers as "signposts," not "traffic lights," in risk assessment. Environ Health Perspect 114:106–114
114. Clements WH, Carlisle DM, Courtney LA, Harrahy EA (2002) Integrating observational and experimental approaches to demonstrate causation in stream biomonitoring studies. Environ Toxicol Chem 21:1138–1146
115. Lam PKS, Gray JS (2003) The use of biomarkers in environmental monitoring programmes. Mar Pollut Bull 46:182–186
116. Sandstrom O, Larsson Å, Andersson J, Appelberg M, Bignert A, Ek H, Förlin L, Olsson M (2005) Three decades of Swedish experience demonstrates the need for integrated long-term monitoring of fish in marine coastal areas. Water Qual Res J Canada 40:233–250
117. Morgan AJ, Kille P, Sturzenbaum SR (2007) Microevolution and ecotoxicology of metals in invertebrates. Environ Sci Technol 41:1085–1096
118. Cattaneo A, Kerimian T, Roberge M, Marty J (1997) Periphyton distribution and abundance on substrata of different size along a gradient of stream trophy. Hydrobiologia 354:101–110
119. Dorigo U, Berard A, Rimet F, Bouchez A, Montuelle B (2010) In situ assessment of periphyton recovery in a river contaminated by pesticides. Aquat Toxicol 98:396–406
120. Morin S, Pesce S, Tlili A, Coste M, Montuelle B (2010) Recovery potential of periphytic communities in a river impacted by a vineyard watershed. Ecol Indic 10:419–426

121. Barata C, Damasio J, López MA, Kuster M, De Alda ML, Barceló D, Riva MC, Raldúa D (2007) Combined use of biomarkers and in situ bioassays in *Daphnia magna* to monitor environmental hazards of pesticides in the field. Environ Toxicol Chem 26:370–379
122. Rotter S, Sans-Piché F, Streck G, Altenburger R, Schmitt-Jansen M (2011) Active biomonitoring of contamination in aquatic systems–An in situ translocation experiment applying the PICT concept. Aquat Toxicol 101:228-236
123. Guasch H, Sabater S (1998) Light history influences the sensitivity to atrazine in periphytic algae. J Phycol 34:233–241
124. Laviale M, Prygiel J, Créach A (2010) Light modulated toxicity of isoproturon toward natural stream periphyton photosynthesis: a comparison between constant and dynamic light conditions. Aquat Toxicol 97:334–342
125. Bonnineau C (2011) Contribution of antioxidant enzymes to toxicity assessment in fluvial biofilms. PhD thesis
126. Guasch H, Ivorra N, Lehmann V, Paulsson M, Real M, Sabater S (1998) Community composition and sensitivity of periphyton to atrazine in flowing waters: the role of environmental factors. J Appl Phycol 10:203–213
127. Lee DM, Guckert JB, Belanger SE, Feijtel TCJ (1997) Seasonal temperature declines do not decrease periphytic surfactant biodegradation or increase algal species sensitivity. Chemosphere 35:1143–1160
128. Serra A, Guasch H, Admiraal W, Van der Geest HG, Van Beusekom SAM (2010) Influence of phosphorus on copper sensitivity of fluvial periphyton: the role of chemical, physiological and community-related factors. Ecotoxicology 19:770–780
129. Villeneuve A, Montuelle B, Bouchez A (2011) Effects of flow regime and pesticides on periphytic communities: evolution and role of biodiversity. Aquat Toxicol 102:123–133
130. Damásio J, Tauler R, Teixidó E, Rieradevall M, Prat N, Riva MC, Soares AMVM, Barata C (2008) Combined use of *Daphnia magna* in situ bioassays, biomarkers and biological indices to diagnose and identify environmental pressures on invertebrate communities in two Mediterranean urbanized and industrialized rivers (NE Spain). Aquat Toxicol 87:310–320
131. Damásio J, Navarro-Ortega A, Tauler R, Lacorte S, Barceló D, Soares AMVM, López MA, Riva MC, Barata C (2010) Identifying major pesticides affecting bivalve species exposed to agricultural pollution using multi-biomarker and multivariate methods. Ecotoxicology 19:1084–1094
132. Damásio J, Fernández-Sanjuan M, Sánchez-Avila J, Lacorte S, Prat N, Rieradevall M, Soares AMVM, Barata C (2011) Multi-biochemical responses of benthic macroinvertebrate species as a complementary tool to diagnose the cause of community impairment in polluted rivers. Water Res 45:3599–3613
133. Behrens A, Segner H (2005) Cytochrome P4501A induction in brown trout exposed to small streams of an urbanised area: results of a five-year-study. Environ Pollut 136:231–242
134. Reynoldson TB, Norris RH, Resh VH, Day KE, Rosenberg DM (1997) The reference condition: a comparison of multimetric and multivariate approaches to assess water-quality impairment using benthic macroinvertebrates. J N Am Benthol Soc 16:833–852
135. De Kock WC, Kramer KJM (1994) Active biomonitoring (ABM) by translocation of bivalve molluscs. In: Kramer KJM (ed) Biomonitoring of coastal waters and estuaries. CRC, Boca Raton, FL
136. Hirst H, Chaud F, Delabie C, Jüttner I, Ormerod SJ (2004) Assessing the short-term response of stream diatoms to acidity using inter-basin transplantations and chemical diffusing substrates. Freshwater Biol 49:1072–1088
137. Bradac P, Wagner B, Kistler D, Traber J, Behra R, Sigg L (2010) Cadmium speciation and accumulation in periphyton in a small stream with dynamic concentration variations. Environ Pollut 158:641–648
138. Gold C, Feurtet-Mazel A, Coste M, Boudou A (2002) Field transfer of periphytic diatom communities to assess short-term structural effects of metals (Cd, Zn) in rivers. Water Res 36:3654–3664

139. Ivorra N, Hettelaar J, Tubbing GMJ, Kraak MHS, Sabater S, Admiraal W (1999) Translocation of microbenthic algal assemblages used for in situ analysis of metal pollution in rivers. Arch Environ Contam Toxicol 37:19–28
140. Rotter S, Sans-Piché F, Streck G, Altenburger R, Schmitt-Jansen M (2011) Active biomonitoring of contamination in aquatic systems – an in situ translocation experiment applying the PICT concept. Aquat Toxicol 101:228–236
141. Berard A, Dorigo U, Humbert JF, Leboulanger C, Seguin F (2002) Application of the Pollution-Induced Community Tolerance(PICT) method to algal communities: its values as a diagnostic tool for ecotoxicological risk assessment in the aquatic environment. Ann Limnol-Int J Lim 38:247–261
142. Dorigo U, Leboulanger C, Berard A, Bouchez A, Humbert J, Montuelle B (2007) Lotic biofilm community structure and pesticide tolerance along a contamination gradient in a vineyard area. Aquat Microb Ecol 50:91–102
143. Schmitt-Jansen M, Altenburger R (2005) Toxic effects of isoproturon on periphyton communities-a microcosm study. Estuar Coast Shelf Sci 62:539–545
144. Blanck H, Wänkberg SA, Molander S (1988) Pollution-induced community tolerance—a new ecotoxicological tool. In: Cairs J, Pratt JR (eds) Functional testing of aquatic biota for estimating hazards of chemicals. Philadelphia
145. Baird DJ, Brown SS, Lagadic L, Liess M, Maltby L, Moreira-Santos M, Schulz R, Scott GI (2007) In situ-based effects measures: determining the ecological relevance of measured responses. Integr Environ Assess Manag 3:259–267
146. Faria M, Huertas D, Soto DX, Grimalt JO, Catalan J, Riva MC, Barata C (2009) Contaminant accumulation and multi-biomarker responses in field collected zebra mussels (*Dreissena polymorpha*) and crayfish (*Procambarus clarkii*), to evaluate toxicological effects of industrial hazardous dumps in the Ebro river (NE Spain). Chemosphere 78:232–240
147. Bocchetti R, Fattorini D, Pisanelli B, Macchia S, Oliviero L, Pilato F, Pellegrini D, Regoli F (2008) Contaminant accumulation and biomarker responses in caged mussels, *Mytilus galloprovincialis*, to evaluate bioavailability and toxicological effects of remobilized chemicals during dredging and disposal operations in harbour areas. Aquat Toxicol 89:257–266
148. Ricart M, Guasch H, Barceló D, Brix R, Conceição MH, Geiszinger A, de Alda MJL, López-Doval JC, Muñoz I, Postigo C, Romaní AM, Villagrasa M, Sabater S (2010) Primary and complex stressors in polluted mediterranean rivers: pesticide effects on biological communities. J Hydrol 383:52–61
149. Puertolas L, Damásio J, Barata C, Soares AMVM, Prat N (2010) Evaluation of side-effects of glyphosate mediated control of giant reed (*Arundo donax*) on the structure and function of a nearby Mediterranean river ecosystem. Environ Res 110:556–564
150. Barata C, Lekumberri I, Vila-Escalé M, Prat N, Porte C (2005) Trace metal concentration, antioxidant enzyme activities and susceptibility to oxidative stress in the tricoptera larvae *Hydropsyche exocellata* from the Llobregat river basin (NE Spain). Aquat Toxicol 74:3–19
151. Barata C, Varo I, Navarro JC, Arun S, Porte C (2005) Antioxidant enzyme activities and lipid peroxidation in the freshwater cladoceran *Daphnia magna* exposed to redox cycling compounds. Comp Biochem Physiol C Toxicol Pharmacol 140:175–186
152. Zhou J, Thompson DK (2002) Challenges in applying microarrays to environmental studies. Curr Opin Biotechnol 13:204–207
153. Wu L, Thompson DK, Li G, Hurt RA, Tiedje JM, Zhou J (2001) Development and evaluation of functional gene arrays for detection of selected genes in the environment. Appl Environ Microbiol 67:5780–5790
154. He ZL, Van Nostrand JD, Wu LY, Zhou JZ (2008) Development and application of functional gene arrays for microbial community analysis. Trans Nonferr Met Soc China 18:1319–1327

155. He Z, Gentry TJ, Schadt CW, Wu L, Liebich J, Chong SC, Huang Z, Wu W, Gu B, Jardine P, Criddle C, Zhou J (2007) GeoChip: a comprehensive microarray for investigating biogeochemical, ecological and environmental processes. ISME J 1:67–77
156. Soloway MS, Briggman V, Carpinito GA, Chodak GW, Church PA, Lamm DL, Lange P, Messing E, Pasciak RM, Reservitz GB, Rukstalis DB, Sarosdy MF, Stadler WM, Thiel RP, Hayden CL (1996) Use of a new tumor marker, urinary NMP22, in the detection of occult or rapidly recurring transitional cell carcinoma of the urinary tract following surgical treatment. J Urol 156:363–367
157. Icarus Allen J, Moore MN (2004) Environmental prognostics: is the current use of biomarkers appropriate for environmental risk evaluation? Mar Environ Res 58:227–232
158. Dorak MT (2006) Real-time PCR. Taylor & Francis

How to Link Field Observations with Causality? Field and Experimental Approaches Linking Chemical Pollution with Ecological Alterations

Helena Guasch, Berta Bonet, Chloé Bonnineau, Natàlia Corcoll, Júlio C. López-Doval, Isabel Muñoz, Marta Ricart, Alexandra Serra, and William Clements

Abstract This chapter summarizes field and laboratory investigations dealing with metals and pesticides (90) and emerging compounds' (10) effects on fluvial communities. The Arkansas River case study is a good example showing how field observations, together with long-term natural experiments and microcosm experiments, provide consistent evidence of metals effects on macroinvertebrate communities. In the case of biofilms, microcosm and mesocosm experiments confirm that metals and pesticides are responsible for the loss of sensitive species in the community, and that this influence is modulated by several biological and environmental factors. Information about the effects of emerging pollutants is very scarce, highlighting the existence of a missing gap requiring future investigations. The examples provided and the recommendations given are proposed as a general guide for studies aiming to link chemical pollution with ecological alterations.

H. Guasch (✉) • B. Bonet • C. Bonnineau • N. Corcoll
Institute of Aquatic Ecology, University of Girona, Campus de Montilivi, 17071 Girona, Spain
e-mail: helena.guasch@udg.edu

J.C. López-Doval • I. Muñoz
Department of Ecology, University of Barcelona, Av. Diagonal, 643, 08028 Barcelona, Spain

M. Ricart
Institute of Aquatic Ecology, University of Girona, Campus de Montilivi, 17071 Girona, Spain

Catalan Institute for Water Research (ICRA), Scientific and Technologic Park of the University of Girona, 17003 Girona, Spain

A. Serra
Centre d'Estudis Avançats de Blanes (CEAB), Consejo Superior de Investigaciones Científicas (CSIC), Accés a la Cala Sant Francesc 14, 17300 Blanes, Girona, Spain

W. Clements
Department of Fish, Wildlife and Conservation Biology, Colorado State University, Fort Collins, CO 80523, USA

H. Guasch et al. (eds.), *Emerging and Priority Pollutants in Rivers*,
Hdb Env Chem (2012) 19: 181–218, DOI 10.1007/978-3-642-25722-3_7,

Keywords Biofilm • Causality • Community ecotoxicology • Macroinvertebrate • Multiple-stress

Contents

1 General Introduction

Among the long list of compounds included in the Priority Pollutants (PP) list (European Water Framework Directive 2000/60/EC [1]) and those not included but considered also of environmental concern, few studies demonstrate their real impact on fluvial ecosystems. Discriminating the influence of environmental variability from chemical pressure on the biota is a challenging issue. It is necessary to detect joint and specific effects of simultaneously co-occurring stressors. This difficult task requires an interdisciplinary approach including environmental chemistry, toxicology, ecotoxicology, ecology, etc., and also a multiscale perspective from field investigations to laboratory experiments, microcosm, mesocosm studies, and back to the field in order to establish cause-and-effect relationships.

The change of scale is easily achieved working with fluvial communities. Exposure might be done under controlled experimental conditions in microcosms and mesocosms or in the field where the response can be evaluated under real exposure conditions. It allows defining non-effect concentrations, and also assessing the effects of different types of stress alone or in conjunction (multiple stressors). In contrast to single-species tests, ecotoxicological studies with communities involve a higher degree of ecological realism, allowing the simultaneous exposure of many species. Furthermore, it is possible to investigate community responses after both acute and chronic exposure, including the evaluation of direct and indirect toxic effects on entire communities.

In particular, fluvial biofilm and benthic macroinvertebrate communities fulfill these conditions and have been used to investigate single- and multiple-stress situations at different spatial and temporal scales.

Fluvial biofilms (also known as phytobenthos or periphyton) are attached communities consisting of cyanobacteria, algae, bacteria, protozoa, and fungi embedded within a polysaccharide matrix [2]. In rivers, these communities are the first to interact with dissolved substances such as nutrients, organic matter, and toxicants. Biofilms can actively influence the sorption, desorption, and decomposition of pollutants. Fluvial biofilms are relatively simple and easy to investigate compared to other communities (i.e., macroinvertebrate or fish communities).

Biological monitoring programs employed by state and federal agencies to assess effects of contaminants have routinely focused on benthic macroinvertebrate communities. Benthic macroinvertebrates are exposed to contaminants in water, sediment, and biofilm, providing a direct pathway to higher trophic levels. Because of considerable variation in sensitivity among species, community composition and the distribution and abundance of benthic macroinvertebrates are useful measures of ecological integrity.

The main aim of this chapter is to present key studies dealing with priority and emerging pollutants that provided clues to link field observations with causality following a community ecotoxicology approach.

While not being an exhaustive review, 100 different investigations have been analyzed, including field and laboratory investigations, most of them dealing with metals and pesticides (90), but several investigations focused on emerging compounds (10) are also provided (Fig. 1).

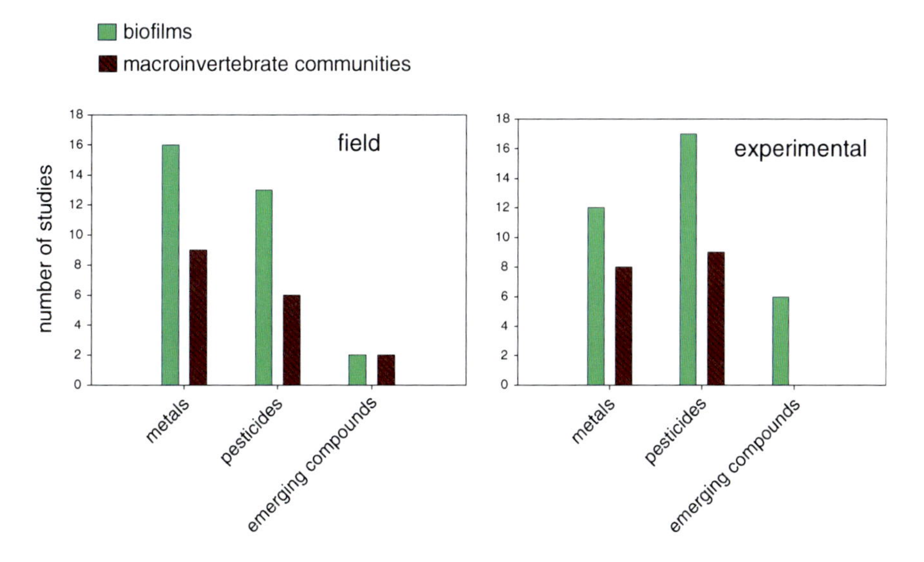

Fig. 1 Summary of field and experimental investigations addressing the effect of toxic exposure on biofilm and macroinvertebrate communities. Studies are grouped on the basis of the type of chemical investigated: metals, pesticides, and emerging compounds

2 Field Investigations Used to Generate Hypotheses

Field investigations, although being the basis of ecotoxicological research, have a high uncertainty linked to environmental variability. Fluvial ecosystems are very dynamic and complex systems, requiring the compilation and statistical analysis of extensive sets of samples and environmental data to characterize the exposure and response of their living organisms. This holistic approach, commonly used in ecology, is less extended in ecotoxicology (see [3] for more details).

Uncertainty might be reduced depending on the scale, type of pollution, and set of response variables investigated (Fig. 2). Large-scale studies are influenced by regional scale variability that may conceal the effects of human impacts. Therefore, human impacts may be better detected in small-scale studies, including reference and impacted situations, within a restricted range of environmental variability [124]. The presence of a specific or dominating type of pollution may also contribute to link causes and effects. The effects of toxic exposure, although being present, are more difficult to detect under multiple-stress situations with many chemical compounds occurring at low concentration and environmental stress factors affecting the biota. Choosing an appropriate set of response variables may allow identifying different types of stress and their ecological consequences (Fig. 2).

The link between chemical pollution and ecosystem damage has been addressed in mining areas (e.g., [4, 5]). Metal pollution in other environments is less

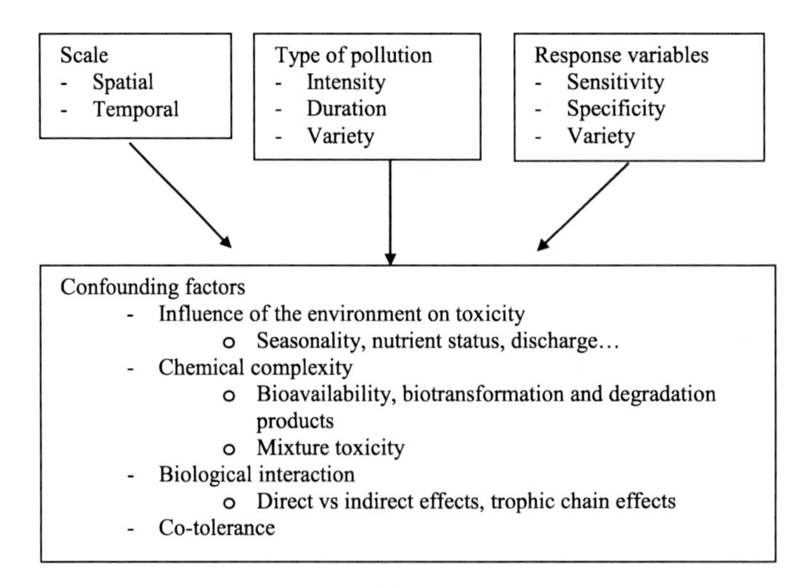

Fig. 2 Scheme showing the main factors influencing the degree of uncertainty in field studies aiming to assess the impact of chemical pollution on natural communities: scale, type of pollution, and set of response variables investigated. Reducing uncertainty will allow to distinguish biological effects caused by toxic exposure from other confounding factors caused by environmental variability, chemical complexity, biological interactions, or co-tolerance

documented. Influences of pesticide pollution or other contaminants received less attention. Emerging pollutants (a group where pharmaceuticals, nanomaterials, the family of perfluorinated compounds, hormones, endocrine disrupting compounds, pesticide degradation products, and newly synthesized pesticides and others are included) are unregulated pollutants, but may be candidates for future regulation depending on research on their potential health effects and monitoring data regarding their occurrence in ecosystems [6]. Their occurrence in aquatic ecosystems has been recently detected and/or their effects on biota recently investigated [7–9].

Many field investigations on metal, pesticides, herbicides, and emerging compounds's toxicity were focused on biofilms, the basis of the food chain, and consumers, mainly macroinvertebrate communities.

2.1 Biofilms

Different types of field studies have been performed to assess the effects of metal and organic contamination on fluvial biofilms (Table 1) such as seasonal monitoring [10], biofilm translocation experiments [11, 12], sampling before and after impact [13, 14], and the study of tolerance induction [15–20].

A key aspect in many ecotoxicological studies is to link water concentration with real exposure conditions due to the variety of factors influencing bioavailabilty and/or the stability of a compound. In the case of biofilms, the analysis of the chemical concentration within the biofilm matrix may contribute to partially overcome this constrain. Evidence of the link between metal exposure (water concentration) and metal contents in biofilms is provided in Behra et al. [21]; Meylan et al. [22]; Le Faucher et al. [23], and also between sediment and biofilm metal concentrations [24] (Table 1), highlighting their possible effects through the trophic chain [25]. Internal concentrations are more difficult to assess and less reported for organic toxic compounds [26].

The link between metal pollution and community changes has been shown in many field studies (see [27] for more details). In particular, the effects of metals on diatom communities have been well documented in mining areas. Hill et al. [13] evaluated the effects of elevated concentrations of metals (Cd, Cu, and Zn) on stream periphyton in the Eagle River, which is a mining impacted river in central Colorado. They found differences in diatom taxa richness, community similarity, biomass, chlorophyll *a* (chl-*a*), and the autotrophic index (AI) that were able to separate metal-contaminated sites from reference or less impacted sites. Sabater [14] evaluated the impact of a mine tailings spill on the diatom communities of the Guadiamar River (SW Spain) showing a clear decrease of diversity (Shannon–Wiener index) and water quality diatom indices. Other variables such as the size of individuals were also attributed to metal pollution in the Riou-Mort [5, 28].

Biomass reduction due to metal toxicity was suggested by Hill et al. [13]; however, these effects were not confirmed in other highly metal-polluted sites

Table 1 Summary of field studies approaching the effects of chemical pollution on fluvial biofilms

Site/sites	Metal/s in water (μg/L) or [a]sediments (μg/g)	Effects	Remarks	Reference
Metals				
High				
Dommel (Belgium)	Zn:6,675; Cd:765 (max)	Bacterial communities in metal-polluted sites more tolerant to Zn Tolerance induction in phototrophs less clear	Zn bioavailability expected to be low due to Fe hydroxide and organic matter	Admiraal et al. [15]
The Coeur d'Alene (Idaho, USA)	[a]As:149; Cd:884; Cu:1,175; Hg:3	Metal contents in polluted sites 10–100× greater than in reference sites Metal concentration of food-web components ranks: biofilm and sediments >invertebrates >whole fish	Bioaccumulated metals may cause physiological effects in indigenous fish	Farag et al. [110]
Boulder River Watershed, Montana (USA)	[a]As:17; Cd:9; Cu:69; Zn:1,800 (max)	Metal concentration of food-web components ranks: biofilm>macroinvertebrates≥sediment>fish tissues>water and colloids, due to entrapment of colloid-bound metals by the biofilm	Metal concentrations higher in low-flow condition	Farag et al. [111]
Guadiamar (after mine tailing spill)	Cd:37 /31[a] Cu:224/1,470[a] Zn:1,080/ 12,000[a] Pb: 30/1,910[a] As: 1,530/200[a] (max)	Changes in diatom species composition and marked reduction of diversity and diatom indices	Slight signs of recovery 14 months after the spill	Sabater [14]
Dommel and Eindergatloop streams (Belgium)	Cd: 330, Zn: 2,864 in Dommel Cd: 282, Zn: 3,147 in Eindergatloop (max)	Changes in diatom species after translocation to metal-polluted sites	Metal tolerance was not the only selective factor	Ivorra et al. [16]
Churnet and Manifold Rivers (England)	Cu: 3,030[a]; Zn: 4,705[a]; Cd: 53[a]; Pb: 3,236[a]	Biofilm metal accumulation more sensitive than water contents (Cd) Marked seasonal variability but identification of metal-tolerant diatom species	Specific diatoms species highlighted metal-tolerant indicator diatom	Holding et al. [24]
Fifteen rivers across Europe	Zn: 2,263 (max)	Regional difference in Zn tolerance in the biofilm (physical and chemical differences)	Large-scale patterns did not follow PICT prediction	Blanck et al. [30]
Riou-Mort River, France	Zn: 2,305, Cd: 27 (max)	The size distribution of diatoms changing from larger to smaller individuals, changes in the taxonomic composition of the assemblages, higher amounts of abnormal frustules Cd in biofilms: 250–1,200 μg/g	No effects of metal pollution on algal biomass due to high nutrients and organic pollution Metal accumulation and diatoms different between seasons	Morin et al. [28], Duong et al. [5]

(continued)

Table 1 (continued)

Metals

Site/sites	Metal/s in water (μg/L) or [a]sediments (μg/g)	Effects	Remarks	Reference
Moderate				
Eagle River, Colorado	Zn: 800, Cu: 18, Fe: 600	Diatom species composition different in metal-polluted sites; increase of valve deformities; lower chl-*a* and AI	Results in agreement with macroinvertebrate studies	Hill et al. [13]
Catalan (NE Spain) $N = 25$	Cd: 0.8, Cr: 6.6, Cu: 50, Pb: 10, Ni: 80, Zn: 63, As: 10, Hg: 4 (max)	CCU (cumulative Criterion Unit) indicated potential toxicity in many sites. Poor relation between concentration and diatom species composition	Scale effects masked metals effects	Guasch et al. [31]
Low				
Birs and Thur River (Switzerland)	Cu: 0.4–10, Zn: 0.7–37 (ref and metal polluted)	Increases in water Cu and Zn reflected by increased metal contents in biofilm and sediments Biofilm contents 35–173× higher in polluted than i n reference sites	Influence of speciation on adsorption on bioavailability	Behra et al. [21]
Furtbach stream (Switzerland)	Cu: 2.5–7.5 (8.95[a], 16[b] max), Zn: 2.9–9.6 (17[a], 25[b] max).	Dissolved and adsorbed metal concentrations increased steeply during rain events	Rain events influenced the metal speciation, and subsequently the metal accumulation	Meylan et al. [22]; Le Faucher et al. [23]
East Fork Poplar Creek (Oak Ridge, Tennessee)	Cu: 3.1, Cd: 0.25, Zn 25 (max)	In comparison to reference sites: Cu, Zn, and Cd biofilm metal contents were 4, 15, and 50× greater (respectively) in 1995. Increase in chl-*a* over time attributed to metal toxicity reduction	Metal reduction downstream attributed to biofilms retention	Hill et al. [25]
Fluvia River (NE Spain)	Cu: 4.2, Cd: 0.08, Zn: 80, Pb: 0.75 (max)	Water and biofilm metal contents in human-impacted sites 3–15× greater than in reference sites. Biofilm metal contents explained a small part of diatom species composition variance	Biofilm metal contents more sensitive than water contents	Guasch et al. [31]

Pesticides

Site/sites	Toxicants in water (μg/L)	Effects	Remarks	Reference
Morcille River (France)	Diuron and its metabolites (DCPMU and 3,4-DCA)	Individual and combined effects of diuron and DCPMU had both short-term effects and long-term effects on phototrophic biofilms, whereas environmental concentrations of 3,4-DCA did not affect biofilm photosynthetic activity	Diuron was more toxic for algae than its metabolites	Pesce et al. [36]
Ozane River (France)	77 pesticides (atrazine, isorpoturon)	Chronic exposure induced community tolerance to atrazine and isoproturon	PICT was confirmed	Dorigo and Leboulanger et al. (2001) [33]
	Diuron		PICT was confirmed	Pesce et al. [35]

(continued)

Table 1 (continued)

Pesticides				
Site/sites	Toxicants in water (μg/L)	Effects	Remarks	Reference
Morcille River (France)		Diuron exposure during biofilm colonization caused an increase in community tolerance with little effects of other environmental variables (nutrients, conductivity, and temperature)		
Avencó River (Spain)	Reference site	Atrazine toxicity on biofilms was influenced by biofilm maturity and light history	Highlights the influence of the environment on toxicity	Guasch et al. [18]
20 rivers across Europe	Pesticides (atrazine) and nutrients	Diatom taxa indicators of different degrees of pollution	Difficulty to separate effects of nutrients from atrazine toxicity	Guasch et al. [19]
7 rivers (Catalunya area, Spain)	Atrazine	Light influenced atrazine toxicity. Biofilms from open-canopy sites were more sensitive to atrazine than biofilms from shaded sites. Differences in the proportion of algal groups and pigment composition were also observed	The influence of environmental light conditions on atrazine toxicity was confirmed	Guasch and Sabater [32]
Ter River (Spain)	Atrazine, Cu	Atrazine and Cu toxicity on biofilms were related to algal abundance and species composition. Seasonal differences in sensitivity were observed	Difficulty to elucidate the causes of the observed temporal and spatial variability in community tolerance	Navarro et al. [17]
Llobregat River (NE Spain)	22 pesticides (mixture)	Organophosphates and phenylureas related with functional and structural changes in biofilms No relationship with the macroinvertebrate community	Biofilms were more sensitive than invertebrates to contamination. Difficult to separate effects of nutrients from phenylureas toxicity [10, 78]. The use of taxonomic distinctness indices recommended for toxicity assessment on algae and benthic invertebrates (Ricciardi et al. 2010)	Ricart et al. [10, 78] Ricciardi et al. (2009)

(continued)

Table 1 (continued)

Pesticides

Site/sites	Toxicants in water (μg/L)	Effects	Remarks	Reference
Morcille River (France)	Pesticides	The upstream–downstream gradient of chemical and nutrients pollution caused changes in diatoms species composition and an increase in biofilm biomass. Biofilm from impacted sites had high capacity to recover when they were translocated to non-impacted sites	Shows biofilm capacity to recover their structure if the impact disappears Difficulty to elucidate the causes of the observed changes	Morin et al. [12]
Don Carlos stream, Argentina[a]	Pb: 13, Cu: 23, phthalate: 20	Changes in community composition, increase of abnormal diatom frustules, drop in Net Primary Production and increase of bacterial density	Difficulty to elucidate specific effects of different toxicants	Victoria and Gómez [38]
Elbe river, (Germany)	Prometryn	Prometryn tolerance increased when biofilms were translocated from reference-R to polluted-P sites and decreased when the translocation was opposite (from P to R)	The exposure history of communities defined the time-response of recovery and adaptation	Rotter et al. [11]

Emerging pollutants

Site/sites	Toxicants in water (μg/L) or sediments[a] (μg/g)	Effects	Remarks	Reference
Llobregat River (NE Spain)	Pharmaceutical products: 0.01–18.7	Pharmaceutical products influenced the distribution of the invertebrate community. No effects on diatom community. High concentrations of anti-inflammatories and β-blockers and high temperatures were related to greater abundance and biomass of *Chironomus* spp. and *Tubifex tubifex*	Multivariate analyses revealed a potential causal relation between stressors and effects along gradients and allowed the generation of hypothesis to be tested in laboratory studies	Muñoz et al. [59]
Don Carlos stream (Argentine Pampean plain)	Phthalates (plastiziers) (max: *n*-butyl phthalate: 20[a])	Shift in the composition of the translocated biofilms, presence of abnormal frustules on diatoms, drop in net primary production	Taxonomic and metabolic variables responded differently to translocation	Victoria and Gómez [38]

Water concentration in μg/L

[a]Indicate sediment conc in μg/g

such as the Riou-Mort where metal pollution co-occurs with high levels of nutrients and organic matter (Table 1).

Some investigations reported the induction of metal tolerance after chronic metal exposure [15] using the Pollution Induced Community Tolerance (PICT) approach. Based on the PICT concept, natural communities colonizing different river sites can be used to investigate their exposure history by comparing their physiological responses to a sudden exposure (short-term toxicity tests referred to as PICT tests in this paper) to high doses of the toxicant investigated [29].

Other publications [16] highlight the presence of many selective factors in addition to metal exposure. It may explain the lack of relationship between metal exposure and community tolerance in large-scale studies [30].

Obtaining clear causal relations is difficult in situations of low but chronic metal pollution due to the co-occurrence of many stress factors (Table 1). A low portion of the diatom species composition variance was explained by biofilm metal contents in a small-scale study [31]. Le Faucher et al. [23] found phytochelatines' (PC) production as a response to low metal exposure, but could not identify the specific metal responsible for that.

Most field studies evaluating the effects of pesticides on biofilms were focused on herbicides (atrazine and its residues, diuron and its residues, prometryn or isoproturon) targeting phototrophic organisms (Table 1). Some investigations (Table 1) evaluated the influence of environmental variability (e.g., light conditions, nutrients, maturity, seasonality, temperature, and hydrology) on biofilm's tolerance to pesticides (e.g., [17–20, 32–36]). Other studies investigated the capacity of biofilms to be adapted to sites presenting pesticides' contamination or their capacity to recover from chemical exposure [11, 12]. Monitoring studies aiming to discriminate the effects of pesticides from other stressors are also reported [10, 19]. In the Llobregat River, organophosphates and phenylureas were related with a significant but low (6%) fraction of variance of several biofilm metrics including algal biomass and the photosynthetic efficiency of the community [10]. In the same case study, a significant correlation between the taxonomic distinctness index of diatoms and diuron concentration was obtained [37]. In many other field studies (Table 1), discerning the specific effects of pesticides toxicity on biofilms was complicated by the co-occurrence of pesticide pollution with other types of pollution such as nutrients, organic pollution, or metals (e.g., [12, 19, 38]). On the other hand, the sensitivity of biofilms to pesticides has also been shown to be influenced by biofilm's age [11, 18] or light conditions [32].

In the case of organic pesticides acting as PSII inhibitors, the PICT approach has been demonstrated to be a valuable tool to assess the effects of chronic exposure [29, 39]. For example, Dorigo et al. [34] showed that the response of algal communities is likely to reflect past selection pressures and suggest that the function and structure of a community could be modified by the persistent or repeated presence of atrazine and isoproturon in the natural environment.

Other pesticides or PPs and emerging contaminants have been poorly investigated (Table 1), probably due to their occurrence at low dose in complex mixtures and the lack of sensitive methods applicable to complex field samples [40].

Overall most ecotoxicological field studies have been focused on structural changes, clearly demonstrated under high pollution conditions. Effects on functional attributes (e.g., photosynthesis) are less clear, probably due to functional redundancy between non-impacted communities and those adapted to chemical exposure (see [41] for more details). Besides, the link between moderate toxic exposure and biological damage is less documented. In these cases, a clear increase of chemical concentration in the biofilm matrix is commonly described, mainly in cases of metal pollution; however, the link with persistent effects on the community is, in most cases, less evident. Moreover, almost all the field studies reported point out the influence of environmental variables on biofilm toxicity. In particular biofilm age, nutrients' availability, and light conditions have been shown to influence biofilms' tolerance to various contaminants. This observation highlights the importance of an extensive monitoring, including a multidisciplinary approach to better assess the effects of pollution in complex fluvial ecosystems.

2.2 Macroinvertebrate Communities

Biomonitoring studies conducted with macroinvertebrate communities have been performed to assess effects of metals and organic contaminants on streams (Table 2). Some of these studies have been conducted at relatively large spatial scales [42, 43], allowing investigators to quantify effects of landscape-level characteristics on responses to stressors. These investigators have also developed new approaches based on the biotic ligand model (BLM) to quantify metal bioavailability and effects in the field [43, 44]. Other studies have employed a more traditional longitudinal study design, in which upstream reference sites are compared to downstream contaminated sites [45–47]. Although these single watershed studies are more restricted spatially, they often include investigations that provide a long-term perspective. For example, Clements et al. [47] documented changes in macroinvertebrates and water quality over a 17-year period and related changes in community composition to long-term improvements in water quality.

One of the most consistent observations from these descriptive studies of metals and organic pollution is that certain groups of macroinvertebrates, especially mayflies (Ephemeroptera), are highly sensitive to chemical stressors [42, 45, 48–50]. Interspecific variation in sensitivity to organic and inorganic contaminants forms the basis for the development of new community-level measures to quantify toxicological effects, such as the species at risk (SPEAR) model developed for pesticides and other organics [46, 51]; see also [3]. Recent studies conducted with macroinvertebrates have identified specific morphological and physiological characteristics that are likely responsible for interspecific variation in sensitivity to toxic chemicals [52].

The study of the effects of organic pollutants such as PCBs (polychlorinated biphenyls) and PAHs (polycyclic aromatic hydrocarbons) on macroinvertebrate communities is scarce in the literature, as has been reviewed by Heiskanen and

Table 2 Summary of field studies approaching the effects of chemical pollution on macroinvertebrates

Metals				
Site/sites	Metal/s in water (μg/L)	Effects	Remarks	Reference
Basento River, Italy	As 10–17, Cd 0.3–8, Cr 17–56, Cu 15–156, Pb 10–1,680, Zn 58–1,048	Benthic communities strongly influenced by metals; metal uptake varied significantly among taxa	Uptake of metals from sediment was more important than bioconcentration from water	Sontoro et al. (2009) [125]
Numerous (>150) streams in Colorado, USA	Cd 0.01–8, Cu 0.15–935, Zn 0.25–1,940	Negative effects were observed at concentrations below the chronic toxicity threshold	A new toxic unit model based on a modification of the biotic ligand model was developed to quantify effects	Schmidt et al. [43]
Numerous (412) streams from England, Japan, Scotland, and USA	Numerous metal cations ranging from bd to >100 × toxic concentrations	Threshold relationships established between metal mixtures and species richness of mayflies, stoneflies, and caddisflies	Related concentrations of cationic metallic species and protons to field effects	Stockdale et al. [44]
Numerous (95) sites in Colorado, USA	Cd, Cu, Zn	Heavy metal concentration was the most important predictor of benthic communities across 78 randomly selected sites	Ability to detect effects dependent on the taxonomic resolution; total abundance of mayflies and abundance of heptageniids were better indicators than abundance of dominant mayfly taxa	Clements et al. [42]
River Avoca, Ireland	Cd, Cu, Fe, Pb, Zn (concentrations not reported)	Acid mine drainage (AMD) sites characterized by lower abundance of EPT taxa and increased abundance of chironomids	Considerable variation in sensitivity among macroinvertebrate metrics	Gray and Delaney [45]
Several tributaries to the Hasama River, Japan	Cd 0.003–4.9, Cu 0.12–5.2, Pb 0.1–19, Zn 5–812	Ephemeroptera abundance and diversity were highly sensitive to metals; orthoclad chironomids were positively associated with metals	Biomass of important prey items for drift-feeding fishes was reduced, demonstrating potential indirect effects on predators	Iwasaki et al. [48]
	Cd, Cu, Zn			

Site/sites	Toxicants in water (μg/L)	Effects	Remarks	Reference
Arkansas River, Colorado, USA		Spatial, seasonal, and annual variation of macroinvertebrates associated with metal concentrations	Macroinvertebrates quickly responded to improvements in water quality after remediation	Clements et al. [47]
Ten streams in northern Idaho, USA	As bd-228, Cd bd-3.2, Pb bd-419, Zn 12–523	Elevated levels of Cd and Zn were correlated with lower macroinvertebrate diversity	Community-level effects persisted 75 years after mining cessation	Lefcort et al. (2010) [126]
Several Peruvian streams	Al 13,000, As 3,490, Mn 19,650, Pb 876, Zn 16,080 (8–3,500 × greater at contaminated sites than at reference sites)	Polluted sites dominated by dipterans and coleopterans; reference sites dominated by crustaceans, ephemeropterans, plecopterans, and trichopterans	Unique challenges associated with high altitude streams; despite elevated metals, diverse communities were observed at polluted sites	Loayza-Muro et al. [50]
Pesticides				
Site/sites	Toxicants in water (μg/L)	Effects	Remarks	Reference
Lourens River, South Africa	Organophosphates (azinphos-methyl and chlorpyrifos): bd (below detection)–0.038 total OP (suspended particles)	Particle-associated OPs affected community structure; greatest effects on mayflies and caddisflies	Study linked population dynamics of sensitive taxa to changes in community structure	Bollmohr and Schulz [49]
29 streams from Finland and France	Mixture of pesticides from agriculture activity in adjacent crop fields	Test the hypothesis that community structure and function (leaf-litter breakdown) can be impaired by pesticides	Reduction of sensitive species in the community due to the presence of pesticides and correlation between elimination of sensitive species and diminution of leaf-litter breakdown	Schäfer et al. [112]
19 sites in the Ob River Basin, southwestern Siberia	Mixture of synthetic surfactants, petrochemicals, and nutrients	Test the hypothesis that a community measure of sensitivity to organic contamination ($SPEAR_{organic}$) was independent of natural variation	($SPEAR_{organic}$) was independent of natural longitudinal variation and strongly dependent on organics	Beketov and Liess [46]

(continued)

Pesticides

Site/sites	Toxicants in water (μg/L)	Effects	Remarks	Reference
Several small streams in Finland, France, and Germany	Mixtures of pesticides expressed using a toxic unit approach	Compare the use of species- and family-level data to show pesticide effects using a species at risk ($SPEAR_{pesticide}$) model	Family-level model had adequate explanatory power to quantify pesticide effects across a broad geographical region	Beketov et al. [51]
150 medium-sized mountain rivers in France	PAH, PCB, and metals in sediments	Some species traits are sensitive to the presence of those pollutants	Species traits is a good tool to predict pollution impacts	Archaimbault et al. [54]
Llobregat River, Spain	Pesticides	Abundance and biomass distribution of macroinvertebrates were not explained by the presence of pesticides but by environmental conditions	Different responses of biofilms and invertebrates communities to pesticides	Ricart et al. [10, 78]

Emerging pollutants

Site/sites	Toxicants in water (μg/L)	Effects	Remarks	Reference
Llobregat River, Spain	Several therapeutic families in water	Potential causal association between the concentration of some anti-inflammatories and β-blockers and the abundance and biomass of several benthic invertebrates	Statistical methodologies were useful tools in primary approaches of stressor's effects. Concentrations of pharmaceuticals in rivers are susceptible of generate changes in communities	Muñoz et al [59]
Llobregat River, Spain	Pharmaceuticals	Calculation of hazard quotients for a mixture of these emerging pollutants. Contrast with macroinvertebrate diversity	A clear negative relationship between diversity indices and higher pharmaceutical concentrations	Ginebreda et al. [113]

Solimini [53]. Archaimbault et al. [54] used invertebrates' species traits in order to relate the structure of the community to the presence of PCBs, PAHs, and metals and predict impacted sites by these compounds. As in the case of the SPEAR approach the use of species traits was also useful in order to predict impacted zones. Beasley and Kneale [55] studied the effects of PAHs and heavy metals by multivariate statistical analysis and they found a clear response of the community to the presence of these pollutants with a decrease in diversity.

The effects of pharmaceuticals on aquatic biota have been recently taken into account and have generated concern [56–58]. Despite the existence of experimental evidence of their effects on aquatic invertebrates, the potential effects of these substances on invertebrates' communities have been poorly studied in the field. Muñoz et al. [59] indicated possible relationships between the presence of some families of pharmaceuticals and the shift in the abundance and biomass of different macroinvertebrate species in freshwater communities.

2.3 Summary of the Hypotheses Generated

Differing in the degree of uncertainty, several hypotheses can be derived from the set of community ecotoxicology field investigations reviewed (Table 3).

3 Experimental Studies Searching for Causality

Although biomonitoring studies and other field approaches provide important insights into the effects of contaminants on aquatic ecosystems, these descriptive studies are limited because of their inability to demonstrate cause-and-effect

Table 3 List of hypotheses derived from field community ecotoxicology investigations

Since chemical pollution is first observed as an increased concentration in the biota, further effects on biological integrity are expected
Chemical exposure is expected to cause morphological and physiological alterations such as:
Photosynthesis reduction if PSII inhibitors are present
Morphological and /or physiological characteristics linked to species sensitivity (e.g., macroinvertebrate life traits)
Chemical exposure is expected to affect the community structure in terms of:
Loss in abundance (biomass)
Loss of sensitive species
Loss in species richness
Community adaptation
Environmental and biological factors expected to influence toxicity effects on biofilms are:
Light conditions during growth
Nutrient status
Age of the community (biofilm's thickness)
Emerging pollutant and especially pharmaceuticals (molecules designed to be biologically actives) present in very low concentrations in rivers are expected to provoke sublethal effects on individuals and community changes after long-term exposure

relationships between stressors and ecological responses. Establishing cause-and-effect relationships between stressors and responses is widely regarded as one of the most challenging problems in applied ecology [60]. Because biological assessments of water quality rely almost exclusively on observational data, causal inferences are necessarily weak [61]. In addition, most descriptive studies are unable to identify underlying mechanisms responsible for changes in aquatic communities. Because contaminants often exert complex, indirect effects on aquatic communities, an understanding of underlying mechanisms is often critical. In general, descriptive approaches such as biomonitoring studies provide support for hypotheses rather than direct tests of hypotheses. Equivocal results of biomonitoring studies result from the lack of adequate controls, nonrandom assignment of treatments, and inadequate replication [62].

Various approaches have been developed to address the lack of causal evidence in ecotoxicological investigations. Several investigators have provided useful advice on how to strengthen causal relationships in descriptive studies [63–65]. Weight of evidence approaches, such as the sediment quality triad [66], combines chemical analyses with laboratory toxicity tests and field assessments. These integrated approaches can provide support for the hypothesis that sediment contaminants are responsible for alterations in community structure. Recent approaches developed by the U.S. Environmental Protection Agency (U.S. EPA) employ formal methods of stressor identification, analogous to those used in human epidemiological studies [63], to determine causes of biological impairment in aquatic systems (http://cfpub.epa.gov/caddis/). Hill's [63] nine criteria and modifications of these guidelines [64, 65, 67] have been employed to strengthen causal relationships between stressors and ecological responses:

- Strength of the association between stressors and responses
- Consistency of the association between stressors and responses
- Specificity of responses to contaminants
- Temporal association between stressors and responses
- Plausibility of an underlying mechanistic explanation
- Coherence with our fundamental understanding of stressor characteristics
- Analogous responses are observed to similar classes of stressors
- A gradient of ecological responses are observed as stressor levels increase
- Experimentation support for the relationship between stressor and responses

Perhaps the most important of these nine criteria is the availability of experimental data to support a relationship between stressors and responses. Regardless of the established strength, consistency, specificity, coherence, plausibility, etc., of the relationship between stressors and responses, there is no substitute for experimentation to demonstrate causality. Experimental approaches provide an opportunity to demonstrate causal relationships between stressors and ecologically relevant responses across levels of biological organization. These experiments are a practical alternative to single-species toxicity tests and address the statistical problems associated with field biomonitoring studies. The maturity of a science such as ecotoxicology is often defined by the transition from purely descriptive to manipulative approaches. The ability to test hypotheses with ecologically realistic experiments represents a major shift in the quality of the questions that can be addressed [68].

Microcosm and mesocosm experiments are often designed to manipulate single or multiple environmental variables, providing opportunities to quantify stressor interactions and identify underlying mechanisms. It is the ability of an investigator to manipulate and isolate individual factors that makes the application of microcosm and mesocosm experiments particularly powerful in ecotoxicological research. Microcosm and mesocosm experiments address two of the key limitations associated with environmental assessment of contaminant effects: the lack of ecological realism associated with traditional laboratory experiments and the inability to demonstrate causation using biomonitoring. In addition, these experimental approaches provide the opportunity to investigate potential interactions among stressors.

Because of the limited spatiotemporal scale, measuring certain responses in microcosm and mesocosm experiments presents significant challenges. The duration of microcosm and mesocosm experiments is of critical importance when assessing effects of contaminants. An important consideration in the development of microcosm and mesocosm approaches is whether greater statistical power and ability to demonstrate causation outweigh their limited spatiotemporal scale. Thus, combining these experimental approaches with results of field studies conducted at a larger spatiotemporal scale is the most reliable way to demonstrate causation.

3.1 Experimental Studies Addressing Cause-and-Effect Relations Between Toxicant Exposure and Biofilm Communities Responses

Different experimental designs have been used to investigate, under controlled conditions, the effects of toxic substances on biofilm communities (Table 4). The scale of exposure ranges from hours to weeks, either continuous (the most common experimental design) or pulsed. After controlled toxic exposure, the PICT approach is often used to assess differences in sensitivity at community level with the same procedure applied to natural biofilms (e.g., see Sect. 2.1).

Many authors have investigated metal and pesticides' effects in microcosms and mesocosms by adding toxicants to natural water (Table 4). This experimental approach was used to investigate the single effect of a toxicant, the effect of mixtures, the bioaccumulation (only for metals), the tolerance and co-tolerance induction (using the PICT approach), and also the interaction between toxicity and environmental and biological factors such as nutrients, light conditions (including UV radiation), or grazing (Table 4). Effects of the most commonly reported metals in the environment: Cu, Zn, Ni, Ag, Cu, Pb, and Cd were investigated at environmental realistic concentrations. As in the field studies reported above, most of the

Table 4 Experimental studies addressing cause–effect relations between toxic exposure and their effects on biofilms

Metals					
Hypothesis	Experimental design			Conclusions and remarks	References
	Conditions	Toxicant (µg/L)	Duration		
Bioaccumulation (bioac) driven by speciation and the presence of ligands	Growth in lab, river water (Glatt River)	Cu:16, Zn: 33 (max)	24 h	Zn bioac controlled by Zn^{2+}. Cu-bioac controlled by weakly complexed Cu due to very low free Cu^{2+} in natural environments	Meylan et al. [81]
Chronic Cu exposure increases community tolerance to Cu and co-tolerance to other metals	Outdoor glass aquaria system	Cu: 317 (max)	16 weeks	Hypothesis confirmed with changes in the community composition (taxonomic changes) and co-tolerance to Zn, Ni, and Ag	Soldo and Behra [85]
Biofilm maturity influences their sensitivity to metals	Growth in river and exposure in indoor channels	Cd: 100	4 and 6 weeks	Cd exposure affected the whole biofilm and diatom assemblages' structure. Mature biofilm less affected	Duong et al. [83]
The biofilm structure offers protection to the communities which live embedded	Laboratory-grown monospecific biofilms[a]. Growth on artificial subtrata in the Meuse river[a] and in indoor aquaria[b]	Cu: 0 < 445 < 953 < 1,905 and 572[b] (max)	7 and 24 h[a] Chronic[b]	Physical structure of the biofilm and not the species composition influenced Cu toxicity during short- and long-term exposures. Cu reduced P uptake	Barranguet et al [114[a], 115[b]]
EPS production as metal tolerance mechanisms in metal-adapted communities	Aliquota from Valencia mine (Mexico) used to colonize artificial substrata in the lab	Cu: 636–6,350, Zn: 654–65,270	1 and 5 days	Metal exposure increased the EPS production and the immobilization of metals. Metals' effects on biomass and the relative abundance of the dominant taxa	Garcia-Meza et al. [88]
Zn (400 µg/L) causes structural and functional damage	Indoor microcosms. Biofilms exposed to Zn and Zn+Cd	Zn: 400, Cd: 20	5 weeks	The effects of the metal mixture could be attributed to Zn toxicity. Photosynthesis was inhibited during the first hours of exposure, and algal growth, mainly diatoms, decreased after 2 weeks of exposure	Corcoll et al. [116]
Zn exposure reduces P-availability	Growth in Gota Alv river	Zn: from 3 to 2,000	4 weeks	Low levels of Zn affected biomass production in biofilms by inducing phosphorus depletion	Paulsson et al. [82]
Phosphate reduces metal toxicity (Zn and Cd)	Growth in the field and exposure in indoor aquaria	Zn: 1,000, Cd: 64, P: 284	3 weeks	Hypothesis confirmed for each metal alone but not for the combination. Sensitivity also influenced by the species composition and their autoecology	Ivorra et al. [84]

Hypothesis	Conditions	Toxicant (μg/L)	Duration	Conclusions and remarks	Reference
Nutrient (P) concentration reduces Cu toxicity	Growth in river and exposure in indoor channels[ab]. Growth in aquaria using algal inocula from the river[c]	Cu: 15[a], 4.2[b] (max in the river), 30[c]	12 days[ab] 3 weeks series[c]	Hypothesis confirmed. The effect of P on Cu toxicity changed depending on the parameter[c]. Biofilms' results confirmed with algal monocultures[b]. Highlights the importance of using a range of parameters that target all biofilm components	Guasch et al. [117] Serra et al. [89], Tlili et al. [39]
UV radiation influences Cd toxicity	Indoor microcosms	Cd: 225–6,800	Long term	UVR: biomass reduction, change in species composition, and an increase in the ratio of UVR-absorbing compounds/physical structure of the biofilm and not the species composition *a*. Induced community tolerance to high-UVR and co-tolerance to Cd	Navarro et al. [90]

Pesticides

Hypothesis	Experimental design			Conclusions and remarks	Reference
	Conditions	Toxicant (μg/L)	Duration		
Isoproturon produces alterations on the diatom species composition	Indoor microcosm system based on large water tanks containing: sediment, macrophytes, mollusks, and periphyton	Isoproturon:5 and 20 Isoproturon in sediment: 100 and 400 μg/kg	34–71 days	Isoproturon produced a reduction of diatoms density. After 71 days, periphyton was adapted to isoproturon exposure by dominant heterotrophic organisms and smaller diatoms species	Pérès et al [75]
Oxyfluorfen exposure produces oxidative stress in biofilms and reduce photosynthetic efficiency, chla	Growth in artificial indoor streams	Oxyfluorfen: $0 < 3 < 7.5 < 15 < 30 < 75 < 150$	40 days	Hypotheses not confirmed. At the end of exposure, biofilms pre-exposed to high concentrations of oxyfluorfen presented a higher capacity to answer to oxidative stress (catalase) linked with changes in algal community (RNA 18S).	Bonnineau PhD
Linuron causes direct and indirect effects in a simple freshwater trophic chain	Outdoor microcosms containing water, phytoplankton, biofilms, zooplankton	Linuron: 15–500	2–8 weeks	Linuron affected the community composition causing a decrease in the amount of algal species digestive for zooplankton altering the energy transfer in the studied trophic chain	Daam et al. [118]

(continued)

Pesticides					
Hypothesis	Experimental design			Conclusions and remarks	Reference
	Conditions	Toxicant (μg/L)	Duration		
Diuron produces direct and indirect effects on biofilms.	Growth and exposure in artificial streams	Diuron: 0.07–7	29 days	Hypothesis confirmed. Diuron produced direct inhibitory effects on algae (photosynthesis inhibition, reduction of biovolume diatoms, and increase of chl-*a*) but also indirect effects on bacteria	Ricart et al. [73]
Long-term exposure to isoproturon, atrazine and prometryn increases community tolerance (PICT	Indoor aquaria	Isoproturon: 2.4–3,120 Atrazine: 7.5–2,000 Prometryn: 2.5–320	10–129 days PICT tests: 1 h	Hypothesis confirmed but the sensitivity of biofilms decreased with increasing age and biomass, respectively	Schmitt-Jansen and Altenburger [87, 119–121]
Long-term diuron exposure increases community tolerance (PICT)	Growth in microcosms from aliquota from Mulde river (Germany)[a] or from Morcille river (France)[b]	Diuron: 0.4–100[a] Chronic: 1[b] Pulses: 7 and 14[b]	3–12 weeks[a] 28 days[b] PICT tests: 1h[a], 3h[b]	PICT confirmed[ab]. The tolerance increased by a factor of 2–3 (based on EC_{50} values)[a]. PICT enhanced after pulse exposure[b]	McClellan et al. [86] Tlili et al [69]
Phosphate influences the diuron-induced community tolerance (PICT)	Growth in aquaria from an aliquota from the Morcille river (France)	Diuron: 10	3 weeks PICT tests: 2–3 h	PICT confirmed but the phosphate gradient affected the response of biofilm to diuron. For PICT approaches it is necessary to dissociate the real impact of toxicants from environmental factors	Tlili et al. [39]
Roundup exposure has an effect on biofilm colonization in "clear" and "turbid" waters	Outdoor mesocosms: "clear" (aquatic macrophytes/ metaphyton) "turbid" (phytoplankton/suspended inorganic matter)	Glyphosate (Roundup®): 8,000	42 days	Hypothesis confirmed. Roundup exposure provoked a delay in biofilm colonization, an increase in eutrophication (due to Roundup degradation), and a shift from"clear" to "turbid" mescosms	Vera et al. (2010) [127]
The exposure duration influences herbicides toxicity towards biofilms and their recovery potential	Growth in Esrum Mollea strem (Denmark). Exposure in 10 mL vials	Isoproturon, metribuzin: $0 < 0.4 < 2 < 10 < 50 < 250 < 1,250$ Pendimethalin, hexazinone: $0 < 0.4 < 2 < 10 < 50$	1 h and 24 h for all 2, 6, 18, 23, and 48 h for metribuzin	Hypothesis confirmed for isoproturon, metribuzin, and pendimethalin. Isoproturon and metribuzin toxicity increased throughout exposure time. Pendimethalin hysteresis effect only visible after 1–2 h of exposure. No effect of exposure duration on biofilm recovery from metribuzin exposure	Gustavson et al. [122]

Grazing enhances atrazine toxicity	Growth in mesocosms	Atrazine: 14 µg/L	18 days	Grazing increased atrazine toxicity on biofilms and affected the metabolism and structure of the algal community	Muñoz et al. [71]
Grazing enhances diuron toxicity to biofilms	Growth in artificial channels	Diuron: 2	29 days	Hypothesis not confirmed. Bacterial survival and photosynthetic activity were affected by diuron but no interaction between toxicant exposure and grazers	López-Doval et al. [74]
Variation in light intensity, in comparison to constant light influences the sensitivity of phototrophic biofilms to isoproturon	Growth in Sensée river (France). Exposure in laboratory under constant or dynamic light conditions	Isoproturon: 0–2,000 (constant light) 2.6–20 (dynamic light)	7 h	A dynamic light regime increased biofilm sensitivity to isoproturon by challenging its photoprotective mechanisms such as the xanthophyll cycle because both stressors (isoproturon and light) affected the photosynthetic activity	Laviale et al. [70]
Light history modulates biofilm response to herbicides exposure	Growth in artificial indoor streams under different light conditions Exposure in 10 mL vials	Glyphosate: 10–1,000,000 AMPA: 10–500,000 Oxyfluorfen: 1.5–1,000 Cu: 20–2,000	6 h	Hypothesis confirmed for glyphosate. Co-tolerance between high-light intensity and glyphosate in terms of photosynthetic efficiency. No effect of light history on oxyfluorfen (increase in ascorbate peroxidase activity) and Cu (decrease in protein content) toxicity	Bonnineau PhD

Emerging pollutants

Hypothesis	Experimental design			Conclusions and remarks	Reference
	Conditions	Toxicant (µg/L)	Duration		
Impacts of pharmaceuticals and personal care products depend on their pre-exposure history	Growth of natural biofilms upstream and downstream of a WWTP. Exposure in 20-mL glass screw-capped test tubes	Triclosan: 0.012–1.2 Ciprofloxacin: 0.015–1.5 Tergitol NP10: 0.005–0.50	13 days	Hypothesis not confirmed. All three compounds caused marked shifts in the community structure at both the upstream and downstream sites and a decline in algal genus richness (at high concentration)	Wilson et al. [79]
Pharmaceuticals at environmental relevant concentrations alter the structure and activity of fluvial biofilms	Growth and exposure in rotating annular bioreactors	Ibuprofen, carbamazepine, furosemide, caffeine: 10	8 weeks	Hypothesis confirmed for the algal and bacterial component of the biofilm. Pharmaceuticals exhibited both nutrient-like and toxic effects on fluvial biofilms	Lawrence et al. [77]

(continued)

Emerging pollutants					
Hypothesis	Experimental design			Conclusions and remarks	Reference
	Conditions	Toxicant (μg/L)	Duration		
Triclosan (TCS) bioavailability is lower in biofilm-associated cell than in suspended ones	Growth on glass discs. Exposure in standard test vessels	TCS: 270–17,300	24 h	Biofilm-associated diatoms were less affected (in terms of photosynthetic efficiency) than suspended cells. Differences in sensitivity could not be attributed to the life form, suggesting that TCS may address multiple target sites	Franz et al. [76]
TCS and triclocarban (TCC) cause effects on the structure and function of river biofilms	Growth in rotating annular bioreactors	TCS, TCC: 10	8 weeks	TCS and TCC affected community structure, architecture, and functioning of the biofilm communities. TCS also produced changes in bacterial community	Lawrence et al. [123]
TCS directly affects the bacterial community (due to the mode of action), while effects on algae are indirect	Growth and exposure in a flow-through system of indoor channels	TCS: 0.05–500	48 h	TCS toxicity was higher for bacteria than algae. The increase in bacterial mortality can be attributed to a direct mode of action of triclosan. Direct effects on algae cannot be disregarded	Ricart et al. [78]
3 β-blockers have toxic effects on fluvial biofilms	Growth from river inocula in crystallizing dishes	Propranolol: 0–531 Metoprolol: 0–522 Atenolol: 0–707,000	24 h	Hypothesis confirmed. Toxicity: Propanolol (photosynthesis inhibition) > metroprolol (bacterial mortality) > atenolol (affected both the algal and bacterial components)	Bonnineau et al. [80]

pesticides studied were herbicides (atrazine, diuron and its residues, isoproturon, prometryn) targeting photosynthetic processes and hence the sensitivity of the phototrophic component of biofilms to these compounds (Table 4).

Investigating pesticides' effects at different temporal scales from hours, days [69, 70], to weeks [39, 71–74] or months [75] allowed to assess the physiological and structural biofilm responses. Herbicides' indirect effects on invertebrates [71, 74] or bacterial communities [73] were also evaluated. Among the long list of the so-called emerging pollutants that are commonly found in environmental samples, few pharmaceuticals and personal care products were selected to investigate their potential effects on biofilms (Table 4), the bactericide triclosan (TCS) being the most investigated compound [76–79]. These studies addressed the protective role of biofilms and the effects of TCS on structural and functional attributes of their heterotrophic and autotrophic component. Short-term effects of three similarly acting β-blockers were investigated by Bonnineau et al [80] using a multibiomarker approach (Table 4).

Overall, the experiments reported tested an important number of hypotheses on toxicant availability, their effects on biofilm communities, and the potential modulating factors (Fig. 3). It was shown that metal bioaccumulation was driven by speciation and the presence of ligands [81]. Negative effects on growth were also reported (Table 3) as a reduction in biomass production (i.e., [82]), diatoms density [75, 83], and in diatoms biovolume [73]. It was also shown that prolonged exposure to a toxic compound alters the community composition causing a selection of tolerant species in most cases (e.g., [83, 84]), an increase in the tolerance of the community in many cases (e.g., [85–87]), and the production of EPS in some occasions [88]. Other investigations also demonstrated that these community responses were modulated by different environmental and biological factors (Fig. 3).

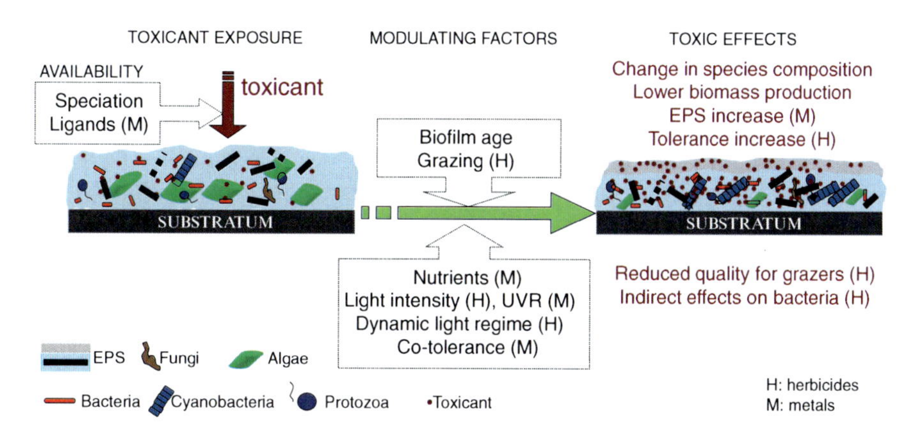

Fig. 3 Figure summarizing experimentally tested hypotheses on (1) toxicants' availability; (2) the biological and environmental factors modulating toxicity; and (3) the toxic effects after prolonged exposure of biofilm communities to different types of toxicants. M indicates hypotheses mainly tested with toxic metals; H for those tested with herbicides. No label for those tested for at least a metal, an herbicide, and an emerging compound

That biofilm's maturity and/or thickness may reduce toxicity is a general rule demonstrated for metals (i.e., [83]), herbicides (e.g., [87]), and emerging compounds such as TCS [76]. On the other hand, increased herbicide toxicity was demonstrated under grazing pressure [71], linked with the observation that grazing was responsible for maintaining the biofilm younger (Fig. 3).

Nutrient concentration may also influence the effects of toxic exposure (Fig. 3). It was shown for toxic compounds like Zn or Cu directly affecting nutrients availability (e.g., [89]).

Light history as well as light conditions during toxicant exposure affected toxicity. It was shown by the increase in sensitivity to Cd observed in biofilms adapted to UVR [90] or the increase in toxicity of isoproturon under a dynamic light regime [70]. In the case of Cu, co-tolerance to other metals was also demonstrated [85].

The toxicity of three similarly acting β-blockers was confirmed, but at very high concentration. In this study, differences in toxicity between the three compounds highlight the need to increase ecological realisms in toxicity testing used to derive environmental quality standards [80].

3.2 Experimental Studies Addressing Cause-and-Effect Relations Between Toxicant Exposure and Macroinvertebrates

Experimental studies have been conducted in aquatic microcosms and mesocosms to examine the effects of heavy metals, pesticides, and other organic chemicals on the structure and function of macroinvertebrate communities (Table 5). Although the duration of most studies was relatively short (1–4 weeks), the duration of several mesocosm and field experiments was considerable longer [91–93]. Endpoints measured in these studies included responses across several levels of organization, from physiological effects to alterations in community structure and ecosystem function. Some of these studies were conducted to provide additional support for field studies, thereby strengthening arguments for causation [60], while others were conducted to verify results of laboratory toxicity tests [46, 94]. Several studies were conducted to identify direct sources of toxicological effects in systems receiving multiple stressors [95, 96] or to quantify the influence of natural habitat characteristics on contaminant effects [97, 98]. Also experiments to examine competition/predation interactions between populations after exposure to pollutants have been developed [99]. A common goal of many experiments was to examine interactions among stressors [91] or to quantify the potential cost of tolerance associated with contaminant acclimation or adaptation [100–102]. A consistent justification for the application of these ecologically realistic approaches in risk assessment was to integrate responses across multiple levels of biological organization [93, 103].

Table 5 Experimental studies addressing cause–effect relations between toxic exposure and their effects on macroinvertebrate communities

Hypothesis	Experimental design			Conclusions and remarks	Reference
	Conditions	Toxicant	Duration		
Metals					
Field experiments to assess the influence of water quality and substratum quality on benthic macroinvertebrate communities	Colonization of clean and metal-contaminated substrate		30 days	Water was acutely toxic to most taxa; however, responses to metals in contaminated biofilms were species-specific	Courtney and Clements [95]
Microcosm experiments conducted to support results of field studies	Communities transferred from field to microcosms	Zn alone, Zn + Cd, and Zn + Cu + Cd	10 days	Experiments confirmed predicted "safe" concentrations of metals; some evidence of synergistic effects due to metal mixtures	Clements [60]
Seasonal variation in metal effects	Communities transferred from field to microcosms	Cd, Cu, Zn	7 days	Effects of metals greater in summer because populations dominated by smaller, early instars	Clark and Clements (2006) [128]
Cost associated with increased tolerance to metals	Communities transferred from reference and metal-contaminated sites to microcosms	Cd, Cu, Zn	10 days	Communities from metal-contaminated sites were tolerant of metals but more susceptible to other stressors (predation, acidification, UV-B radiation)	Clements [100], Courtney and Clements [101], Kashian et al. [102]
Heavy metals and UV-B radiation interact to structure communities	Field experiment excluded UV-B; microcosm experiment quantified UV-B effects on reference and contaminated communities	UV-B + metals (Cd, Cu, Zn)	Field: 60 days microcosm: 7 days	Effects of UV-B treatments are consistently greater in metal-contaminated streams compared to reference streams	Zuellig et al. [91]
Separate relative effects of acidity and metals in acid mine drainage (AMD)	Measured colonization of AMD-contaminated substrate in the field	Acidity and metals	5 weeks	Most toxicity associated with acidity leaching from sediment rather than metals	Dsa et al. [96]

Hypothesis	Experimental design			Conclusions and remarks	Reference
	Conditions	Toxicant (μg/L)	Duration		
Pesticides					
Determine no observed effect concentrations for several structural and functional measures	Small, synthetic microcosms seeded with zooplankton, phytoplankton, and snails from a pond	Chlorpyrifos, carbendazim, and linuron	2–3 weeks	Responses and effect similar to those observed in larger systems; absence of sediment and macrophytes accounted for greater persistence of pesticides in these systems	Daam and Van den Brink [97]

(continued)

Pesticides

Hypothesis	Experimental design			Conclusions and remarks	Reference
	Conditions	Toxicant (μg/L)	Duration		
Quantify the drift responses of benthic invertebrates	Mayflies, blackflies and amphipods exposed to pesticides in 1.2 m glass channels	11 different pesticides	48 h	Drift initiated by 6 of 11 pesticides at levels 7–22 times lower than LC_{50} values; greatest effects from neurotoxic insecticides	Beketov and Liess [46]
Quantify effects of a single pulse of insecticide	Mesocosm study designed to compare community LOEC with LC_{50} for individual organism	Thiacloprid	7 months	Long-term community LOEC comparable to LC_{50} for sensitive species; rate of recovery more dependent on life history characteristics than pesticide concentrations	Beketov et al. [92]
Demonstrate causal relationship between pesticide and responses	Field experiment conducted in tributary of the River Arrow, UK	Cypermerthrin, chlorpyrifos	20 months	Effects of pesticides mitigated by no-spray buffer zones; individual-level effects occurred at low concentrations, but were not translated to populations or communities	Maltby and Hills [93]
Compared structural and functional measures	Outdoor stream channels with 4 species of benthic invertebrates	Pyrethroid	10 d	No effects on macroinvertebrate density; greater algal biomass and reduced decomposition in treated mesocosms	Rasmussen et al. (2008) [129]
	Field deployed stream mesocosms containing natural benthic communities	Imidacloprid	20 days	Significant effects on macroinvertebrate abundance and diversity; demonstrated the importance of integrating studies of individual species with model mesocosm studies	Pestana et al. [103]
Measure effects of a mixture of insecticide and herbicide	Indoor microcosms exposed to a range of concentrations	Atrazine and lindane	14 weeks	Direct and indirect effects observed; macroinvertebrate community was affected at all but the lowest levels (0.01 TU)	Van den Brink et al. [94]

Test direct and indirect effects on community structure	Outdoor microcosms treated after 6 weeks of initial colonization	Carbendazim	8 weeks	Both direct toxic effects and indirect effects due to reduced competition and altered food chain	Daam et al. (2010) [130]
A mixture of pesticide produces effects on the ecology of aquatic microcosms	Indoor freshwater plankton-dominated microcosms	Atrazine: 55 Lindane: 28	28 days	Lindane seriously affected macroinvertebrate community. Atrazine produced fewer effects than expected, probably due to decreased grazer stress on the algae as a result of the lindane application	Van den Brink et al. [94]
Test changes in species composition and in competence/predation relations	Experimental pond	Atrazine and endosulfan	4 weeks	The pesticides changed the intensity in the competitive or predatory interactions	Rohr and Crumrine [99]
Test if refuge zones are important in the recuperation of communities from short-term exposure to insecticide	Experimental ditches	Lufenuron	161 days	Existence of refuge zones and species' vital traits are important factors enhancing recovery	Brock et al. [98]

3.3 Integrating Descriptive and Experimental Approaches to Demonstrate Causation in Stream Bioassessment Studies: Case Study of a Metal-Polluted Stream

To demonstrate how descriptive and experimental approaches can be integrated to demonstrate causation, we present results of a case study conducted in a metal-polluted stream in Colorado, USA. A long-term monitoring program of water quality and benthic macroinvertebrates was initiated in the Arkansas River in 1989 [47]. Concentrations of heavy metals (Cd, Cu, and Zn) are greatly elevated downstream, and often exceed acutely toxic levels. Over the past 17 years, we measured physicochemical characteristics, habitat quality, heavy metal concentrations, and macroinvertebrate community structure seasonally (spring and fall) at locations upstream and downstream from several sources of metal contamination. Three years after we began this research, a large-scale restoration program was initiated to reduce metal concentrations and reestablish trout populations in the upper Arkansas River basin. Because these data were collected before and after remediation, this long-term research provided a unique opportunity to quantify ecological responses to improvements in water quality. For the purposes of this case study, we present macroinvertebrate and metals data collected from one upstream and one downstream station collected from 1989 to 2006.

To support this descriptive study, microcosm experiments were conducted to develop concentration–response relationships between heavy metals and measures of macroinvertebrate community structure. Benthic macroinvertebrate communities for these experiments were obtained from an uncontaminated reference stream with no history of metal contamination using a technique previously described [104]. Communities were transferred to stream microcosms and exposed to combinations of Cu and Zn at concentrations that bracketed those measured at metal-contaminated sites in the field. Because these experiments involved a mixture of heavy metals, an additive measure of toxicity was used to express metal concentrations relative to the U.S. EPA chronic criterion values. The cumulative criterion unit (CCU) was defined as the ratio of the measured metal concentration to the hardness-adjusted criterion value and summed for each metal.

Significant concentration–response relationships were developed for total abundance and species richness in stream microcosms (Fig. 4). Total macroinvertebrate abundance was more sensitive to metals than species richness, a finding previously reported from stream microcosm experiments [105]. The LC_{20} concentrations for macroinvertebrate abundance and species richness, defined as the CCU levels that caused a 20% reduction, were approximately 2.3 and 9.0, respectively. These experimental data support the hypothesis that macroinvertebrate communities were highly sensitive to heavy metals and that relatively low concentrations resulted in significant alterations in community composition.

Heavy metal concentrations in the Arkansas River were seasonally variable, but decreased significantly after the remediation program was initiated (Fig. 5). Metal

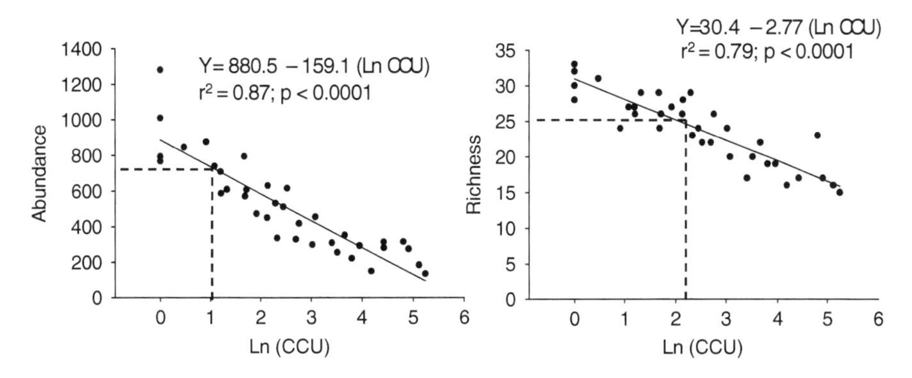

Fig. 4 Results of microcosm experiments showing concentration–response relationships between metal concentration (CCU), total macroinvertebrate abundance, and species richness. *Dashed lines* show the estimated LC_{20} concentrations for each metric

concentrations at station EF5 rapidly decreased below the estimated EC_{20} values in 1992. In contrast, metal concentrations at station AR3 remained elevated until about 1999, fluctuated between the estimated EC_{20} values for several years, and then decreased below these levels in 2003.

Macroinvertebrate communities in the Arkansas River responded to these improvements in water quality after remediation (Fig. 5). Macroinvertebrate communities at station EF5 quickly recovered after metal levels were reduced below the EC_{20} values. Abundance and species richness recovered more slowly at station AR3, consistent with expectations based on long-term changes in metal concentrations and estimated EC_{20} values for these metrics. Total macroinvertebrate abundance recovered in approximately 2003, whereas species richness recovered in 2000.

Although the biomonitoring results from the Arkansas River were consistent with the hypothesis that heavy metals are responsible for reduced abundance and richness, these data are not sufficient to demonstrate causation. Long-term improvements in water quality and the associated increases in abundance and species richness after remediation represent a natural experiment that allowed a more rigorous test of this hypothesis. Finally, highly significant concentration–response relationships between macroinvertebrate community metrics and heavy metal concentration from stream microcosm experiments provided estimates of metal levels that likely impact benthic communities. The consistency of these LC_{20} estimates with concentrations measured in the field where abundance and richness recovered greatly strengthened the argument that heavy metals were responsible for alterations in benthic communities. These results demonstrate the importance of employing ecologically realistic experimental techniques to support descriptive studies for developing causal arguments.

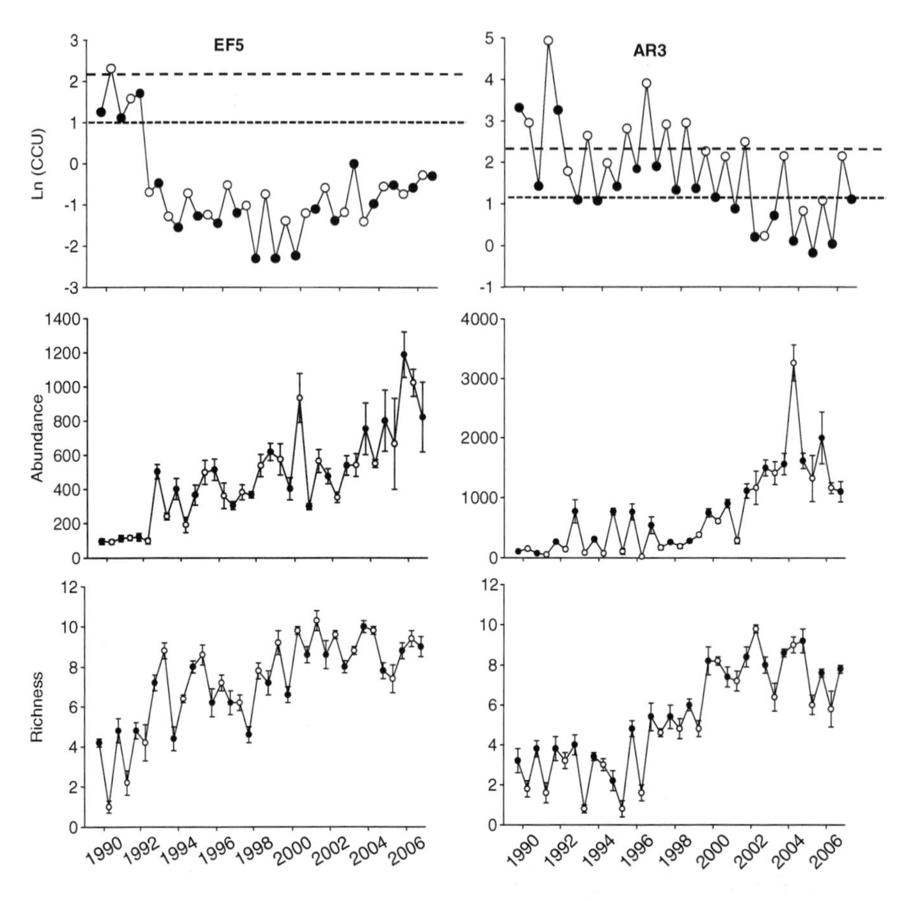

Fig. 5 Long-term changes in metal concentration (CCU), abundance, and species richness at Arkansas River stations EF5 and AR3. *Horizontal lines* correspond to the EC20 values for species richness (*dashed*) and abundance (*dotted*) estimated from microcosm experiments. Alternating *solid* and *open* symbols refer to samples collected in spring and fall, respectively. Macroinvertebrate data are means ±s.e.

4 General Discussion and Prospects

The papers found in the literature show how field and laboratory investigations evolved from the end of the 1990s, when community ecotoxicology studies focused on metals and pesticides began, to present days, when the first emerging compounds investigations have appeared (Fig. 1). To date, quite a lot of information has been generated about the effects of emerging pollutants (mixtures or simple compounds) on single organisms but community approaches are very scarce, highlighting the existence of a missing gap requiring future investigations.

The set of studies presented in this chapter illustrate how field and laboratory investigations complement to provide causality between toxicant exposure in running waters and benthic communities' responses. As a general scheme (Fig. 6), hypotheses based on field observations cannot be confirmed without experimentation and thus needed to provide causal evidence. If the formulated hypothesis is not confirmed, new field observations may be required and the results should be analyzed again in order to evaluate the role of different environmental factors (confounding factors) influencing the biological responses observed. The newly generated hypothesis should also be experimentally tested including possible interactions by simulating multiple-stress situations (Fig. 6).

Many of the hypotheses derived from field investigations have been validated following this general scheme. The Arkansas River case study is a good example showing how field observations, together with long-term natural experiments and microcosm experiments, provide consistent arguments and evidence of metals' effects on the biota. More precisely, metal pollution levels were responsible for reduced abundance and taxa richness of the macroinvertebrate community. Long-term monitoring studies following temporal trends in both chemical pollution and the biological responses are very informative, but unfortunately very scarce. In the case of biofilms, microcosm and mesocosm experiments confirmed that metals and pesticides are responsible for the loss of sensitive species in the community, and that this influence is modulated by several biological and environmental factors including the successional stage (biofilms age), the trophic status of the river or stream (nutrient concentration), and the light regime (whether is it an open canopy or shaded reach). These concluding findings should be included, in the future, in

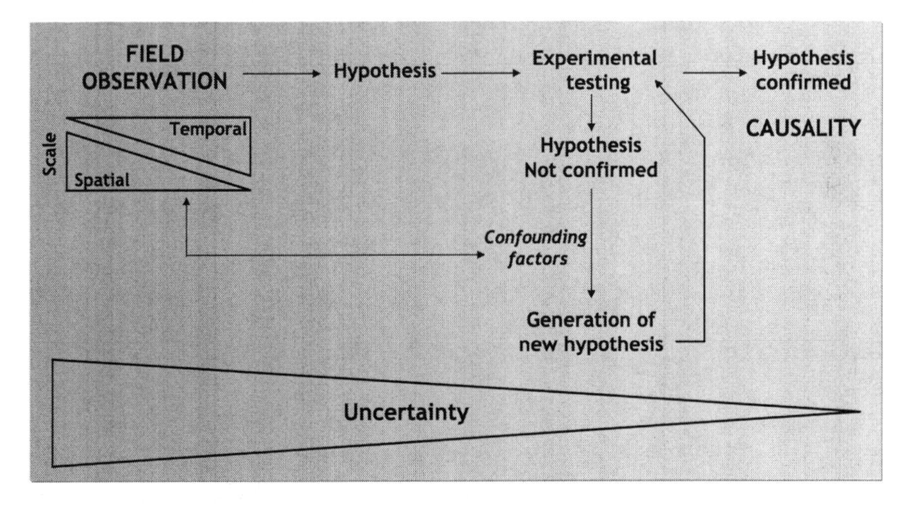

Fig. 6 Graph illustrating the steps required to derive causality. Hypothesis formulated on the basis of field observations should be experimentally tested for confirmation. If the formulated hypothesis is not confirmed, new field observations may be required and the results analyzed again in order to evaluate the role of different environmental factors (confounding factors) influencing the biological responses observed. The newly generated hypothesis should also be experimentally tested including possible interactions simulating multiple-stress situations

risk assessment models in order to account for the influence that the environment exerts on the effects of chemicals on the biota.

In contrast to the results obtained with metals and pesticides already included in the list of priority pollutants, emerging pollutants and especially pharmaceuticals (molecules designed to be biologically actives) are not expected to cause changes on natural communities easily to detect. Given that the levels reported in rivers for these substances in waters and sediments are generally low, no lethal effects on the species are expected at concentrations found in the environment [56, 106]; hence, new approaches should be used in order to know the effects of these substances on natural communities. Experimental investigations on communities are difficult because of the long-term studies required and because of the inconspicuous endpoints that need to be studied, whereas in field studies the difficulties in predicting the effects of emerging pollutants and changes in community arise from the nature of the effects caused by these substances, such as feminization or changes in behavior or emergence time, not studied enough in invertebrates in wildlife. Effects in communities can be detected by examining other mechanisms than the direct effect in species' density (like in the case of pesticides) such as sublethal or long-term effects on physiology, on reproductive traits or hormone-mediated processes (Endocrine Disrupting Compounds as explained in Lagadic et al. [107] and Soin and Smagghe [108]), or on other less obvious traits such as behavior [109].

Overall, the examples provided in this chapter, together with the recommendations given, are proposed as a general guide for studies aiming to link chemical pollution with ecological alterations. The proposed approach, although being complex and probably expensive in terms of dedication, is strongly recommended for investigative monitoring, situations where routine monitoring may fail in the detection of the causes accounting for a poor ecological status.

Acknowledgments The research was funded by the Spanish Ministry of Science and Innovation (CTM2009-14111-CO2-01 and SCARCE-Consolider 2010 CSD2009-00065) and the EC Sixth Framework Program (KEYBIOEFFECTS MRTN-CT-2006-035695 and MODELKEY GOCE-511237).

References

1. European Commission (2000) Directive 2000/60/EC of the European Parliament and of the Council – Establishing a Framework for Community Action in the Field of Water Policy. European Commission, Belgium
2. Sabater S, Guasch H, Ricart M, Romaní AM, Vidal G, Klünder C, Schmitt-Jansen M (2007) Monitoring the effect of chemicals on biological communities. The biofilm as an interface. Anal Bioanal Chem 387:1425–1434
3. Muñoz I, Sabater S, Barata C (2012) Evaluating ecological integrity in multi-stressed rivers: from the currently used biotic indices to newly developed approaches using biofilms and invertebrates. In: Guasch H, Ginebreda A, Geiszinger A (eds) Emerging and priority pollutants in rivers. The handbook of environmental chemistry, vol 19. Springer, Heidelberg

4. Li XD, Thornton I (2001) Chemical partitioning of trace and major elements in soils contaminated by mining and smelting activities. Appl Geochem 16:1693–706
5. Duong TT, Morin S, Herlory O et al (2008) Seasonal effects of cadmium accumulation in periphytic diatom communities of freshwater biofilms. Aquat Toxicol 90:19–28
6. Barceló D (2003) Emerging pollutants in water analysis. TrAC 22(10):xiv–xvi
7. Daughton CG, Ternes TA (1999) Pharmaceuticals and personal care products in the environment: agents of subtle change? Environ Health Perspect 107:907–938
8. Daughton CG (2004) Non-regulated water contaminants: emerging research. Environ Impact Assess Rev 24:711–732
9. Chapman PM (2006) Emerging substances-emerging problems? Environ Toxicol Chem 25:1445–1447
10. Ricart M, Guasch H, Barceló D et al (2010) Primary and complex stressors in polluted Mediterranean rivers: pesticide effects on biological communities. J Hydrol 383:52–61
11. Rotter S, Sans-Piché F, Streck G et al (2011) Active bio-monitoring of contamination in aquatic systems—An in situ translocation experiment applying the PICT concept. Aquat Toxicol 101:228–236
12. Morin S, Pesce S, Tlili A et al (2010) Recovery potential of periphytic communities in a river impacted by a vineyard watershed. Ecol Indic 10:419–426
13. Hill BH, Willingham WT, Parrish LP et al (2000) Periphyton community responses to elevated metal concentrations in a Rocky Mountain stream. Hydrobiologia 428:161–169
14. Sabater S (2000) Diatom communities as indicators of environmental stress in the Guadiamar River, S-W. Spain, following a major mine tailings spill. J Appl Phycol 12:113–124
15. Admiraal W, Blanck H, Buckert De Jong M et al (1999) Short-term toxicity of zinc to microbenthic algae and bacteria in a metal polluted stream. Wat Res 33:1989–1996
16. Ivorra N, Hettelaar J, Tubbing GMJ et al (1999) Translocation of microbenthic algal assemblages used for in situ analysis of metal pollution in rivers. Arch Environ Contam Toxicol 37:19–28
17. Navarro E, Guasch H, Sabater S (2002) Use of microbenthic algal communities in ecotoxicological tests for the assessment of water quality: the Ter river case study. J Appl Phycol 14:41–48
18. Guasch H, Muñoz I, Rosés N et al (1997) Changes in atrazine toxicity throughout succession of stream periphyton communities. J Appl Phycol 9:137–146
19. Guasch H, Ivorra N, Lehmann V et al (1998) Community composition and sensitivity of periphyton to atrazine in flowing waters: the role of environmental factors. J Appl Phycol 10:203–213
20. Guasch H, Admiraal W, Sabater S (2003) Contrasting effects of organic and inorganic toxicants on freshwater periphyton. Aquat Toxicol 64:165
21. Behra R, Landwehrjohann R, Vogel K et al (2002) Copper and zinc content of periphyton from two rivers as a function of dissolved metal concentration. Aquat Sci 64:300–306
22. Meylan S, Behra R, Sigg L (2003) Accumulation of copper and zinc in periphyton in response to dynamic variations of metal speciation in freshwater. Environ Sci Technol 37:5204–5212
23. Le Faucher S, Behra R, Sigg L (2005) Thiol and metal contents in periphyton exposed to elevated copper and zinc concentrations: a field and microcosm study. Environ Sci Technol 39:8099–8107
24. Holding KL, Gill RA, Carter J (2003) The relationship between epilithic periphyton (biofilm) bound metals and metals bound to sediments in freshwater systems. Environ Geochem Health 25:87–93
25. Hill WR, Ryon MG, Smith JG et al (2010) The role of periphyton in mediating the effects of pollution in a stream ecosystem. Environ Manage 45:563–576
26. Akkanen J, Slootweg T, Mäenpää K, Leppänen MT, Agbo S, Gallampois C, Kukkonen JVK (2012) Bioavailability of organic contaminants in freshwater environments. In: Guasch H, Ginebreda A, Geiszinger A (eds) Emerging and priority pollutants in rivers, The handbook of environmental chemistry, vol 19. Springer, Heidelberg

27. Morin S, Cordonier A, Lavoie I, Arini A, Blanco S, Duong TT, Tornés E, Bonet B, Corcoll N, Faggiano L, Laviale M, Pérès F, Becares E, Coste M, Feurtet-Mazel A, Fortin C, Guasch H, Sabater S (2012) Consistency in diatom response to metal-contaminated environments. In: Guasch H, Ginebreda A, Geiszinger A (eds) Emerging and priority pollutants in rivers, The handbook of environmental chemistry, vol 19. Springer, Heidelberg
28. Morin S, Vivas-Nogues M, Duong TT et al (2007) Dynamics of benthic diatom colonization in a cadmium/zinc-polluted river (Riou-Mort, France). Arch Hydrobiol 168:179–187
29. Blanck H, Wänkberg SÅ, Molander S (1988) Pollution-induced community tolerance-a new ecotoxicological tool. In: Cairs J Jr, Pratt JR (eds) Functional testing of aquatic biota for estimating hazards of chemicals, 988. ASTM STP, Philadelphia, PA, pp 219–230
30. Blanck H, Admiraal W, Cleven RFMJ et al (2003) Variability in zinc tolerance, measured as incorporation of radio-labeled carbon dioxide and thymidine, in periphyton communities sampled from 15 European river stretches. Arch Environ Contam Toxicol 44:17–29
31. Guasch H, Leira M, Montuelle B et al (2009) Use of multivariate analyses to investigate the contribution of metal pollution to diatom species composition: search for the most appropriate cases and explanatory variables. Hydrobiologia 627:143–158
32. Guasch H, Sabater S (1998) Light history influences the sensitivity to atrazine in periphytic algae. J Phycol 34:233–241
33. Dorigo U, Leboulanger C (2001) A PAM fluorescence-based method for assessing the effects of photosystem II herbicides on freshwater periphyton. J Appl Phycol 13:509–515
34. Dorigo U, Bourrain X, Bérard A et al (2004) Seasonal changes in the sensitivity of river microalgae to atrazine and isoproturon along a contamination gradient. Sci Total Environ 318:101–114
35. Pesce S, Margoum C, Montuelle B (2010) In situ relationship between spatio-temporal variations in diuron concentrations and phtotrophic biofilm tolerance in a contaminated river. Wat Res 44:1941–1949
36. Pesce S, Lissalde S, Lavieille D et al (2010) Evaluation of single and joint toxic effects of diuron and its main metabolites on natural phototrophic biofilms using a pollution-induced community tolerance (PICT) approach. Aquat Toxicol 99:492–499
37. Ricciardi F, Bonnineau C, Faggiano L et al (2009) Is chemical contamination linked to the diversity of biological communities in rivers? TrAC 28:592–602
38. Victoria SM, Gómez N (2010) Assessing the disturbance caused by an industrial discharge using field transfer of epipelic biofilm. Sci Total Environ 408:2696–2705
39. Tlili A, Bérard A, Roulier JL et al (2010) PO_4^{3-} dependence of the tolerance of autotrophic and heterotrophic biofilm communities to copper and diuron. Aquat Toxicol 98:165–177
40. Jelić A, Gros M, Petrović M, Ginebreda A, Barceló D (2012) Occurrence and elimination of pharmaceuticals during conventional wastewater treatment. In: Guasch H, Ginebreda A, Geiszinger A (eds) Emerging and priority pollutants in rivers. The handbook of environmental chemistry, vol 19. Springer, Heidelberg
41. Corcoll N, Ricart M, Franz S, Sans-Piché F, Schmitt-Jansen M, Guasch H (2012) The use of photosynthetic fluorescence parameters from autotrophic biofilms for monitoring the effect of chemicals in river ecosystems. In: Guasch H, Ginebreda A, Geiszinger A (eds) Emerging and priority pollutants in rivers, The handbook of environmental chemistry, vol 19. Springer, Heidelberg
42. Clements WH, Carlisle DM, Lazorchak JM et al (2000) Heavy metals structure benthic communities in Colorado mountain streams. Ecol Appl 10:626–638
43. Schmidt TS, Clements WH, Mitchell KA et al (2010) Development of a new toxic-unit model for the bioassessment of metals in streams. Environ Toxicol Chem 29:2432–2442
44. Stockdale A, Tipping E, Lofts S et al (2010) Toxicity of proton–metal mixtures in the field: Linking stream macroinvertebrate species diversity to chemical speciation and bioavailability. Aquat Toxicol 100:112–119
45. Gray NF, Delaney E (2008) Comparison of benthic macroinvertebrate indices for the assessment of the impact of acid mine drainage on an Irish River below an abandoned Cu-S mine. Environ Pollut 155:31–40
46. Beketov MA, Liess M (2008) Potential of 11 pesticides to initiate downstream drift of stream macroinvertebrates. Arch Environ Contam Toxicol 55:247–253

47. Clements WH, Vieira NKM, Church SE (2010) Quantifying restoration success and recovery in a metal-polluted stream: a 17 year assessment of physicochemical and biological responses. J Appl Ecol 47:899–910
48. Iwasaki Y, Kagaya T, Miyamoto K et al (2009) Effects of heavy metals on riverine benthic macroinvertebrate assemblages with reference to potential food availability for drift-feeding fishes. Environ Toxicol Chem 28:354–363
49. Bollmohr S, Schulz R (2009) Seasonal changes of macroinvertebrate communities in a Western Cape River, South Africa, receiving nonpoint-source insecticide pollution. Environ Toxicol Chem 28:809–817
50. Loayza-Muro RA, Elias-Letts R, Marticorena-Ruiz JK et al (2010) Metal-induced shifts in benthic macroinvertebrate community composition in Andean high altitude streams. Environ Toxicol Chem 29:2761–2768
51. Beketov MA, Foit K, Schafer RB et al (2009) SPEAR indicates pesticide effects in streams – comparative use of species – and family-level biomonitoring data. Environ Pollut 157: 1841–1848
52. Buchwalter DB, Cain DJ, Martin CA et al (2008) Aquatic insect ecophysiological traits reveal phylogenetically based differences in dissolved cadmium susceptibility. Proc Natl Acad Sci USA 105:8321–8326
53. Heiskanen AS, Solimini AG (2005) Analysis of the current knowledge gaps for the implementation of the water framework directive. European Commission, Joint Research Center, Ispra, Italy
54. Archaimbault V, Usseglio-Polatera P, Garric J et al (2010) Assessing pollution of toxic sediment in streams using bio-ecological traits of benthic macroinvertebrates. Freshwater Biol 55:1430–1446
55. Beasley G, Kneale P (2002) Reviewing the impact of metals and PAHs on macronvertebrates in urban watercourses. Prog Phys Geogr 26:236–270
56. Cunningham VL, Buzby M, Hutchinson T et al (2006) Effects of human pharmaceuticals on aquatic life: next steps. Environ Sci Technol 40:3456–3462
57. Fent K, Weston AA, Caminada D (2006) Ecotoxicology of human pharmaceuticals. Aquat Toxicol 76:122–159
58. Jjemba PK (2006) Excretion and ecotoxicity of pharmaceutical and personal care products in the environment. Ecotoxicol Environ Saf 63:113–130
59. Muñoz I, López-Doval JC, Ricart M et al (2009) Bridging levels of pharmaceuticals in river water with biological community structure in the Llobregat river basin (NE Spain). Environ Toxicol Chem 28:2706–2714
60. Clements WH (2004) Small-scale experiments support causal relationships between metal contamination and macroinvertebrate community responses. Ecol Appl 14:954–967
61. Platt JR (1964) Strong inference. Science 146:347–353
62. Hurlbert SH (1984) Pseudoreplication and the design of ecological field experiments. Ecol Monogr 54:187–211
63. Hill AB (1965) The environment and disease: association or causation. Proc Roy Soc Med 58:295–300
64. Suter GW II (1993) A critique of ecosystem health concepts and indexes. Environ Toxicol Chem 12:1533–1539
65. Beyers DW (1998) Causal inference in environmental impact studies. J N Am Bentholl Soc 17:367–373
66. Chapman PM (1986) Sediment quality criteria from the sediment quality triad: an example. Environ Toxicol Chem 5:957–964
67. Clements WH, Newman MC (2002) Community ecotoxicology. Wiley, Chichester, UK, p 336
68. Popper KR (1972) The logic of scientific discovery, 3rd edn. Hutchinson, London, England

69. Tlili A, Dorigo U, Montuelle B et al (2008) Responses of chronically contaminated biofilms to short pulses of diuron an experimental study simulating flooding events in a small river. Aquat Toxicol 87:252–263
70. Laviale M, Prygiel J, Créach A (2010) Light modulated toxicity of isoproturon toward natural stream periphyton photosynthesis: a comparison between constant and dynamic light conditions. Aquat Toxicol 97:334–342
71. Muñoz I, Real M, Guasch H et al (2001) Effects of atrazine on periphyton under grazing pressure. Aquat Toxicol 55:239–249
72. Guasch H, Lehmann V, van Beusekom B et al (2007) Influence of phosphate on the response of periphyton to atrazine exposure. Arch Environ Contam Toxicol 52:32–37
73. Ricart M, Barceló D, Geszinger A et al (2009) Effects of low concentrations of the phenylurea herbicide diuron on biofilm algae and bacteria. Chemosphere 76:1392–1401
74. López-Doval JC, Ricart M, Guasch H et al (2010) Does grazing pressure modify diuron toxicity in a biofilm community? Arch Environ Contam Toxicol 58:955–962
75. Pérès F, Florin D, Grollier T et al (1996) Effects of the phenylurea herbicide isoprotruron on periphytic diatom communities in freshwater indoor microcosms. Environ Pollut 94:141–152
76. Franz S, Altenburger R, Heilmeier H et al (2008) What contributes to the sensitivity of microalgae to triclosan? Aquat Toxicol 90:102–108
77. Lawrence JR, Swerhone GDW, Wassenaar LI et al (2005) Effects of selected pharmaceuticals on riverine biofilm communities. Can J Microbiol 51:655–669
78. Ricart M, Guasch H, Alberch M et al (2010) Triclosan persistence through wastewater treatment plant and its potential toxic effects on river biofilms. Aquat Toxicol 100:346–353
79. Wilson BA, Smith VH, Denoyelles F et al (2003) Effects of three pharmaceutical and personal care products on natural freshwater algal assemblages. Environ Sci Technol 37:1713–1719
80. Bonnineau C, Guasch H, Proia L et al (2010) Fluvial biofilms: a pertinent tool to assess β-blockers toxicity. Aquat Toxicol 96:225–233
81. Meylan S, Behra R, Sigg L (2004) Influence of metal speciation in natural freshwater on bioaccumulation of copper and Zinc in periphyton: a microcosm study. Environ Sci Technol 38:3104–3111
82. Paulsson M, Mansson V, Blanck H (2002) Effects of zinc on the phosphorus availability to periphyton communities from the river Gota Alv. Aquat Toxicol 56:103–113
83. Duong TT, Morin S, Coste M et al (2010) Experimental toxicity and bioaccumulation of cadmium in freshwater periphytic diatoms in relation with biofilm maturity. Sci Total Environ 408:552–562
84. Ivorra N, Hettelaar J, Kraak MHS et al (2002) Responses of biofilms to combined nutrient and metal exposure. Environ Toxicol Chem 21:626–632
85. Soldo D, Behra R (2000) Long-term effects of copper on the structure of freshwater periphyton communities and their tolerance to copper, zinc, nickel and silver. Aquat Toxicol 47:181–189
86. McClellan K, Altenburger R, Schmitt-Jansen M (2008) Pollution-induced community tolerance as a measure of species interaction in toxicity assessment. J Appl Ecol 45:1514–1522
87. Schmitt-Jansen M, Altenburger R (2008) Community-level microalgal toxicity assessment by multiwavelength-excitation PAM fluorometry. Aquat Toxicol 86:49–58
88. Garcia-Meza JV, Barranguet C, Admiral W (2005) Biofilm formation by algae as a mechanism for surviving on mine tailings. Environ Toxicol Chem 24:573–581
89. Serra A, Guasch H, Admiraal W, Van der Geest HG, Van Beusekom SAM (2010) Influence of phosphorus on copper sensitivity of fluvial periphyton: the role of chemical, physiological and community-related factors. Ecotoxicology 19:770–780
90. Navarro E, Robinson CT, Behra R (2008) Increased tolerance to ultraviolet radiation (UVR) and cotolerance to cadmium in UVR-acclimatized freshwater periphyton. Limnol Oceanogr 53:1149–1158

91. Zuellig RE, Kashian DR, Brooks ML et al (2008) The influence of metal exposure history and ultraviolet-B radiation on benthic communities in Colorado Rocky Mountain streams. J N Am Bentholl Soc 27:120–134
92. Beketov MA, Schafer RB, Marwitz A et al (2008) Long-term stream invertebrate community alterations induced by the insecticide thiacloprid: effect concentrations and recovery dynamics. Sci Tot Environ 405:96–108
93. Maltby L, Hills L (2008) Spray drift of pesticides and stream macroinvertebrates: experimental evidence of impacts and effectiveness of mitigation measures. Environ Pollut 156:1112–1120
94. Van den Brink PJ, Crum SJH, Gylstra R et al (2009) Effects of a herbicide–insecticide mixture in freshwater microcosms: risk assessment and ecological effect chain. Environ Pollut 157:237–249
95. Courtney LA, Clements WH (2002) Assessing the influence of water quality and substratum quality on benthic macroinvertebrate communities in a metal-polluted stream: an experimental approach. Freshwat Biol 47:1766–1778
96. Dsa JV, Johnson KS, Lopez D et al (2008) Residual toxicity of acid mine drainage-contaminated sediment to stream macroinvertebrates: relative contribution of acidity vs. metals. Water Air Soil Pollut 194:185–197
97. Daam MA, Van den Brink PJ (2007) Effects of chlorpyrifos, carbendazim, and linuron on the ecology of a small indoor aquatic microcosm. Arch Environ Contam Toxicol 53:22–35
98. Brock TCM, Roessink I, Dick J et al (2009) Impact of a benzoyl urea insecticide on aquatic macroinvertebrates in ditch mesocosms with and without non-sprayed sections. Environ Toxicol Chem 28(10):2191–2205
99. Rohr JR, Crumrine PW (2005) Effects of an herbicide and an insecticide on pond community structure and processes. Ecol Appl 15:1135–1147
100. Clements WH (1999) Metal tolerance and predator–prey interactions in benthic macroinvertebrate stream communities. Ecol Appl 9:1073–1084
101. Courtney LA, Clements WH (2000) Sensitivity to acidic pH in benthic invertebrate assemblages with different histories of exposure to metals. J N Am Bentholl Soc 19:112–127
102. Kashian DR, Zuellig RE, Mitchell KA et al (2007) The cost of tolerance: sensitivity of stream benthic communities to UV-B and metals. Ecol Appl 17:365–375
103. Pestana JLT, Alexander AC, Culp JM et al (2009) Structural and functional responses of benthic invertebrates to imidacloprid in outdoor stream mesocosms. Environ Pollut 157:2328–2334
104. Clements WH, Cherry DS, Cairns J Jr (1988) Structural alterations in aquatic insect communities exposed to copper in laboratory streams. Environ Toxicol Chem 7:715–722
105. Carlisle DM, Clements WH (1999) Sensitivity and variability of metrics used in biological assessment of running waters. Environ Toxicol Chem 18:285–291
106. Farré M, Pérez S, Kantiani L et al (2008) Fate and toxicity of emerging pollutants, their metabolites and transformation products in the aquatic environment. TrAC 27:991–1007
107. Lagadic L, Cotellec MA, Caquet T (2007) Endocrine disruption in aquatic pulmonate molluscs: few evidences, many challenges. Ecotoxicology 16:45–59
108. Soin T, Smagghe G (2007) Endocrine disruption in aquatic insects: a review. Ecotoxicology 16:83–93
109. De Lange HJ, Noordoven W, Murkc AJ et al (2006) Behavioural responses of *Gammarus pulex* (Crustacea, Amphipoda) to low concentrations of pharmaceuticals. Aquat Toxicol 78:209–216
110. Farag AM, Woodward DF, Goldstein JN, Brumbaugh W, Meyer JS (1998) Concentrations of metals associated with mining waste in sediments, biofilm, benthic macroinvertebrates, and fish from the Coeur d'Alene River Basin, Idaho. Arch Environ Contam Toxicol 34:119–127
111. Farag AM, Nimick DA, Kimball BA, Church SE, Harper DD, Brumbaugh WD (2007) Concentrations of metals in water, sediment, biofilm, benthic macroinvertebrates, and fish in the Boulder River Watershed, Montana, and the role of colloids in metal uptake. Arch Environ Contam Toxicol 52:397–409

112. Schäfer RB, Caquet T, Siimes K, Mueller R, Lagadic L, Liess M (2007) Effects of pesticides on community structure and ecosystem functions in agricultural streams of three biogeographical regions in Europe. Sci Total Environ 382:272–285
113. Ginebreda A, Muñoz I, López de Alda M, Brix R, López-Doval JC, Barceló D (2010) Environmental risk assessment of pharmaceuticals in rivers: relationships between hazard indexes and aquatic macroinvertebrate diversity indexes in the Llobregat River (NE Spain). Environ Int 36:153–162
114. Barrangeut C, Charantoni E, Plans M et al (2000) Short-term response of monospecific and natural algal biofilms to copper exposure. Eur J Phycol 35:397–406
115. Barranguet C, Plans M, Van Der Brinten E et al (2002) Development of photosynthetic biofilms affected by dissolved and sorbed copper in a eutrophic river. Environ Toxicol Chem 21:1955–1965
116. Corcoll N, Bonet B, Leira M, Guasch H (2011) Chl-*a* fluorescence parameters as biomarkers of metal toxicity in fluvial biofilms: an experimental study. Hydrobiologia 673:119–136
117. Guasch H, Navarro E, Serra A et al (2004) Phosphate limitation influences the sensitivity to copper in periphyton algae. Freshwat Biol 49:463–473
118. Daam MA, Rodrigues AMF, Van den Brink PJ et al (2009) Ecological effects of the herbicide linuron in tropical freshwater microcosms. Ecotoxicol Environ Saf 72:410–423
119. Schmitt-Jansen M, Altenburger R (2005) Toxic effects of isoproturon on periphyton communities – a microcosm study. Estuar Coast Shelf Sci 62:539–545
120. Schmitt-Jansen M, Altenburger R (2005) Predicting and observing responses of algal communities to photosystem II-herbicide exposure using pollution-induced community tolerance and species-sensitivity distributions. Environ Toxicol Chem 24:304–312
121. Schmitt-Jansen M, Altenburger R (2007) The use of pulse-amplitude modulated (PAM) fluorescence-based methods to evaluate effectgs of herbicides in microalgal systems of different complexity. Toxicol Environ Chem 89:665–681
122. Gustavson K, Mohlenberg F, Schlüter L (2003) Effects of exposure duration of herbicides on natural stream periphyton communities and recovery. Arch Environ Contam Toxicol 45:48–58
123. Lawrence JR, Zhu B, Swerhone GDW, Roy J, Wassenaar LI, Topp E, Korber DR (2009) Comparative microscale analysis of the effects of triclosan and triclocarban on the structure and function of river biofilm communities. Sci Total Environ 407:3307–3316
124. Guasch H, Serra A, Bonet B, et al. (2010) Metal ecotoxicology in fluvial biofilms. Potential influence of water scarcity. In: Sabater S, Barceló D (eds) Water scarcity in the Mediterranean: perspectives under global change. Springer, Hdb Env Chem 8:41–53
125. Sontoro A, Blo G, Mastrolitti, Fagioli F (2009) Bioaccumulation of heavy metals by aqua tic invertebrates along the Basento river in the south of Italy. Water, air and soil pollution. 201 (14)19–31
126. Lefcort H, Vancura J, Lider EL (2010) 75 years after mining ends streams insect diversity is still affeceted by heavy metals. Ecotoxicology 19(8)1416–25
127. Vera MS, Lagomarsino L, Sylvester M et al. (2010) New evidences of Roundup (glyphosate formulation) impact on the periphyton community and the water quality of freshwater ecosystems. Ecotoxicology 19(4):710–21
128. Clark JL, Clements WH (2006) The use of in situ and stream microcosm experiments to assess population- and community-level responses to metals. Environ. Toxicol. Chem. 25(9) 2306–12
129. Rasmussen JB, Gunn JM, Sherwood, GD et al. (2008) Direct and indirect (foodweb mediated) effects of metal exposure on the growth of yellow perch (Perca flavescens): implications for ecological risk assessment. Human and Ecological Risk Assessment 14:317–350.
130. Daam MA, Satapornvanit K, Van den Brink PJ et al. (2010) Direct and indirect effects of the fungicide Carbendazim in tropical freshwater microcosms. Arch. Env. Cont. Toxicol. 58 (2) 315–324.

Evaluating Ecological Integrity in Multistressed Rivers: From the Currently Used Biotic Indices to Newly Developed Approaches Using Biofilms and Invertebrates

Isabel Muñoz, Sergi Sabater, and Carlos Barata

Abstract This chapter reviews the different approaches to evaluate ecological integrity in rivers affected by multiple stressors, with emphasis on community responses and using both functional and structural descriptors. Special attention is given to the effect of chemical substances, both with respect to multiple effects and of single toxicants and mixtures in combination with natural and chemical disturbances. Most current bioassessment methods are not designed to detect specific impacts of chemical pollutants and can hardly be used to discern between multiple impacts. Combining well-designed field surveys with multivariate data analysis, in situ experiments using community functional responses, and laboratory analysis may be an inclusive approach to understand the effects of multiple stressors on biological communities. The chapter describes examples of integrated models that combine different environmental and biological variables to outline the mechanisms behind interactions effects. Maintaining the ecological integrity of rivers needs of an interdisciplinary work to generate complete knowledge of cause and effects, and should integrate this knowledge in the decision-making processes.

I. Muñoz (✉)
Department of Ecology, University of Barcelona, Av. Diagonal, 643, 08028 Barcelona, Spain
e-mail: imunoz@ub.edu

S. Sabater
Institute of Aquatic Ecology, University of Girona, Campus Montilivi, 17071 Girona, Spain

Catalan Institute for Water Research (ICRA), Parc Científic i Tecnològic de la Universitat de Girona, Edifici Jaume Casademont, C/Pic de Peguera, 15, 17003 Girona, Spain

C. Barata
Department of Environmental Chemistry, CSIC, C/Jordi Girona 18-26, 08034 Barcelona, Spain

H. Guasch et al. (eds.), *Emerging and Priority Pollutants in Rivers*,
Hdb Env Chem (2012) 19: 219–242, DOI 10.1007/978-3-642-25722-3_8,

Keywords Ecological integrity • Ecotoxicology • In situ bioassays • Integrated models • Multiple stressors

Contents

Abbreviations

AFDW	Ash-free dry weight
ANN	Artificial neural network
BDI	Biological diatom index
BMWP	Biological monitoring working party
BQE	Biological quality element
CA	Concentration addition model
DW	Dry weight
EPS	Extracellular polymeric substances
EQS	Environmental quality standard
HPLC	High performance liquid chromatography
HQ	Hazard quotients
IA	Independent action model
LC50	Lethal concentration 50%
MDS	Multidimensional scaling
msPAF	Multi-species potentially affected fraction of species
PAM	Pulse amplitude modulated technique
PICT	Pollution-induced community tolerance
PSI	Pollution-sensitivity index
RDA	Redundancy analysis
SOM	Self-organizing map
SPEAR	Species at risk index
SSD	Species sensitivity distribution
TU	Toxic unit
WFD	European water framework directive

1 Introduction

Well-preserved fluvial ecosystems house particular habitats and diverse biota, conferring well-structured, resilient systems that maintain native biodiversity and natural processes. These systems allow the connectivity of the different river compartments, allowing to consider them single functional entities. River systems offer several essential functions with interest for humans. Among these, water provisioning, water quality amelioration, food production, and leisure opportunities can be considered immediate services to human societies. However, in industrialized countries rivers are subject to a wide range of pressures and are heavily exploited and managed. The co-occurrence of multiple stressors from both point and diffuse sources in urban, agricultural, and industrial sectors compromise the quality of water resources, and the evaluation of ecological integrity. Major environmental stressors include, among others, contamination with toxic chemicals, habitat degradation and fragmentation, changes in hydromorphology, nutrient pollution, and the presence of invasive species. Combinations of chemical, physical, and biological stressors are reflected in water quality and ecosystem impairment ([1] and references therein). Further, impacts of climate change are also direct drivers of biodiversity loss and ecosystem change [2], adding to the direct human-related effects to ecosystems. Global climate change is expected to exacerbate the degradation of aquatic ecosystems through a decrease in precipitation [3] that will increase the competition for water resources between human uses and those required for natural ecosystem. The predicted increase in the frequency and intensity of storms and floods, or prolonged periods of drought, will create additional stressors on ecosystems that are often already subject to chronic impacts [4].

The potential effects of stressors extend through several organization levels from molecules to communities throughout individuals and species populations. There is not a linear translation of disturbances to effects at the different biological levels. As an example, the inability of a single species to tolerate a toxicant can be compensated by the adaptation of other species within the community and therefore to an unexpected lack of response of the whole community to the stressor [5]. Alternatively, low sublethal levels of a stressor may not cause observational effects on sensitive species in the short term, but indeed can produce long-term changes in the whole community (e.g., increase of nutrient inputs, presence of endocrine disrupter compounds, [6]). The existence of multiple and scale-dependent mechanisms, potentially nonlinear responses of biota to disturbance, and the difficulties of separating current from historical effects can make it difficult to establish definite relationships between disturbance and ecosystem integrity [7]. Additionally, with the presence of a high number of chemicals (some of them emerging compounds) in the natural freshwater systems, Holmstrup et al. [8] underline that synergistic interactions between the natural stressors and toxicants are common phenomena in river ecosystems in industrialized countries.

Much of the information available on the effects of stressors on organisms is obtained through tests under controlled conditions, or separately derived from observations in field studies where only a limited set of stressors are considered.

The simultaneous exposure to several stressors requires careful consideration of the possible interaction between the stressors themselves and the effect on species populations, communities, and ecosystems. This is essential to identify and prioritize the major management issues and to seek the means to identify, diagnose, and tackle multiple-stressor effects. The growing interest in understanding multiple-stressor effects and how to manage them is underlined by several recent reviews, among them a special issue on multiple stressors in freshwater ecosystems [9] and other specialized papers [8, 10].

Understanding the effects of multiple human-derived stressors on ecosystems requires considering a large array of response variables. This chapter will review which are the different approaches to evaluate ecological integrity in rivers being affected by multiple stressors. The main focus of this review is more on community than individual responses and in both functional and structural descriptors. A special attention is given to the effect of chemical substances either with respect to multiple effects or of toxicant mixtures and the combination of natural and chemical disturbances. The chapter is organized in three sections. A first one describes those approaches based on field bioassesment, where the use of biological metrics and their link with ecological and chemical status are discussed. The second section illustrates the experimental field approach as a tool to understand the multiple toxicant effects on communities. Finally, the third section drives us to the new modeling approach developed to improve knowledge of mechanisms explaining the occurrence of interactions effects.

2 Field Bioassessment for Ecological Status

The European Water Framework Directive (WFD) defines different Biological Quality Elements (BQEs) with respect to their composition and abundance. The declared BQEs are benthic algae (including macrophytes), phytoplankton, macroinvertebrates, and fish. These BQE data can be matched with data relating to chemical pressures, the latter grouped into three types: those arising from general water chemistry problems (e.g., pH, oxygen), those arising from a lack or excess of nutrients (mainly N and P), and those arising from exposure to priority substances exceeding their Environmental Quality Standard (EQS) values, which are ultimately specified in Annex X of the WFD. The information provided by the biological community can be summarized through several metrics, potentially useful as descriptors of multistress.

2.1 Biofilms

Benthic algae are a major component of biofilms in rivers. Biofilms respond to altered chemical, physical, and biological parameters. Because of the small size and rapid growth of the organisms composing the biofilm, they are potentially useful as "early warning system" for disturbances. The changes in biofilms are, however, rarely linear to the degree of change and their responses show a feedback on external parameters. Disturbances trigger direct responses in biofilms, but also indirect responses including adaptive changes, or altered exploitation by grazing fauna. This applies to their use in multistress situations where a variety of influences affect the outcome of the biofilm.

One potential parameter to be used as response to stressors is biomass. The biomass of phototrophic biofilms may be informative of the trophic state of the system (Sabater 2008) [11], but its use in other situations is less common. Chlorophyll-*a*, growth rate (differences in algal cell densities), or directly biomass [expressed as dry weight (DW) or ash-free dry weight (AFDW)] have been commonly used as descriptors and may express the long-lasting effect of toxicants on these communities. Long-term effects of herbicides and heavy metals were of significant reductions of chlorophyll-*a* concentrations, dry mass, and AFDM [12]. The effects of a chronic Cu exposure on biofilms were of lower algal biomass, higher proportion of green algae, lower proportion of brown algae, and higher EPS content per unit of biomass than the unexposed community [13]. Effects of herbicides on cell density and biomass in periphytic diatom communities have also been detected. Debenest et al. [14] determined that herbicide (isoproturon and s-metolachlor) exposure inhibited the biomass natural increase of benthic algae.

Sometimes the specific pigments of the different groups could be used as biomarkers, since the taxonomic groups of microalgae contain characteristic sets of accessory pigments. This approach requires the use of HPLC or other chromatographic techniques. Both HPLC pigment analysis and in vivo chlorophyll fluorescence (Pulse Amplitude Modulated, PAM technique) have been used for site-specific risk assessment in rivers. The PAM has been proposed as an alternative method for biomass estimation. The measurement of fluorescence (F_0) under dark conditions correlates well with chlorophyll-*a* content [15]. The use of F_0 as a surrogate of chlorophyll-*a* has the advantage to allow a continuous monitoring without the need of performing the chemical extraction of the sample. Changes in fluorescence detected by PAM methodology have been used extensively as an ecotoxicological tool for the detection of toxicants on benthic algae biomass [16]. Multiwavelength excitation has become available on recent PAM fluorometers and the deconvolution of the fluorescent signal of mixed samples potentially reveals the contribution of algal groups with different sorption spectra [17]. Pesce et al. [18] determined the adaptation of the biological community to chronic exposition to diuron by means of PAM fluorometry.

Stressors exert a selection pressure toward a community better adapted to the new situation associated with the disturbance. This general ecological principle was developed and applied to ecotoxicology. Assessment of structural and physiological parameters after exposing a community to increasing amounts of the toxicant may be used to demonstrate the effects of the exposure, but also provides an overall impression of the level of toxicant contamination between various sites or situations. This theoretical framework provides the basis of the concept of pollution-induced community tolerance (PICT). The concept has been tested and applied as a prospective as well as retrospective assessment tool for toxicity. Pollution-induced community tolerance (PICT) concept is based on the assumption that the toxicant exerts selection pressure on the biological communities when exposure reaches a critical level for a sufficient period of time and therefore sensitive species are eliminated. However, induced tolerance of microbial biofilm communities cannot be usually attributed only to one stressor, but to the existence of several of them acting as multiple stressors ([19]; see more references in Guasch et al.). Another constraint of the PICT method is that it does not allow for a quantifiable and comparable response.

Community diversity and composition provide complementary information to that of the density or biomass estimation to the response of the algal community to the stressor. Diversity has been effectively used as an indicator of change in community structure when comparing impacted and references sites [20]. However, its use as a measure of organic pollution across systems is doubtful. Some community diversity indices consider the richness (species or taxa numbers) and frequency of their relative abundance. A number of indices have been developed to describe the diversity of biological communities, for algae (mostly diatoms), invertebrate, and fish. The response may be dependent on the temporal status of a community, and it is complicated to make a general prediction about how diversity may be affected by pollution in a natural community. As diversity patterns vary along environmental gradients (e.g., the size of a stream or river, its water flow, and chemical contamination), changes in diversity can be analyzed only by comparing sampling sites along a spatial contamination gradient (e.g., in a polluted site downstream of a related reference site) or by assessing historical data series at a given site [21]. A comparative analysis in the Llobregat River (Spain) used a phylogenetic-derived index (the taxonomic distinctness; Ricciardi et al. 2009 [22]) to reflect the water-borne concentration of the herbicide diuron on the diatom community. The taxonomic distinctness index performed better than the usually used Shannon–Wiener or Simpson indices. There was—according to the diversity distinctness index—a significant decrease of diatom diversity with increasing diuron concentrations in the river. This example illustrates that chemical pollution can be correlated with integrated community indices and therewith provide a first hint on community-level effects. Other approaches consider the proportion of taxonomic groups in the community, which can be also effectively related to a prevalent toxicant. Diatom communities were more tolerant than green algae and Chrysophyceae to the herbicide atrazine in a multisite comparison across Europe [23].

Other indices do not rely on diversity estimates and are more commonly used to link the ecological status of the site of investigation to the chemical pollution. Most of these indices were derived previously to the implementation of the WFD, and others have been refined afterward. Irrelevant to this, they were designed to reflect directly the organic pollution that was occurring in the systems, and not so much the potential influence of others co-occurring with them. Though used for many classes of algae as well as for cyanobacteria, diatom communities are the most commonly used as indicators of general water pollution, as well as of the integrity of biological habitats [24, 25]. The widespread use of diatoms as indicators for water quality derives from several facts. Diatom taxa may account up to 80% of those present in streams, rivers, littoral of lakes, and wetlands. Their structural elements in their siliceous cell wall allow a reliable taxonomical identification at the specific or subspecific level. Diatom communities react to changes in nutrient content, pH, habitat, salinity, and light in terms of both composition and relative contribution of the taxa. The most tolerant taxa are favored while the sensitive taxa are depleted or eliminated, following the rules of interspecific competition. Diatom indices have been developed and widely used in European waters (e.g., [26, 27]). However, the ability of these indices to reflect the effect of toxicants is hampered by the interference between the responses of toxicants and variations in organic matter or nutrients. This difficulty is less obvious when indices are applied to extremely polluted sites. Sabater [28] detected that several diatom indices reflected the effects of strong heavy metal pollution. Dorigo et al. [29] used the BDI (biological diatom index) and PSI (pollution-sensitivity index) to relate community structure of diatoms to pollution of river sites and noticed that the BDI on average was lower at herbicide-impacted sites. In spite of these positive results, the use of diatom indices for toxicity detection calls for caution because it also reflects the trophic state of the system.

Regarding the heterotrophic component of the biofilm, various methods exist to evaluate the effects of stressors on the heterotrophic activity of biofilms. Carbon utilization assays provide a metabolic profile that can be altered by the presence of pollutants or other stressors [30]. The most widely used approach is the response of extracellular enzyme activities to stressor factors. Sensitivity and direct relationship with organic matter use and, therefore, microbial growth make extracellular enzyme activities a relevant tool to assess the interference of toxicants [20]. The exposure of a microbial community under a toxic compound may induce either an increase or a decrease of enzyme activity depending on its possible relevance (or its degradation products) as a carbon, nitrogen, or phosphorus source for microbial growth. Beta-glucosidase activity of natural freshwater biofilms collected in situ was related to heavy metal pollution [31].

2.2 Macroinvertebrates

The commonly used metrics of biological assessment for rivers and streams using macroinvertebrates are similar to those described for biofilms and are based on (1) taxonomic richness and composition (number of species, diversity indices, number of individuals, percentage of some taxa, etc.); or (2) biological information on ecological functions (e.g., habits and species traits of the aquatic fauna, such as feeding types, habitat preferences, tolerance/intolerance measures, and others) [32–34]. The first type of metric depends only on the presence/absence and the proportion of taxa lists, whereas the second needs a profound knowledge of species ecological features. However, the capability of the currently applied bioassessment methods to detect effects of specific pollutants on invertebrates is also small, as they were not designed to detect effects of specific stressors (e.g., [35]). Recent analysis of available large-scale pesticide and biomonitoring data has shown that conventional bioassessment indices, such as BMWP [36], Saprobic Index [37], percentage of selected families, and species number, do not correlate well with pesticide contamination and vary significantly with environmental variables such as pH, current velocity, and temperature [38]. Damasio et al. [39] and Barata et al. [40] also observed poor correlation between biotic indexes (PSI, BMWP) and toxicological biomarkers in macroinvertebrates, considering this metric as insufficient to determine the effects of chemical pollution. In general, correlation analyses between the indices (using algal and invertebrate data) and particular stressors [41] cannot be used as demonstrative of their applicability to detect multistress situations.

Combination of several measures and indices addressing different stressors (multimetric indices) has also been extensively tested for algae, invertebrates, macrophytes, and fish (e.g., [42–44]).

A more functional community approach than that of biological indices is the use of species traits. The biological traits (e.g., size, growth form, morphology, life cycles, and behavior) are variables describing biological characteristics of organisms. The effects of environmental stressors determine the presence of either biological trait. In other words, biological traits provide simple explanations on how organisms respond to environmental constraints. As an example, the presence of species with resistance eggs or diapause is common in systems with seasonal drought periods, as a strategy to maintain a non-active period during the low water flow. The use of multiple species traits in diverse groups of organisms to indicate human-caused stressors has increased recently (e.g., [45–47]).

Some studies approximate these community characteristics to the effects of chemical pollution. In this way, the "species at risk" index (SPEAR) is recommended with respect to the chemical stress. This index represents the ratio of physiologically sensitive species in the macroinvertebrate community, which assesses the effects from organic toxicants [48, 49]. The main advantage of the SPEAR system is that it is based on biological traits of stream invertebrates, and not on taxonomic units or abundance parameters like many other conventional bioassessment indices. Therefore, the application of SPEAR is relatively independent from confounding factors, and its application is not constrained by geographical and geomorphological influences on biological communities. Species are classified and grouped according to their vulnerability to pesticides. Ecological traits used to define these groups include sensitivity to toxicants, generation time, migration ability, and the presence of aquatic stages during the time of maximum pesticide application. Species with high generation time and low migration ability will be considered at risk due to their low capacity to avoid the chemical exposure. Currently, three SPEAR indicators exist: $SPEAR_{pesticides}$ $SPEAR_{organic}$, and $SPEAR_{metals}$ designed to detect and quantify effects of pesticides, general organic toxicants (e.g., petrochemicals, synthetic surfactants, [50]), and metals, respectively.

As mentioned previously, several harmful substances co-occur in significant concentrations in polluted waters and the biota exposition to these substances requires consideration. The toxic unit (TU) is a good approximation [51] that quantifies the toxic stress associated with a mixture of pollutants. TU is defined as the ratio of a chemical concentration and its observed LC_{50} (1). In a mixture of chemicals, TU_{sum} will be the sum of the concentrations (C) of n individual compounds expressed as a fraction of their respective LC_{50}, assuming an additive behavior of all components and representing the maximal expected effect of a mixture. The TU_{sum} allows comparing the toxic risk of sites with differing chemical exposure profiles. To derive respective toxic units (TU), the measured compound concentrations can be scaled to the toxicity of each compound toward standard test organisms (such as *Daphnia magna,* invertebrate; *Selenastrum capricornutum,* algae; and *Pimephales promelas,* fish), covering all trophic levels. The resulting three values indicate the respective risks for the aquatic biota and could be used for prioritization purposes.

$$\text{Toxic Units} = \log \sum_{i=1}^{n} \frac{C_i}{\text{LC50}_i}. \tag{1}$$

High values of TU for invertebrates in monitoring sites from different European rivers were negatively correlated with the SPEAR index [48, 52, 53]. This relationship indicated that potential effects from chemical exposure were higher than expected. Environmental risk assessment of emerging compounds (specifically pharmaceuticals) has been recently estimated with the use of the hazard quotients (HQ). As for TU, this quotient is calculated as the ratio between predicted or measured environmental concentrations and their chronic toxicity [54]. With this approximation, Ginebreda et al. [55] describe inverse linear correlation between HQ for pharmaceuticals and diversity based on invertebrate community. Consequently, toxic stress should be considered equally for the assessment of ecological status, by applying basic mixture toxicity concepts with regard to different BQE.

2.3 *Multivariate Approaches*

Multivariate statistical techniques are useful to find spatial and temporal correlation of stressors and effects on biological communities along gradients. These techniques combine environmental and biological variables and ordinate samples and species in the space, providing a good illustration from complex ecological data. The question to solve is whether the environmental variable x explains or not a significant proportion of the multivariate variation in the community data. Moreover, the use of different sets of environmental variables (chemicals, nutrients, etc.) allows calculating which part of the variance associated with biological community can be explained by each variable set [56], therefore providing indication of which is the most important stressor.

Redundancy analysis (RDA) is one example of multivariate statistical analysis that finds linear combinations of the predictor variables (environmental) which explain the greatest variation in the species data. Some examples of application are the works of Muñoz et al. [57] and Ricart et al. [22] in the Llobregat River (Fig. 1). In the former, some pharmaceuticals, as well as temperature, explained 71% of the species variance in invertebrate density. The second study focused on pesticides and biofilm, observing that more than 90% of variance of the chlorophyll-*a* response was explained by pesticides, but some bacterial extracellular enzymatic

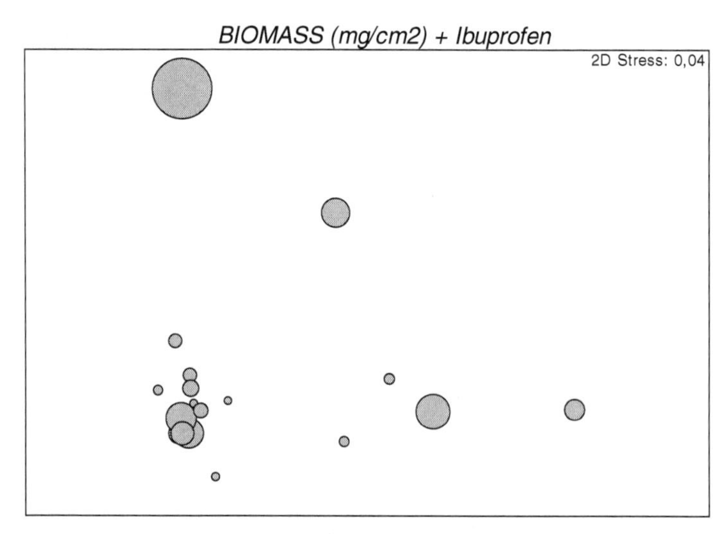

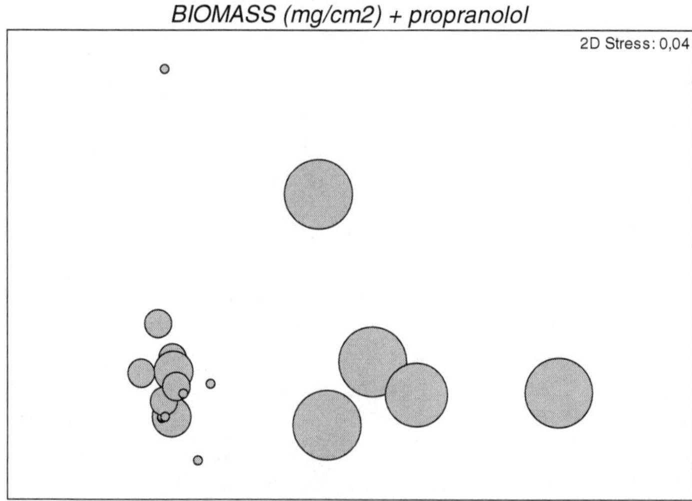

Fig. 1 Biomass distribution of macroinvertebrates in the Llobregat River (NE Spain). The MDS treatment distributes the sampling sites (21 sites) in two dimensions. Superimposed to sites the *circles* relate the presence of two pharmaceuticals, their size corresponding to the respective concentration. Some sites with higher pharmaceutical concentration are grouped and have similar distribution of species biomass

activities were determined mainly by other environmental (temperature and sulfates) variables.

The Nonmetric Multidimensional scaling (MDS, Fig. 2) was successfully applied in marine systems affected by chemical pollution [58]. Other multivariate methods, such as Partial Least Square Projections to Latent Structures regression, were also a complementary tool to identify major contaminants affecting different measured biomarkers in several bivalve species [59]. Structural and functional variables of a community or species population can be analyzed with these

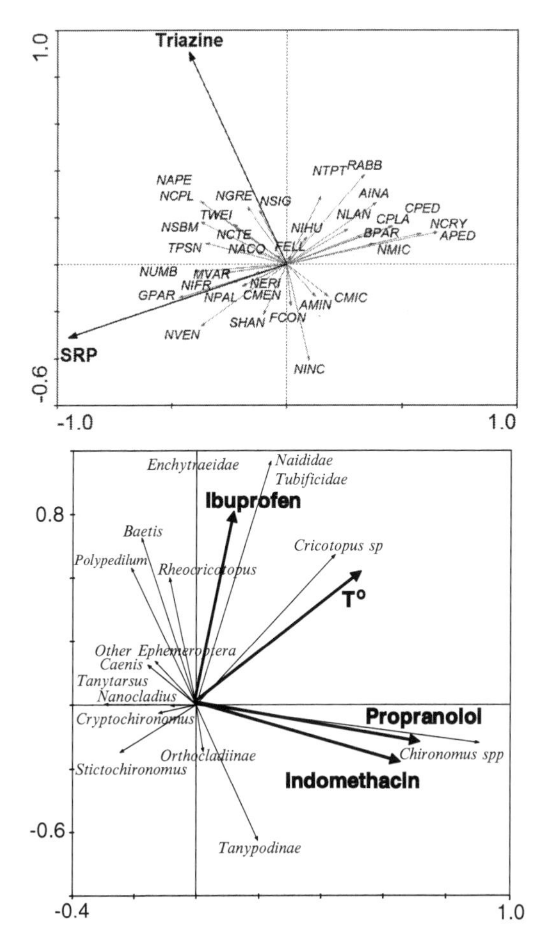

Fig. 2 Two examples of application of the RDA multivariate analysis at the Llobregat River (NE Spain). The environmental variables more significantly correlated with RDA axes are represented by *arrows*, which point in the direction of maximum change in the value of the associated variable. The figure at the *top* shows the relationships between the diatom species composition and the presence of the herbicide atrazine. The figure *below* shows the distribution of macroinvertebrate species in soft sediment and the presence of some pharmaceuticals. Modified from Ricart et al. [22] and Muñoz et al. [57]

statistical techniques. The use of these techniques is limited in the study of environmental risk of chemicals, but has been widely used in ecology. Understanding the field patterns with respect to several stressors will be helpful to define the subsequent monitoring efforts and to identify key taxa suitable for different environmental stressors. This approach might be useful to find spatial and temporal correlations between the stressors and their effects along gradients. Finally, such approach can generate hypotheses to be examined under experimental controlled conditions.

Most of the described approaches derive evidence of causality from associations between variables, but these associations are not evidences for direct causality [60]. Experiments are arguably the best way to distinguish between different causal mechanisms, but manipulative experiments are often difficult to perform. Improvements can be made to survey designs and sampling methods in order to more effectively sort out the effects of multiple stressors caused by human impacts on rivers [61]. Most described tools are not new and come from fundamental inference in ecology. Integration of ecology and ecotoxicology may be a major drive to understand, preserve, and sustainably use the aquatic systems [62].

3 Field Experiments (In Situ Bioassays)

Recently the development of new bioassays with caged, single species or communities has allowed determining pollutant effects in situ. Key advantages of in situ bioassays over whole affluent toxicity tests and biological surveys of riparian communities include a greater relevance to the natural situation, especially with respect to the contamination scenario, and higher ability to detect effects in caged individuals or communities more rapidly (hours to days) than resulting changes in community structure (months to years) [63]. For primary producers in situ bioassays have been developed basically for phytoplankton and biofilms. While the former involves immobilization of microalgae into solidified algal-growth medium that substantially modifies the bioavailability of pollutants in the field [64], the latter involves the use of biofilms grown on top of transplantable surfaces, and hence in direct contact with local water [65]. Furthermore, contrary to most in situ bioassays, biofilms as a complex community of bacteria, fungi, and algae offer the possibility to assess effects at individual and community levels [10]. In situ assays using macrophytes are surprisingly scanty, despite that several species are widely used in aquatic toxicology. In situ assays have been developed for planktonic and benthic invertebrate species, in both hard and soft sediment environments. These bioassays basically consist in exposing invertebrate species in specially designed cages that allow water or/and suspended solids and/or sediment to pass throughout, but not the organisms. For soft sediments it is possible to evaluate the contribution of surface water, pore water, and sediment itself using different chambers and species and/or by deploying the same chamber on top, in between, or inside the sediment. An active research in this field has been conducted by Allen

Burton from the Institute of Environmental Quality, Dayton, USA. The most reliable measured endpoints apart from mortality varied across species. These endpoints include leaf decomposition in shredders like *Gammarus* [63], body mass growth for *Chironomus* [66], and post-exposure microalgae clearance rates for *Daphnia* [67, 68]. Post-exposure responses such as prey consumption rates have also been used in several omnivorous macroinvertebrate species, but their ecological relevance is questionable since artificial (*Artemia* nauplii) rather than natural (detritus) diets have been used [69]. When selecting an array of in situ assays it is important to use ecological relevant species either sensitive to the studied pollution source or inhabiting different environmental compartments. Cladoceran species can be selected as surrogate species for water column algae filter feeders, chironomids despite being quite insensitive to pollutants are good candidates for soft sediment filter feeders on detritus, and amphipods like *Gammarus* are quite sensitive and good surrogates for macroinvertebrate grazing on gross (leaf) material. Other in situ bioassays have included snails, caddisfly, and stonefly larvae [70–75] or have been developed in tropical ecoregions using surrogated European species ([76–78]; Domingues et al. 2008; [79]). Several of these bioassays have permitted detecting lethal and sublethal responses that are biologically linked with key ecological processes such as primary production, detritus processing and algal grazing rates, reproduction, responses in the field of the same species or of the whole community, and of specific toxicological mechanisms (Schulz and Liess 2001; [63, 67, 80–82]). Nevertheless, it is important to consider that most of these studies have been focused to detect major point contaminant impacts such as the effects of pesticides after rainfall runoffs or application events, of metals from mine drainage, the presence of endocrine disruption compounds, and of industrial effluent discharges. Therefore, the usefulness of in situ bioassay responses in detecting sublethal effects on river biota exposed chronically to multiple environmental factors or/and to low levels of contaminants still needs further study [83]. This is especially important for rivers from heavily urbanized and industrialized areas within developed countries, whose riparian aquatic communities are affected by the physical degradation of their habitats and the physicochemical alteration of river water (i.e., an excess of salts, nutrients, and contaminants coming from sewage treatment plant effluent discharges). In those situations, it is thus vital to develop indicators of adverse change in ecological systems able to discriminate and differentiate causal agents. For example, by combining biomarker and toxicological in situ responses with multimetric indices based on community species assemblages, it should be possible to identify causal agents impairing river biota exposed chronically to multiple environmental factors. Recently, combining biomarker, post-exposure feeding *Daphnia magna* in situ bioassay, and community-based indexes, Damasio et al. [83] and Puertolas et al. [84] were able to distinguish stressors affecting species assemblages from those having specific detrimental effects on aquatic invertebrates. Riparian forest and habitat degradation and salinity were the factors affecting most macroinvertebrate riparian assemblages and pesticides had the greatest effects on transplanted organisms. Nevertheless, these results have to be considered with caution. According to Maltby and Hills [85] and Puertolas et al. [84], effects observed in caged organisms not necessarily translate to higher ecological levels due to the buffering capacity of benthic communities. Affected species can be replaced by more tolerant ones or by immigrants from non-exposed populations of the same species.

In summary, most existing published information on in situ bioassays may inform us about functional effects occurring on key ecological processes such as primary production within biofilms, algal grazing, or detritus processing in transplanted invertebrate species, but require to be implemented by additional monitoring tools to assess effects at ecological level. In situ biofilm responses constitute a useful field bioassay since it already considers single- and community-level effects. Nevertheless, in many rivers detritus is the main food source and hence its processing across the food web using bioassays needs to be evaluated. There are, however, several methodological and conceptual procedures that need to be addressed in the future. The first one relies on the reliability of existing in situ bioassays in terms of statistical power, consistence, and replicability in assessing degrees of impairment. Many endpoints such as the *Gammarus* leaf processing or *Chironumus* growth bioassay required long exposure periods (one or more weeks) and hence are greatly affected by water temperature, water flow, and other parameters. Thus, a large number or replicated chambers and modeling approaches are needed to account for undesired environmental factors and to compare effects across control and polluted sites. On the contrary, endpoints obtained under standardized conditions such as post-exposure responses are less variable, require less replicates, and do not need regression approaches for comparing purposes. If possible the use of groups rather than single individuals reduces variability. The use of several control sites differing in environmental factors and of polluted ones with an increasing or decreasing degree of pollution is also recommendable. For species that cannot be cultured in the lab or/and occur naturally in the studied system it is also recommendable to perform reciprocal deployment with populations collected from both control and polluted sites to account for physiological or/and genetic adaptation. There is increasing evidence that communities adapted chronically to pollution respond differently from non-adapted ones being usually more tolerant to pollution [5]. Pollution-induced community tolerance (PICT) can indeed be one of the reasons explaining why in situ responses of transplanted species not always translate into community-level effects.

4 Modelization Approach to Evaluate Joint Effects of Stressors

The effects of a stressor on species are commonly studied using laboratory experiments where conditions are optimal for the organisms and only few factors are changed. In natural systems, organisms are exposed to suboptimal conditions, due in most of the cases for exposure to severe environmental stress. As has been mentioned in this and other chapters of this book, it is difficult to analyze jointly these effects and then an underestimation or sometimes overestimation of the risk can be common with important environmental and economical consequences. An appropriate complement for the study of cumulative stress may be modelization.

Some mechanistic models have been described to make predictions about the interaction between eutrophication processes and contaminants using individual or a group of species that can interact. This mechanistic understanding is important to develop conceptual models in order to predict joint effects of stressors. Kooi et al. [86] study the sublethal effect of toxicants on the functioning (biomass, production, and nutrient recycling) and structure (species composition and complexity) of a simple aquatic ecosystem in a well-mixed environment (chemostat reactor). The dynamic behavior of this ecosystem is described by a set of ordinary differential equations. The system is stressed by a toxicant dissolved in the in-flowing water. In this paper, bifurcation theory analyzes the long-term dynamics of the models. The authors highlighted that the dynamic behavior of the stressed ecosystem can be much more complicated than that of the unstressed system. The effects of pesticides and environmental conditions (oxygen depletion and decreased salt concentration) were modeled for fishes by Sekine et al. [87] based on acute experimental data. Coors and de Meester [88]) find that the presence of a pesticide enhances parasite virulence in invertebrate host model.

To calculate the toxicity of a mixture on the basis of known toxicities of the individual components of the mixture, we can use the concentration addition (CA) and the Independent action (IA) models. The former presumes that each toxicant has similar mode of action. The independent action (IA) model, developed to detect effect interactions between more than one chemical exposure with dissimilar modes of action, has been adapted to study interactions between natural and chemical stressors in soils [89, 90]. A mathematical model is fitted to dose–response data generating a response surface. Although such data only have a phenomenological similarity to mixture toxicity studies, deviations from the model may be interpreted as synergistic or antagonistic interactions, and can be used for describing larger or smaller effects than those expected from single-stressor effects.

Stressors cause direct and indirect effects on organisms. Moreover, in natural systems species interact, through competition or predation for example, and the presence of a stress might affect one species and indirectly these interactions between species (Fig. 3). Relyea and Hoverman [62] propose that the framework of direct (effects on density and traits) and indirect (top-down, bottom-up) effects developed in basic ecology for food webs' interactions is effective for considering how pesticides cause indirect effects on ecological communities. These authors describe some examples highlighting the usefulness of mechanistic studies in determining the underlying causes of community and ecosystem changes. Only when we identify underlying mechanisms can we begin to develop a general understanding of how thousands of different pollutants impact on aquatic ecosystems.

The self-organizing map (SOM) is a type of artificial neural network (ANN) and a tool in exploratory phase of data mining ([91], and references therein). It produces a low-dimensional (typically two-dimensional), discretized representation of the input data that can be effectively utilized to visualize and explore properties of

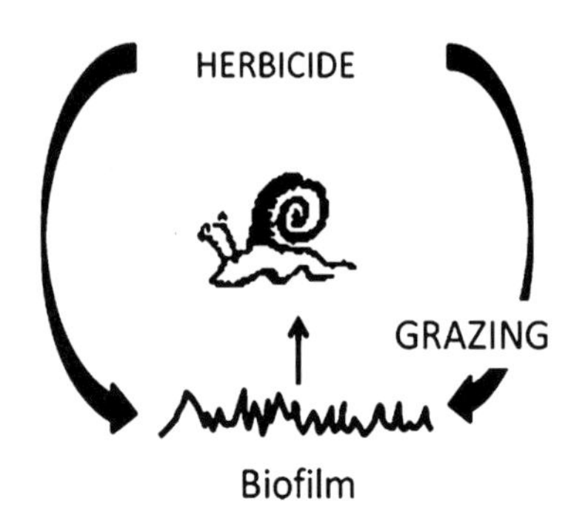

	Control treatment	Atrazine treatment	Grazing treatment	Atrazine + Grazing treatment
μg Chl a cm^{-2}	25	24	6*	3*
μg C cm^{-1} h^{-1}	15	16	6*	2*
Total diatom abundance ($x10^3$ cells cm^{-2})	7451	5218	3289*	1732*
Abundance of filamentous diatom growthform		Significant decrease		
Abundance of diatom prostrate growth form			Significant decrease	Significant decrease

Fig. 3 An herbicide has direct effects on biofilm. The presence of an herbivorous causes direct effects on biofilm, reducing their density, and indirect effects favoring the effects of the herbicide, and decreasing the biofilm adaptation to pesticide exposure. Data on table show changes observed in an experimental study with atrazine. Interaction of pesticide and snails decreases significantly chlorophyll-*a* content, photosynthetic C incorporation, total diatom abundance, and changes algal community structure. Data from Muñoz et al. [102]

these data. The goal of such approaches is to identify, within existing data, species assemblages that can be used retrospectively to identify probable causal agents (i.e., specific chemical pressures) of poor ecological quality. The use of this technique has increased in the last years and we can learn from a number of application examples in different European rivers. These extend from diatoms (e.g., [92]), macroinvertebrates [93, 94], and fish [95] communities. All these works highlight the ecologically relevant results of these models, identifying the relative importance of environmental and toxic stress factors on the patterns observed in the community.

The "multi-species potentially affected fraction of species" (msPAF) directly quantifies the expected loss of species taking into account mixture toxicity [96, 97]. This is a predictive model that attributes impairment in biological condition to multiple causes. The toxic pressure at a site is a modeled value, obtained by combining local monitoring data on chemical concentrations with a train of models subsequently considering bioavailability, Species Sensitivity Distributions (SSD), and mixture toxicity. The resulting measure covers all trophic levels (algae, invertebrates, and fish) and considers both organic and metal compounds and is represented as a site-specific pie diagram of impact magnitudes and probable causes [98]. De Zwart et al. [99]) illustrate the results of a multiple stress diagnostic analysis for the River Scheldt basin using msPAF model compared to WFD Ecological Quality Standards (EQS).

We emphasize that the combination of some of the described approaches can help to obtain an integrative view of the chemical and biological state of a river. The spatial patterns obtained help to detect risk levels and to prioritize management action on reaches or on groups of compounds causing the greatest risk. The studies of Tuikka et al. [100] in three European rivers and Faggiano et al. [101] in the Adour-Garonne basin are recent examples of this procedure.

5 Concluding Remarks

Rivers are increasingly impaired by effects caused by multiple stressors of physical, chemical, and biological nature. To achieve a good ecological quality in fluvial systems requires performing integrated assessments of all available information, and selecting the most effective management options to improve ecosystem quality. Selecting the criteria of future monitoring designs under multiple stress scenarios is a priority.

Some of them have been enumerated along this chapter. These include understanding better the interaction mechanisms between stressors, improvement in predicting the effects of their combination, quantifying the effects on biological elements, and determining the respective impacts of toxic chemical relative to other stressors. The greater and more distinct the variables used, the greater will be our ability to identify the effects of major stressors impairing communities. Combining field surveys and modeling approaches is a desirable way to generate hypothesis on key stressors, sensitive species, and potential community effects. This approach, complemented with in situ bioassays and laboratory (micro or mesocosms) experiments, may establish a mechanistic link between stressors and ecological effects.

The application of different modelization tools that include basic concepts on toxicology (as the concentration addition and independent action) for stressors action, combined with the use of multivariate methods, constitutes a powerful tool to be used in river evaluations. Special mention is for the multi-species potentially affected fraction (msPAF) and the self-organizing maps (SOM), examples of integrated models that combine different variables and that can be applied to the basin scale.

There is also growing recognition on the need to incorporate ecosystem process measurements in assessment programs, since several goals of river management directly relate to the maintenance of natural ecological processes. Such measures are often sensitive to causal factors that are known to affect river health and it is possible to develop simple but powerful predictive models of cause and effect [9]. An example is the relatively recent use of multiple biological traits of organisms as indicators of a biological impairment. Species traits are the result of the community adaptation to the environmental features acting at different temporal and spatial scales.

To be effective, the maintenance of the ecological integrity of river ecosystems needs an interdisciplinary work to generate complete knowledge of cause and effects and, in a further step, integrate this knowledge in the decision-making processes.

Acknowledgements This research was funded by the EC Sixth Framework Program (MODELKEY, contract No. 511237-GOCE) and the Spanish Ministry of Science (Consolider-Ingenio 2010 program, CSD2009-00065).

References

1. Ormerod SJ, Dobson M, Hildrew AG, Townsend CR (2010) Multiple stressors in freshwater ecosystems. Freshwater Biol 55:1–4
2. Millenium Ecosystem Assessment (2005) Ecosystems and human wellbeing: wetlands and water. World Resources Institute, Washington, DC
3. IPCC (2007) Climate change 2007: impacts, adaptation and vulnerability. Working Group II. Fourth Assessment Report, UNEP
4. Sabater S, Tockner K (2010) Effects of hydrologic alterations on the ecological quality of river ecosystems. In: Sabater S, Barcelo D (eds) Water scarcity in the mediterranean: perspectives under global change. Springer, Berlin
5. Blanck H (2002) A critical review of procedures and approaches used for assessing pollution-induced community tolerance (PICT) in biotic communities. Hum Ecol Risk Assess 8:1003–1034
6. Sumpter JP, Johnson AC, Williams RJ (2006) Modeling effects of mixtures of endocrine disrupting chemicals at the River Catchment Scale. Environ Sci Technol 40:5478–5489
7. Allan JD, Castillo MM (2007) Stream ecology: structure and function in running waters. Springer, Dordrecht, The Netherlands
8. Holmstrup M, Bindesbøl AM, Oostingh GJ et al (2010) Interactions between effects of environmental chemicals and natural stressors: a review. Sci Total Environ 408:3746–3762
9. Bunn SE, Abal EG, Smith MJ et al (2010) Integration of science and monitoring of river ecosystem health to guide investments in catchment protection and rehabilitation. Freshwater Biol 55:223–240
10. Sabater S, Guasch H, Ricart M et al (2007) Monitoring the effect of chemicals on biological communities. The biofilm as an interface. Anal Bioanal Chem 387:1425–1434
11. Sabater S (2008) Alterations of the Global Water Cycle and their effects on river structure, function and services. Freshwater Reviews 1, 75-88
12. Atazadeh I, Kelly M, Sharifi M et al (2009) The effects of copper and zinc on biomass and taxonomic composition of algal periphyton communities from the River Gharasou, Western Iran. Oceanol Hydrobiol Stud 38:3–14
13. Serra A, Guasch H (2009) Effects of chronic copper exposure on fluvial systems: linking structural and physiological changes of fluvial biofilms with the in-stream copper retention. Sci Total Environ 407:5274–5282

14. Debenest T, Pinelli E, Coste M et al (2009) Sensitivity of freshwater periphytic diatoms to agricultural herbicides. Aquat Toxicol 93:11–17
15. Guasch H, Admiraal W, Sabater S (2003) Contrasting effects of organic and inorganic toxicants on freshwater periphyton. Aquat Toxicol 64:165–175
16. Brack W, Frank H (1998) Chlorophyll a fluorescence: a tool for the investigation of toxic effects in the photosynthetic apparatus. Ecotox Environ Saf 40:34–41
17. Schreiber U, Gademann R, Bird P et al (2002) Apparent light requirement for activation of photosynthesis upon rehydration of desiccated beachrock microbial mats. J Phycol 38:125–134
18. Pesce S, Margoum C, Montuelle B et al (2010) In situ relationships between spatio-temporal variations in diuron concentrations and phototrophic biofilm tolerance in a contaminated river. Water Res 44:1941–1949
19. Tlili A, Berard A, Lourier JL et al (2010) PO_4^{3-} dependence of the tolerance of autotrophic and heterotrophic biofilm communities to copper and diuron. Aquat Toxicol 98:165–177
20. Patrick R (1977) Ecology of freshwater diatoms–diatom communities. In: Werner D (ed) The biology of diatoms. Blackwell, Oxford
21. Ricart M, Guasch H, Barceló D et al (2010) Primary and complex stressors in polluted mediterranean rivers: pesticide effects on biological communities. J Hydrol 383:52–61
22. Ricciardi F, Bonnineau C, Faggiano L, Geiszinger A, Guasch H, López-Doval J, Muñoz I, Proia L, Ricart M, Romani AM, Sabater S (2009) Is chemical contamination linked to the diversity of biological communities in rivers? Trac-Trends in Analytical Chemistry 28:592–602
23. Guasch H, Ivorra N, Lehmann V et al (1998) Community composition and sensitivity of periphyton to atrazine in flowing waters: the role of environmental factors. J Appl Phycol 10:203–213
24. Passy SI, Pan YD, Lowe RL et al (1999) Ecology of the major periphytic diatom communities from the Mesta River, Bulgaria. Int Rev Hydrobiol 84:129–174
25. Rott E, Duthie HC, Pipp E (1998) Monitoring organic pollution and eutrophication in the Grand River, Ontario, by means of diatoms. Can J Fish Aquat Sci 55:1443–1453
26. Descy JP, Coste M (1991) A test of methods for assessing water quality based on diatoms. Verh Internat Verein Limnol 24:2112–2116
27. Kelly MG, Whitton BA (1995) The trophic diatom index: a new index for monitoring eutrophication in rivers. J Appl Phycol 7:433–444
28. Sabater S (2000) Diatom communities as indicators of environmental stress in the Guadiamar River, S-W Spain, following a major mine tailings spill. J Appl Phycol 12:113–124
29. Dorigo U, Bourrain X, Berard A et al (2004) Seasonal changes in the sensitivity of river microalgae to atrazine and isoproturon along a contamination gradient. Sci Total Environ 318:101–114
30. Lawrence JR, Zhu B, Swerhone GDW et al (2009) Comparative microscale analysis of the effects of triclosan and triclocarban on the structure and function of river biofilm communities. Sci Total Environ 407:3307–3316
31. Fechner LC, Gourlay-France C, Uher E et al (2010) Adapting an enzymatic toxicity test to allow comparative evaluation of natural freshwater biofilms' tolerance to metals. Ecotoxicology 19:1302–1311
32. Barbour MT, Gerritsen J, Snyder BD, Stribling B (1999) Rapid bioassessment protocols for use in streams and wadeable rivers: periphyton, macroinvertebrates and fish, EPA 841-B-99-002, 2nd edn. US Environmental Protection Agency, Washington, DC
33. Hering D, Moog O, Sandin L, Verdonschot PFM (2004) Overview and application of the AQEM assessment system. Hydrobiologia 516:1–20
34. Statzner B, Resh VH, Doledec S (1994) Ecology of the Upper Rhone River: a test of habitat templet theories. Freshwater Biol 31:253–554
35. Bohmer J, Rawer-Jost C, Zenker A (2004) Multimetric assessment of data provided by water managers from Germany: assessment of several different types of stressors with macrozoobenthos communities. Hydrobiologia 516:215–228

36. Armitage PD, Moss D, Wright JF, Furse MT (1983) The performance of a new biological water quality score system based on macroinvertebrates over a wide range of unpolluted running water sites. Water Res 17:333–347
37. Friedrich G (1990) A revision of the saprobic system. J Water Wastewater Res 23:141–152
38. Liess M, Schafer RB, Schriever CA (2008) The footprint of pesticide stress in communities-species traits reveal community effects of toxicants. Sci Total Environ 406:484–490
39. Damasio JB, Barata C, Munne A et al (2007) Comparing the response of biochemical indicators (biomarkers) and biological indices to diagnose the ecological impact of an oil spillage in a Mediterranean river (NE Catalunya, Spain). Chemosphere 66:1206–1216
40. Barata C, Lekumberri I, Vila-Escalé M et al (2005) Trace metal concentration, antioxidant enzyme activities and susceptibility to oxidative stress in the tricoptera larvae *Hydropsyche exocellata* from the Llobregat river basin (NE Spain). Aquat Toxicol 74:3–19
41. Blanco S, Becares E (2010) Are biotic indices sensitive to river toxicants? A comparison of metrics based on diatoms and macroinvertebrates. Chemosphere 79:18–25
42. Furse MT, Hering D, Brabec K et al (2006) The ecological status of European rivers: evaluation and intercalibration of assessment methods. Hydrobiologia 566:299–309
43. Hill BH, Herlihy AT, Kaufmann PR et al (2000) Use of periphyton assemblage data as an index of biotic integrity. J N Am Benthol Soc 19:50–67
44. Karr JR (1991) Biological integrity – a long-neglected aspect of water-resource management. Ecol Appl 1:66–84
45. Archaimbault V, Usseglio-Polatera P, Garric J et al (2010) Assessing pollution of toxic sediment in streams using bio-ecological traits of benthic macroinvertebrates. Freshwater Biol 55:1430–1446
46. Fore LS, Grafe C (2002) Using diatoms to assess the biological condition of large rivers in Idaho (USA). Freshwater Biol 47:2015–2037
47. Statzner B, Bêche LA (2010) Can biological invertebrate traits resolve effects of multiple stressors on running water ecosystems? Freshwater Biol 55:80–119
48. Liess M, Von der Ohe PC (2005) Analizing effects of pesticides on invertebrate communities in streams. Environ Toxicol Chem 24:954–965
49. Von der Ohe PC, Prüß A, Schäfer RB et al (2007) Water quality indices across Europe – a comparison of the good ecological status of five river basins. J Environ Monitor 9:970–978
50. Beketov MA, Liess M (2008) An indicator for effects of organic toxicants on lotic invertebrate communities: independence of confounding environmental factors over an extensive river continuum. Environ Pollut 156:980–987
51. Sprague JB (1970) Measurement of pollutant toxicity to fish, II-utilizing and applying bioassay results. Water Res 4:3–32
52. Schäfer RB, Caquet T, Siimes K et al (2007) Effects of pesticides on community structure and ecosystem functions in agricultural streams of three biogeographical regions in Europe. Sci Total Environ 382:272–285
53. Von der Ohe PC, de Deckere E, Prüß A et al (2009) Towards an integrated assessment of the ecological and chemical status of European River Basins. Integr Environ Assess Manag 5:50–61
54. Cooper ER, Siewicki TC, Phillips K (2008) Preliminary risk assessment database and risk ranking of pharmaceuticals in the environment. Sci Total Environ 398:26–33
55. Ginebreda A, Muñoz I, López de Alda M et al (2010) Environmental risk assessment of pharmaceuticals in rivers: relationships between hazard indexes and aquatic macroinvertebrate diversity indexes in the Llobregat River (NE Spain). Environ Int 36:153–162
56. Borcard D, Legendre P, Drapeau P (1992) Partialling out the spatial component of ecological variation. Ecology 73:1045–1055
57. Muñoz I, López-Doval JC, Ricart M et al (2009) Bridging levels of pharmaceuticals in river water with biological community structure in the Llobregat river basin (NE Spain). Environ Toxicol Chem 2:2706–2714

58. Clarke KR (1999) Nonmetric multivariate analysis in community level ecotoxicology. Environ Toxicol Chem 18:118–127
59. Damasio J, Navarro-Ortega A, Tauler R et al (2010) Identifying major pesticides affecting bivalve species exposed to agricultural pollution using multi-biomarker and multivariate methods. Ecotoxicology 19:1084–1094
60. Sokal RR, Rohlf FJ (1995) Biometry: the principles and practice of statistics in biological research, 3rd edn. Freeman, New York
61. Downes BJ (2010) Back to the future: little-used tools and principles of scientific inference can help disentangle effects of multiple stressors on freshwater ecosystems. Freshwater Biol 55:60–79
62. Relyea R, Hoverman J (2006) Assessing the ecology in ecotoxicology: a review and synthesis in freshwater systems. Ecol Lett 9:1157–1171
63. Maltby L, Clayton SA, Wood RM et al (2002) Evaluation of the *Gammarus pulex* in situ feeding assay as a biomonitor of water quality: Robustness, responsiveness, and relevance. Environ Toxicol Chem 21:361–368
64. Moreira-Santos M, Soares AMVM, Ribeiro S et al (2004) A phytoplankton growth assay for routine in situ environmental assessments. Environ Toxicol Chem 23:1549–1560
65. Courtney LA, Clements WH (2002) Assessing the influence of water and substratum quality on benthic macroinvertebrate communities in a metal-polluted stream: an experimental approach. Freshwater Biol 47:1766–1778
66. Chappie DJ, Burton GA (1997) Optimization of in situ bioassays with Hyalella azteca and Chironomus tentans. Environ Toxicol Chem 16:559–564
67. Mc William RA, Baird DJ (2002) Postexposure feeding depression: a new toxicity endpoint for use in laboratory studies with *Daphnia magna*. Environ Toxicol Chem 21:1198–1205
68. Mc William RA, Baird DJ (2002) Application of postexposure feeding depression bioassays with *Daphnia magna* for assessment of toxic effluents in rivers. Environ Toxicol Chem 21:1462–1468
69. Soares S, Cativa I, Moreira-Santos M et al (2005) A short-term sub-lethal in situ sediment assay with *Chironomus riparius* based on postexposure feeding. Arch Environ Contam Toxicol 49:163–172
70. Crichton CA, Conrad AU, Baird DJ (2004) Assessing stream grazer response to stress: a post-exposure feeding bioassay using the freshwater snail *Lymnaea peregra* (Muller). Bull Environ Contam Toxicol 72:564–570
71. Gust M, Buronfosse T, Geffard O et al (2010) In situ biomonitoring of freshwater quality using the New Zealand mudsnail *Potamopyrgus antipodarum* (Gray) exposed to waste water treatment plant (WWTP) effluent discharges. Water Res 44:4517–4528
72. Pestana JLT, Alexander AC, Culp JM et al (2009) Structural and functional responses of benthic invertebrates to imidacloprid in outdoor stream mesocosms. Environ Pollut 157:2328–2334
73. Sataporavanit K, Baird DJ, Little DC et al (2009) Laboratory toxicity test and post-exposure feeding inhibition using the giant freshwater prawn *Macrobrachium rosenbergii*. Chemosphere 74:1209–1215
74. Schulz R, Liess M (2001) Toxicity of aqueous-phase and suspended particle-associated fenvalerate: chronic effects after pulse-dosed exposure of *Limnephilus lunatus* (Trichoptera). Environ Toxicol Chem 20:185–190
75. Schmitt C, Streck G, Lamoree M et al (2011) Effect directed analysis of riverine sediments-The usefulness of *Potamopyrgus antipodarum* for in vivo effect confirmation of endocrine disruption. Aquat Toxicol 101:237–243
76. Lopes I, Moreira-Santos M, da Silva DM et al (2007) In situ assays with tropical cladocerans to evaluate edge-of-field pesticide runoff toxicity. Chemosphere 67:2250–2256
77. Lopes I, Moreira-Santos M, Rendon-von Osten J et al (2011) Suitability of five cladoceran species from Mexico for in situ experimentation. Ecotox Environ Safe 74:111–116

78. Moreira SM, Moreira-Santos M, Rendon-von Osten J et al (2010) Ecotoxicological tools for the tropics: sublethal assays with fish to evaluate edge-of-field pesticide runoff toxicity. Ecotox Environ Saf 73:893–899
79. Domingues I, Satapornvanit K, Yakupitiyage A et al (2008) In situ assay with the midge Kiefferulus calligaster for contamination evaluation in aquatic agro-systems in central Thailand. Chemosphere 71:1877–1887
80. Barata C, Damasio J, Lopez MA et al (2007) Combined use of biomarkers and in situ bioassays in *Daphnia magna* to monitor environmental hazards of pesticides in the field. Environ Toxicol Chem 26:370–379
81. Maltby L, Clayton SA, Yu HX et al (2000) Using single-species toxicity tests, community-level responses and toxicity identification evaluations to investigate effluent impacts. Environ Toxicol Chem 19:151–157
82. Moreira-Santos M, Fonseca AL, Moreira SM et al (2005) Short-term sublethal (sediment and aquatic roots of floating macrophytes) assays with a tropical chironomid based on postexposure feeding and biomarkers. Environ Toxicol Chem 24:2234–2242
83. Damasio J, Tauler R, Teixido E et al (2008) Combined use of *Daphnia magna* in situ bioassays, biomarkers and biological indices to diagnose and identify environmental pressures on invertebrate communities in two Mediterranean urbanized and industrialized rivers (NE Spain). Aquat Toxicol 87:310–320
84. Puertolas L, Damasio J, Barata C et al (2011) Evaluation of side-effects of glyphosate mediated control of giant reed (*Arundo donax*) on the structure and function of a nearby Mediterranean river ecosystem. Environ Res 110:556–564
85. Maltby L, Hills L (2008) Spray drift of pesticides and stream macroinvertebrates: experimental evidence of impacts and effectiveness of mitigation measures. Environ Pollut 156:1112–1120
86. Kooi BW, Bontje D, Voorn GAK, Kooijman SALM (2008) Sublethal contaminants effects in a simple aquatic food chain. Ecol Model 112:304–318
87. Sekine M, Nakanishi H, Ukita M (1996) Study on fish mortality caused by the combined effects of pesticides and changes in environmental conditions. Ecol Model 86:259–264
88. Coors A, de Meester L (2008) Synergistic, antagonistic and additive effects of multiple stressors: predation threat, parasitism and pesticide exposure in *Daphnia magna*. J Appl Ecol 45:1820–1828
89. Hojer R, Bayley M, Damgaard CF et al (2001) Stress synergy between drought and a common environmental contaminant: studies with the collembolan *Folsomia candida*. Global Change Biol 7:485–494
90. Long SM, Reichenberg F, Lister LJ et al (2009) Combined chemical (fluoranthene) and drought effects on *Lumbricus rubellus* demonstrate the applicability of the independent action model for multiple stressor assessment. Environ Toxicol Chem 28:629–636
91. Lek S, Guegan JF (1999) Artificial neural networks as a tool in ecological modeling, an introduction. Ecol Model 120:65–73
92. Gevrey M, Rimet F, Park YS et al (2004) Water quality assessment using diatom assemblages and advanced modeling techniques. Freshwater Biol 49:208–220
93. Comte L, Lek S, de Deckere E et al (2010) Assessment of stream biological responses under multiple-stress conditions. Environ Sci Pollut R 17:1469–1478
94. Park YS, Chon TS, Kwak IS, Lek S (2004) Hierarchical community classification and assessment of aquatic ecosystems using artificial neural networks. Sci Total Environ 327:105–122
95. Gevrey M, Sans-Fiche F, Grenouillet G et al (2009) Modeling the impact of landscape types on the distribution of stream fish species. Can J Fish Aquat Sci 66:484–495
96. De Zwart D, Posthuma L (2005) Complex mixture toxicity for single and multiple species: proposed methodologies. Environ Toxicol Chem 24:2665–2676
97. Posthuma L, Traas TP, Suter GW (2002) Species sensitivity distributions in ecotoxicology. Lewis, Boca Raton, FL

98. De Zwart D, Dyer SD, Posthuma L, Hawkins CP (2006) Use of predictive models to attribute potential effects of mixture toxicity and habitat alteration on the biological condition of fish assemblages. Ecol Appl 16:1295–1310
99. De Zwart D, Posthuma L et al (2009) Diagnosis of ecosystem impairment in a multiple stress context – how to formulate effective river basin management plans. Integr Environ Assess Manag 5:38–49
100. Tuikka AI, Schmitt C, Hoss S et al (2011) Toxicity assessment of sediments from three European river basins using a sediment contact test battery. Ecotox Environ Saf 74:123–131
101. Faggiano L, de Zwart D, Garcia-Berthou E et al (2010) Patterning ecological risk of pesticide contamination at the river basin scale. Sci Total Environ 408:2319–2326
102. Muñoz I, Real M, Guasch H et al (2001) Effects of atrazine on periphyton under grazing pressure. Aquat Toxicol 55:239–249

Comparing Chemical and Ecological Status in Catalan Rivers: Analysis of River Quality Status Following the Water Framework Directive

Antoni Munné, Lluís Tirapu, Carolina Solà, Lourdes Olivella, Manel Vilanova, Antoni Ginebreda, and Narcís Prat

Abstract In Europe, diverse biological indices and metrics have been developed for ecological status assessment in rivers using macroinvertebrate, diatoms, macrophytes, and fish communities according to the Water Framework Directive (2000/60/EC). Additionally, priority and hazardous substances (pesticides, PAHs, heavy metals, chlorinated and non-chlorinated solvents, endocrine disruptors, etc.) must be analyzed using their environmental quality standards (EQS) according to the 2008/105/EC Directive. Chemical and biological elements have to be properly combined to set the final water quality status. We compare ecological and chemical status outputs in a Mediterranean watershed (the Catalan river basins, NE Spain), in order to provide useful information about the strengths and weaknesses of quality status classification in rivers.

A total of 367 sites with different sampling frequencies along the monitoring program period (for six following years) were used to determine the chemical and the ecological status in Catalan rivers. The results of the monitoring program carried out in Catalan rivers (2007–2009) show a higher percentage of nonfulfillment quality objectives due to ecological status rather than chemical status. A total of 144 river water bodies (39%) do not achieve the good biological quality according to the 2000/60/EC Directive, whereas 68 river water bodies (19%) do not achieve the EQS for priority and hazardous substances provided by the 105/2008/EC Directive (chemical status). Both chlorinated pesticides (mainly endosulfan, trifluralin, and

A. Munné (✉) • L. Tirapu • C. Solà • L. Olivella • M. Vilanova
Catalan Water Agency, c/Provença, 204-208, E-08036 Barcelona, Spain
e-mail: anmunne@gencat.cat

A. Ginebreda
Department of Environmental Chemistry, IDÆA-CSIC, c/Jordi Girona, 18-26, E-08034 Barcelona, Spain

N. Prat
Department of Ecology, University of Barcelona, Av. Diagonal, 654, E-08028 Barcelona, Spain

H. Guasch et al. (eds.), *Emerging and Priority Pollutants in Rivers*,
Hdb Env Chem (2012) 19: 243–266, DOI 10.1007/978-3-642-25722-3_9,

hexachlorocyclohexanes) and endocrine disruptors (nonylphenols and octilfenols) are the main substances responsible for quality standard failures in Catalan rivers.

Some chemical values must be carefully considered, since they are found near the EQS and their threshold detection values. EQS values for some priority substances (mainly heavy metals and organic compounds) are extremely low, up to threshold detection levels, which make chemical results uncertain. Additionally, bad chemical status does not necessarily imply biological community damages, at least in short time. A total of 21 river water bodies (6%) showed priority substance concentrations over the EQS thresholds, whereas biological elements showed good quality. Biological indices based on community structure and composition cannot detect specific chemical alterations at very low concentrations. Complementary analysis for risk assessment using biomarkers, species sensitivity distribution toxicity test, or other emerging tools can provide additional information of possible coming problems, which should be considered for investigative monitoring.

Keywords Biological quality • Catalan rivers • Chemical status • Mediterranean area • Priority substances • Water Framework Directive

Contents

1 Introduction and Objectives

The publication of the Water Framework Directive (2000/60/EC) (WFD) by the European Parliament and Commission at the end of 2000, and the subsequent approved Priority Substances Directive (105/2008/EC), provided significant changes in the process of assessing water quality in the European aquatic ecosystems [1–3]. This challenge must be carried out by using a new monitoring program, which EU Member Estates are bound to apply since 2007, according to the WFD requirements. Water quality measurements using biological elements reach a legal status together with the chemical status. Both biological quality and priority substances must be taken into account to establish a comprehensive water status diagnostics in rivers. That is why, the selection of proper biological elements, and their quality indices and class boundaries, became fundamental elements, which can significantly affect the outcome classification [4]. Moreover, priority and hazardous substances (pesticides, PAHs, heavy metals, chlorinated and non-chlorinated solvents, endocrine disruptors, etc.) must be carefully analyzed using suitable methods for their environmental quality standards (EQS) (2009/90/EC Directive). Chemical and biological analysis must be finally combined to set the final water quality status, and results need to be easily interpretable in terms of both chemical and ecological status to define a suitable program of measures.

In Europe, diverse biological indices and metrics have been developed for water quality assessment in rivers [5–8], using macroinvertebrate (e.g., [9–12]), diatoms (e.g., [13, 14]), macrophytes (e.g., [15]), and fish communities (e.g., [16]). Biological indices have also been applied in Catalan rivers since long time ago, basically developed in research centers [17–20], and more recently applied by Water Authorities following the WFD requirements [21–23]. The assessment of biological quality in rivers has been developed and enriched by contributions of several research centers and water authorities in order to achieve the normative definitions of WFD and their compliance with a gradient stressor [24, 25]. Additionally, quality class comparisons between countries have been carried out in order to harmonize and intercalibrate the information obtained [26, 27]. Thus, biological elements and their indices allow a comprehensive procedure to assess the ecological integrity for quality status classification in rivers. However, the WFD requires additional analysis combining biological indices and the presence of priority and hazardous chemical substances (chemical status). The chemical and ecological combination is required to ensure correct diagnosis in order to protect water ecosystem damages.

Chemical status is essentially defined by the compliance with the established environmental quality standards (EQS). A total of 33 substances or groups of substances, the so-called Priority Substances (PS) and Priority Hazardous Substances (PHS), were firstly set out by 2455/2001/EC Decision, and later by 2008/105/EC Directive. The list of PS and PHS is the outcome of an extensive risk assessment study carried out by the Fraunhofer Institut (Schmallenberg, Germany), according to the so-called COMMPS (Combined Monitoring-based and Modeling-based Priority Setting Scheme) methodology [28, 29]. The COMMPS procedure aims to quantify the risk associated with the exposure of a given chemical making use of two kinds of data, named modeling-based and monitoring-based. This procedure establishes a ranking of chemical substances according to a risk priority index. The exposure index of a chemical substance is calculated using all its measured concentration values in every sampling site from those compiled throughout Europe, whereas the effect index calculation is carried out by direct and indirect effects on aquatic organisms (toxicity and potential bioaccumulation), as well as indirect effects on humans (carcinogenicity, mutagenicity, adverse effects on reproduction, and chronic effects). The ranked list of substances resulting from the COMMPS procedure was updated in two new studies carried out by using both modeling [30] and monitoring procedures [31]. Setting out EQS for the different PS and PHS is based on the consideration of the reported chronic and acute ecotoxicities at different trophic levels, using appropriate safety factors (Annex V of 2000/60/EC

Directive). Regarding some chemical elements, EQS values are often very strict, but they should be understood under a precautionary principle.

For all those aforementioned priority substances, the European Commission shall submit proposals to control the progressive reduction of discharges, and phase out emissions of PHS in the environment within the forthcoming 20 years. As a whole, all the foresaid required information is supported on the data supplied by the corresponding surveillance or operational monitoring programs that must be carried out by their respective water authorities.

This chapter attempts to compare ecological and chemical status outputs in a Mediterranean watershed (the Catalan river basins, NE Spain), in order to provide useful information about the strengths and weaknesses of quality status classification in rivers according to the WFD requirements (2000/60/EC Directive). The biological quality indices provide a final diagnosis based on the analysis of biological community structure and ecosystem function, whereas chemical analysis of priority substances provides an indirect diagnosis from a set of environmental quality standard thresholds (EQS) previously established according to toxicity test on biota. Therefore, there is an obvious need to compare and analyze the relationship between chemical and ecological status to enhance the output interpretation. The whole of the cases determining good ecological status do not necessarily agree with a good chemical status, and vice versa. Hence, differences need to be carefully analyzed to properly classify the river quality status.

2 Monitoring Program Carried Out in Catalan Rivers

2.1 Sampling Sites and Study Area

Catalan basins are located in NE. Spain (Catalonia), by the Mediterranean Sea. Water is managed by the Catalan Water Agency, which is under the authority of the Catalan Autonomous Government (Generalitat de Catalunya). Catalonia contains two main hydrographic areas: on the one hand, the Catalan River Basin District, composed by several small watersheds that completely drain in Catalonia (Llobregat, Ter, Muga, Daró, Fluvià, Francolí, Foix, Besòs, Gaià, Tordera, Riudecanyes, and several small coastal streams), and on the other hand, a part of the Ebro basin, a transboundary big basin which drains through several regions in Spain (Fig. 1). The Catalan Water Agency has full competences for managing water supply and restoring quality status in the Catalan River Basin District, whereas Catalan Water Agency only monitors and makes suggestions for water management in the interregional basin (the part of the Ebro basin located within Catalonia). Transboundary basins among regions are managed by the Spanish central government through water authorities called Hydrographic Confederations, one for each Basin District. Thus, the Ebro watershed is managed by the Ebro Hydrographic Confederation.

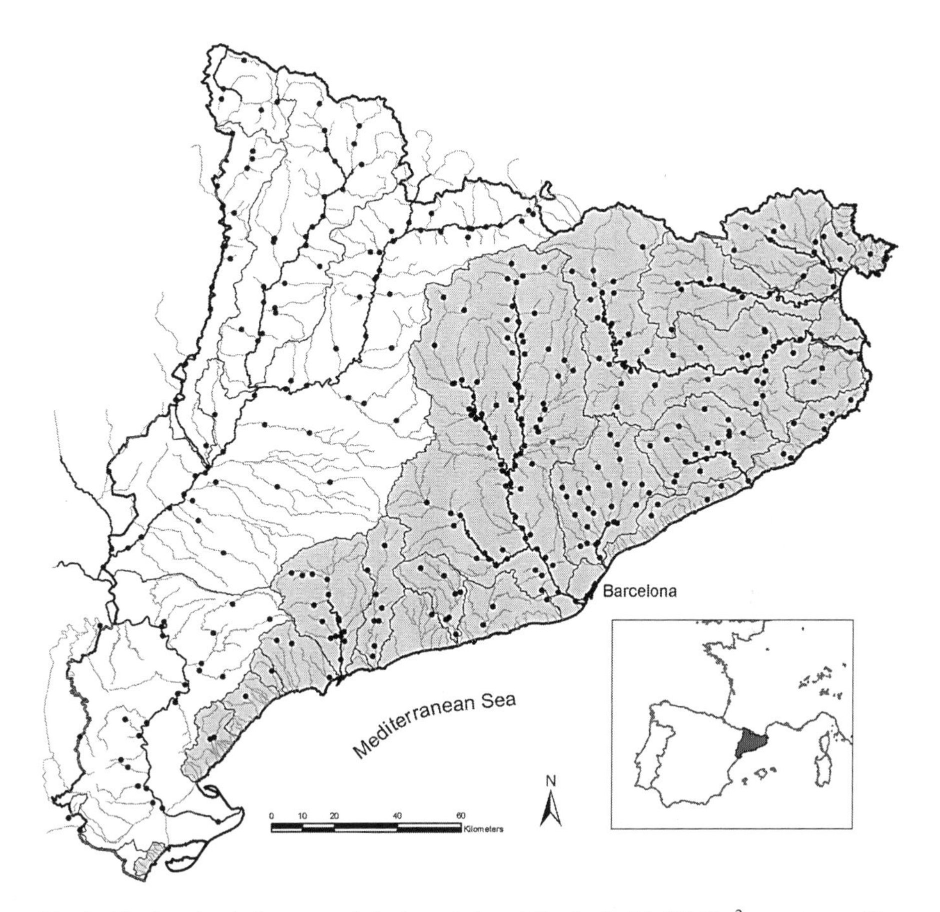

Fig. 1 The location in Europe and the boundaries of Catalonia (31,990 Km^2) are shown. The Catalan River Basin District, made up by several watersheds (in *gray*), is fully located in Catalonia and drains to the Mediterranean Sea. The rest of the Catalan region is part of the Ebro watershed (interregional basin)

The Catalan River Basin District occupies a total area of 16,423 km^2, representing 52% of the territory of Catalonia, and housing 92% of the Catalan population (6.8 million inhabitants). The interregional basins that flow in Catalonia (Ebro river and its tributaries) have a total area of 15,567 km^2, representing 48% of the Catalan territory, and housing the remaining 8% of the Catalan population (0.6 million inhabitants).

A total water of 3,123 hm^3 per year (equivalent to 100 m^3/s) is used in whole Catalonia. From that volume of water, 38% (1,186 hm^3 per year) comes from the Catalan River Basin District, while 62% (1,937 hm^3 per year) is drawn from the interregional basins (part of the Ebro watershed located in Catalonia). Urban usage, comprising household and industrial consumption, accounts for 27.4% of total withdrawals (856 hm^3 per year), while agriculture uses, including irrigation and livestock consumption, accounts for the remaining 72.6% (2,267 hm^3 per year). The percentage of use significantly varies between the Catalan River Basin District and the part of the Ebro basin located in Catalonia. Thus, urban and industrial use makes up for the majority of the Catalan River Basin District, representing 65% of total consumption, whereas agricultural use predominates in the interregional basin, which represents over 95% of the total water consumed [32].

Catalonia, and especially the Catalan River Basin District, is one of the most industrialized areas of Spain. Waste and leak discharges from industrial and urban activities, and specific mining wastes (e.g., salt wastes), severely impact downstream areas in the Catalan River Basin District. Therefore, source-point pollution from industrial areas is mainly found in the lower Llobregat (also affected by salt mine activities), Tordera, Besòs, and Francolí basins [33, 34], whereas rivers close to agricultural and irrigation areas, mostly located in the lower Ebro basin and its main tributaries (e.g., the lower Segre), are mostly affected by diffuse pollution, and meaningful concentrations of pesticides and nitrates can be found [33].

According to the river basin characterization carried out by the Catalan Water Agency [35], a total of 6,639 km of the river network was used to select water bodies in all the Catalan basins. A total of 367 river water bodies were set out (with an average of 18 km per water body), from which a total of 136 (37%) were identified under risk of not fulfilling WFD objectives due to relevant pressures from human activities. River monitoring procedure and its sampling frequencies were defined taking into account water bodies in risk, and at least one sampling site per water body was established. So, a total of 367 sites (Fig. 1), with different sampling frequencies along the monitoring program period (for 6 following years), have been used to determine the chemical and the ecological status in Catalan rivers. Data used in this work have been sampled from 2007 to 2009.

2.2 *Quality Elements and Methods*

Suitable Mediterranean type-specific indices for each biological quality element (BQE) required by the WFD were considered for the biological quality assessment in Catalan rivers (Table 1). Quality indices for each biological element and reference conditions, using diatoms, macroinvertebrates, and fish fauna, were previously selected considering river typology. Quality classes were later combined among BQEs using the "one out, all out" criteria in order to establish the final biological quality class [36]. This is a restrictive procedure since the worst biological quality item is used to set the final quality status.

Biological data (macroinvertebrate, diatom, and fish fauna) was obtained from spring samples (from April to June). Changes in biological community composition are often found along time in Mediterranean rivers [37], and samples need to be collected in the same period to avoid natural disturbances [38]. Samples were gathered following specific sampling protocols for macroinvertebrate, diatoms, and fish fauna (protocols are available in the Catalan Water Agency WEB page). The IBMWP [39] for macroinvertebrate, the IPS [13] for diatoms, and the IBICAT

Table 1 Chemical elements (priority and hazardous substances) analyzed, and biological indexes used by the monitoring program for quality status surveillance in Catalan rivers

Quality assessment	Quality elements or group of chemical substances used	Biological indices or chemical substances used for quality assessment
	Macroinvertebrates	IBMWP
Biological quality	Diatom	IPS
	Fish fauna	IBICAT
Chemical status (priority and hazardous substances)	Heavy metals (16) substances	Mercury, aluminum, antimony, arsenic, cadmium, cobalt, copper, chrome, bari, iron, manganese, molybdenum, nickel, lead, selenium, zinc
	Chlorinated and non-chlorinated solvents (30) individual substances or grouped substances	Benzene, chlorobenzene, ethylbenzene, xylenes, naphthalene, toluene, 1,1,1-trichloroethane, 1,1,2,2-tetrachloroethane, 1,1,2-trichloroethane, 1,1-dichloroethylene, 1,1-dichloroethane, 1,2,3-trichlorobenzene, 1,2,4-trichlorobenzene, 1,2-dichloroethane, 1,2-dichloropropane, 1,2-dichlorobenzene, 1,3,5-trichlorobenzene, 1,3-dichlorobenzene, 1,4-dichlorobenzene, bromodichloromethane, bromoform, chloroform, dibromochloromethane, dichloromethane, tetrachlorethylene, carbon tetrachloride, trichloroethylene, trichlorofluorometane, 1,2-dichloroetilene (*c* + *t*), 1,3-dichloropropene(*c* + *t*)
	Chlorinated pesticides (25) individual substances or grouped substances	Pentachlorophenol, alachlor, endosulfan I, endosulfan II, endosulfan sulfate, heptachlor, heptachlor epoxide A, heptachlor epoxide B, lindane, a-hexachlorocyclohexane, pentachlorobenzene, hexachlorobutadiene, 2.4'-DDD, 4,4'-DDD, 2,4'-DDE, 4,4'-DDE, 2.4'-DDT, 4,4'-DDT, aldrin, dieldrin, endrin, endrin ketone, hexachlorobenzene, isodrin, metolachlor.
	Other pesticides (triazines, organophosphates and miscellaneous) (16) individual substances or grouped substances	Chlorpyrifos, diazinon, azinphosethyl, ethion, fenitrothion, malathion, methylparathion, molinate, parathion, chlorfenvinfos (E + Z), atrazine, propazine, simazine, terbutryne, tertbutilazine, trifluralin.
	Polycyclic aromatic hydrocarbons (PAH) (7) individual substances or grouped substances	Anthracene, benzo(a)pyrene, benzo(b) fluoranthene, benzo(ghi)perilene, benzo (k)fluoranthene, fluoranthene, Indene (1.2.3)pyrene.
	Endocrine disruptors (3) individual substances or grouped substances	4-*tert*-octylphenol, 4-nonylphenol, brominated biphenyl ether

index for fish fauna [23] were used to calculate the quality class for each BQE. These indices mainly came from European projects [16, 25], and are commonly being applied in Spain since long time ago. The IBMWP and IPS indices were also analyzed in the European intercalibration process carried out by Mediterranean countries (Mediterranean Geographical Intercalibration Group: Med-GIG), and later adopted by the European Commission as WFD normative compliant. All these indices have been properly tested and they are well correlated with the stressor gradient for major human activity pressures [21]. Data of quality indices were standardized by calculating the EQR values (Ecological Quality Ratio) [26]. Values of each quality index were divided by reference values to obtain the EQR. Reference values were previously obtained by calculating the average for selected reference sites according to each river type. EQR values allow proper comparisons among quality indices and BQEs, and their later combinations.

A total of 97 priority substances and group of substances (isomers, metabolites, etc.) were analyzed in Catalan rivers (Table 1) using suitable standard procedures. Atomic fluorescence spectroscopy for mercury, inductively coupled plasma mass spectrometry for metals, headspace extraction procedure for solvent substances [40], solvent extraction with simultaneous derivatization for pentachlorophenol [41], and solid-phase stirred bar extraction [42] for the rest of organic compounds were used. All chemicals were also analyzed or confirmed using GC-MS according to the 2009/90/EC Directive. From these 97 substances and group of substances, 42 are included in the Annex I of the 105/2008/EC Directive, whereas 55 substances are not currently included in this Directive (Table 2), but they are required by Spanish-national laws or are likely to be found in Catalan rivers due to industrial or agricultural activities. Only substances and thresholds provided by the 105/2008 Directive (EQSs) were applied for chemical status assessment. Values of heavy metals (lead, cadmium, mercury, and nickel), chlorinated solvents, pesticides (chlorine, phosphorus, and triazine), polycyclic aromatic hydrocarbons, and endocrine disruptors (nonylphenols, octilphenols, and brominated diphenyl ether compounds) were analyzed and compared with the EQS provided by the directive. All other chemical elements are used to analyze their evolution along time and detect possible hot spots.

Table 2 Number of substances or group of substances measured by the Catalan Water Agency in the Monitoring Program carried out in Catalan rivers (2007–2009)

Group of compounds	Chemicals analyzed included in the 105/2008/EC Directive	Chemicals analyzed and not included in the 105/2008/EC Directive	Chemicals not analyzed and included in the 105/2008/EC Directive
Heavy metals	4	12	0
Solvents	8	23	0
Chlorinated pesticides	14	9	0
Others pesticides	6	11	3
PAH	7	0	0
Endocrine disruptors	3	0	2
Total	42	55	5

Five substances or group of substances included in the 105/2008/EC Directive (tributyltin compounds, diuron, isopropuron, Di(2-ethilhexil)phtalate (DEHP), and chloroalkanes) were not analyzed (Table 2). These elements require complex laboratory procedures and their analysis is difficult. Also some detection thresholds are not easy to be achieved and can provide misleading results. Quality thresholds provided by the 105/2008/EC Directive for some compounds (e.g., brominated diphenyl ether and tributyltin compounds) are extremely low.

2.3 *Frequency and Monitoring Campaigns*

The Catalan monitoring program is based on the combination of several frequencies depending on quality element and the risk classification of each water body (Table 3). Sampling sites belong to each risk category, and the number of performed samples in the monitoring program period is defined according to each biological and chemical element. The entire monitoring program is completed after 6 years. Notice that, for management purposes, the differentiation among types of monitoring is not as relevant as the need to optimize and efficiently allocate the available resources on a working calendar or schedule. Therefore, the problem to be solved by the Monitoring Program can be envisaged in a logistic perspective. A specific monitoring program adapted to each typology of water was designed, and it is carrying out by the Catalan Water Agency in Catalan rivers.

3 Data Analysis and Results

3.1 *Biological Quality*

A total of 145 water bodies (66% of water bodies with data available) achieve good quality according to the IPS index based on diatoms (Table 4). Diatoms are very

Table 3 Sampling frequency applied by the Monitoring Program for biological quality and chemical status assessment in Catalan rivers

Water body classification	Biological quality			Chemical status
	Macroinvertebrates	Diatoms	Fish fauna	Priority substances
With risk	6	3	2	72
Without risk	2	1	1	1
Reference	3	2	1	6

Sampling frequencies for each water body are classified according to its risk to not achieve the good status (previously analyzed using human pressures). Numbers in boxes refer to samples required within the 6 year period of whole monitoring program. Therefore, “6” means once a year, “1” once every 6 years, “3” once every 2 years, and “72” monthly samples

Table 4 Number of river water bodies classified in five biological quality classes for each biological quality element (diatoms, macroinvertebrate and fish fauna) in the Catalan River Basin District (RBD), the Ebro watershed located in Catalonia, and in whole Catalan rivers. Some water bodies were not sampled yet (basically headwaters without human pressures), and were considered without data

		Achieve biological quality		Do not achieve biological quality			
		High	Good	Moderate	Poor	Bad	Without data
Diatoms	Catalan RBD	39	43	33	21	10	102
	Ebro watershed	50	13	8	2	0	46
	All Catalan rivers	89	56	41	23	10	148
Macroinvertebrate	Catalan RBD	65	45	42	20	5	71
	Ebro watershed	53	25	8	1	0	32
	All Catalan rivers	118	70	50	21	5	103
Fish	Catalan RBD	52	29	21	17	23	106
	Ebro watershed	72	4	7	5	3	28
	All Catalan rivers	124	33	28	22	26	134
Biological quality	Catalan RBD	19	50	69	36	16	58
	Ebro watershed	44	23	18	4	1	29
	All Catalan rivers	63	73	87	40	17	87

sensitive organisms for organic pollution discharges and they have a rapid response to disturbances [13]. Therefore, low IPS index values are found near urban and industrial areas, mainly located in lower rivers (Fig. 2), where generally continuously high nutrient loads are received. Bad quality using IPS values is found in Besòs river, the lower Anoia, Llobregat, and Muga rivers, and from the middle Tordera until the mouth.

Meaningful differences are found between the Catalan River Basin District and the Ebro watershed located in Catalonia. Whereas 56% of water bodies with data available (82 water bodies) achieve the WFD objectives in the Catalan River Basin District, a total of 86% of water bodies with data available achieve the WFD objectives in Ebro basin.

Similar results, though a little bit more optimistic, are found using macroinvertebrates as a BQE (Fig. 3). Quality objectives (good and high quality) are achieved for 71% of water bodies with data available (188 water bodies) in all Catalan rivers (Table 4). Macroinvertebrates are also sensitive to organic pollution, but they have a slower and sustained response to disturbances than diatoms [7]. However, low IBMWP index values are quite consistent with diatoms results. The mid-Anoia river and lower Llobregat, Foix, Besòs, Tordera, and Muga rivers presented very low quality levels. Rivers located in the Ebro watershed show higher quality values; therefore high and good quality are achieved for a 90% of water bodies with data available (78 water bodies), whereas 62% (110 water bodies) are achieved in the Catalan River Basin District.

Fig. 2 Biological quality results using diatom communities (IPS index) in Catalan rivers (2007–2009). High quality is shown in *blue*, good quality in *green*, moderate quality in *yellow*, poor quality in *orange*, and bad quality in *red* color. Water bodies without data are shown in *gray* color. Reservoirs are shown in *light blue*

Fish fauna show the worst quality scenario from measured biological elements in Catalan rivers. A total of 67% of water bodies with data (157 water bodies) presented consistent fish communities dominated by native species in whole Catalan rivers (Table 4). However, this situation is very different between both water bodies located in the Catalan River Basin District and in the Ebro basins. Whereas 83% water bodies with data available (76 water bodies) achieve high and good quality in the Ebro watershed located in Catalonia, only a 57% of water bodies (81 water bodies) with data available achieve WFD objectives in the Catalan River Basin District. Fish fauna are sensitive to water pollution, but also to morphological and flow alterations, river discontinuity, and habitat lost. Low quality values using fish fauna denote lack of suitable hydrological conditions, and rivers located in the Catalan River Basin District are characterized by water scarcity, whence habitat

Fig. 3 Biological quality results using macroinvertebrate communities (IBMWP index) in Catalan rivers (2007–2009). High quality is shown in *blue*, good quality in *green*, moderate quality in *yellow*, poor quality in *orange*, and bad quality in *red* color. Water bodies without data are shown in *gray* color. Reservoirs are shown in *light blue*

alteration and water abstractions considerably affect native fish community composition. Moreover, also other threats such as exotic alien species invasions, basically due to fishing or other human activities, negatively affect fish communities and prevent from achieving a good biological quality. Llobregat and Ter basins show a high number of water bodies dominated by nonnative fish species (Fig. 4). The lower Ebro and Segre rivers (in the Ebro basin located in Catalonia) also show a high number of nonnative species.

Finally, high and good biological quality, combining macroinvertebrate, diatoms, and fish quality classes, are achieved in 49% of water bodies with data available (136 water bodies) in whole Catalan rivers (Table 4). Biological quality was determined using the worst three biological quality levels measured. Results show high and good quality in high mountain streams, basically rivers draining

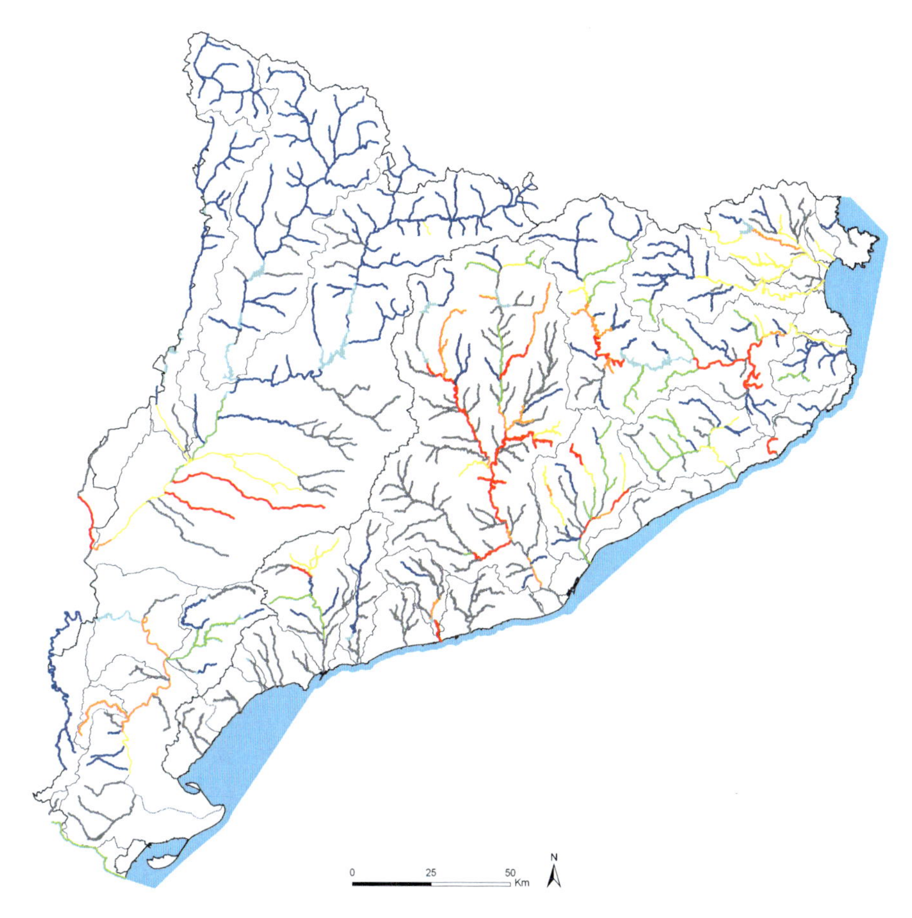

Fig. 4 Biological quality results using fish fauna (IBICAT index) in Catalan rivers (2007–2009). High quality is shown in *blue*, good quality in *green*, moderate quality in *yellow*, poor quality in *orange*, and bad quality in *red* color. Water bodies without data are shown in *gray* color. Reservoirs are shown in *light blue*

from the Pyrenees and protected areas, and small streams far from urban and agricultural areas, mostly tributaries of major rivers. The lower Besòs, Llobregat, Francolí, and Muga rivers, located close to urban and industrial areas, show many water bodies with bad or poor biological quality. Also, lower Segre river and its tributaries that flow between irrigation areas show bad and poor biological quality (Fig. 5). According to the diatoms, macroinvertebrate, and specially due to fish fauna, meaningful differences are found between the Catalan River Basin District and the Ebro watershed located in Catalonia. Whereas only 36% of water bodies with data available (69 water bodies) achieve the WFD objectives in the Catalan River Basin District, a total of 74% of water bodies with data available (67 water bodies) achieve the WFD objectives in the Ebro basin located in Catalonia.

Fig. 5 Biological quality results in Catalan rivers (2007–2009). High quality is shown in *blue*, good quality in *green*, moderate quality in *yellow*, poor quality in *orange*, and bad quality in *red* color. Water bodies without data are shown in *gray* color. Reservoirs are shown in *light blue*

3.2 Chemical Status

Quality standards provided by the 105/2008/EC Directive are achieved in a 75% of water bodies with data available (206 water bodies) in whole Catalan rivers (Table 5). A total of 68 water bodies do not achieve good chemical status in Catalan rivers. Unfulfilled quality standards are close to industrial areas basically located in the Catalan River basin District (lower Llobregat and Besòs rivers), and close to drainage irrigation fields located in the lower Segre river (in the Ebro watershed) (Fig. 6). Also, additional unfulfilled quality standards are found in some small streams located in the upper Segre watershed, draining from Pyrenees, although these values must be tentatively considered and require further validation.

Table 5 Number of river water bodies classified according to chemical status for each group of priority substances and group of substances (chlorinated pesticides, other pesticides, chlorinated and non-chlorinated solvents, endocrine disruptors, heavy metals, and polycyclic aromatic hydrocarbons) in the Catalan River Basin District (Catalan RBD), the Ebro watershed located in Catalonia, and in whole Catalan rivers. Some water bodies were not sampled yet (basically headwaters without human pressures), and were considered without data

		Achieve good status	Do not achieve good status	Without data
Heavy metals	Catalan RBD	170	11	67
	Ebro watershed	95	0	24
	All Catalan rivers	265	11	91
Chlorinated and non-chlorinated solvents	Catalan RBD	180	1	67
	Ebro watershed	95	0	24
	All Catalan rivers	275	1	91
Chlorinated pesticides	Catalan RBD	172	9	67
	Ebro watershed	67	28	24
	All Catalan rivers	239	37	91
Other pesticides	Catalan RBD	174	7	67
	Ebro watershed	86	9	24
	All Catalan rivers	260	16	91
Polycyclic aromatic hydrocarbons	Catalan RBD	181	0	67
	Ebro watershed	95	0	24
	All Catalan rivers	274	2	91
Endocrine disruptors	Catalan RBD	148	33	67
	Ebro watershed	93	2	24
	All Catalan rivers	241	35	91
Chemical status	Catalan RBD	145	36	67
	Ebro watershed	63	32	24
	All Catalan rivers	208	68	91

Chemical status results are analyzed for each different set of compounds (Table 5). Both chlorinated pesticides and endocrine disruptors are the main substances responsible of quality standard failures in Catalan rivers. Chlorinated pesticides do not achieve quality standards in 37 water bodies (a 13% of Catalan water bodies with data available), and they are mainly located in the Ebro watershed (lower Segre river close to irrigation areas). Mostly endosulfan, trifluralin, and also hexachlorocyclohexanes (lindane) are the main hazardous substances found over their EQS. Trifluralin is found in tributaries of lower Llobregat and Segre rivers with an average concentration of 0.05 μg/L (slightly over its EQS value: 0.03 μg/L). Lindane is found in nine water bodies, basically located near urban and industrial areas in Besòs and Llobregat rivers, and also close to agricultural zones in the lower Segre river, with an average value of 0.03 μg/L (EQS: 0.02 μg/L), and with a maximum value of 0.1 μg/L (EQS: 0.07 μg/L), whereas endosulfan (mainly endosulfan II) is broadly detected in some small tributaries draining from high mountains in the upper Segre watershed. Endosulfan is detected in 28 water bodies

Fig. 6 Chemical status results in Catalan rivers (2007–2009). Good quality is shown in *green*, and bad quality in *red* color. Water bodies without data are shown in *gray* color. Reservoirs are shown in *light blue*

with an average value of 0.02 μg/L (EQS: 0.005 μg/L), with a maximum value of 0.3 μg/L (EQS: 0.01 μg/L). The origin of endosulfans found in the upper watershed area of Segre river is not clear. No relevant human activities, neither industrial nor significant agricultural uses, are located in this area, and data should be carefully considered. Similar concentrations and equivalent isomeric relationship have been reported for endosulfan substances in high mountain water bodies by other authors [43, 44] in the same region. Atmospheric deposition is envisaged as a possible source of endosulfan concentration detected in high mountain rivers and lakes, due to their higher proportion of β-isomer compared with the α-isomer found in those samples. The abundance of α-isomer of endosulfan gradually increases with respect to β-isomer in the atmosphere concentration according to the distance from the endosulfan source. So, endosulfan atmospheric composition is normally dominated by α-isomer. However, the highest volatility and lower water solubility of α-isomer can increase the β-isomer composition in rain deposition and its input in rivers [45, 46]. Endosulfan concentrations found in Catalan high mountain rivers need to be afterward monitored to properly set out their origin.

Regarding triazines, organophosphates, and miscellaneous compounds (named other pesticides in this manuscript), they do not achieve quality standards in 16 water bodies (6% of water bodies with data), basically located in the Ebro watershed (9 water bodies), although they can also be found in the Catalan River Basin District (7 water bodies). Chlorpyrifos, chlorfenvinfos, and simazine are the most detected compounds found in Catalan rivers. They are mainly found at low concentrations close to quality standard values provided by the 105/2008/EC Directive. Chlorpyrifos is found close to irrigation areas in the lower Segre river, and also in the lower Llobregat watershed, in 18 water bodies with an average concentration of 0.03–0.09 μg/L (EQS: 0.03 μg/L), whereas chlorfenvinfos and simazine are only located in two water bodies. Levels of chlorfenvinfos are detected at 0.1–0.15 μg/L (EQS: 0.1 μg/L) in lower Llobregat watershed, whereas simazine is also slightly detected at 1 μg/L (EQS: 1 μg/L) in the upper Francoli river. Most pesticides have been detected close to or slightly over the EQS values and their threshold detection. That is the reason why they must be tentatively considered and later evaluated over time to be confirmed.

Endocrine disruptors are mainly found in the Catalan River Basin District, close to industrial areas. A total of 35 water bodies do not achieve the quality standards in whole Catalan rivers, from which 33 are located in the Catalan River Basin District (in lower Llobregat and Besòs rivers). Nonylphenols (EQS: 0.3 μg/L) and octilfenols (EQS: 0.1 μg/L) are present in the wastewater mainly coming from industrial uses with an average concentration of 0.4–0.8 μg/L. Endocrine disruptors are mostly detected in industrialized and high populated areas of Besòs basin, and the lower Llobregat and Anoia rivers. Similar endocrine disruptors' concentrations have been found by other authors close to industrial areas, where several types of dissolvent or similar compounds are mainly used in industrial processes [47].

Regarding heavy metal concentrations, they are also mainly found near industrialized areas (Besòs basin, and the lower Llobregat river), and additionally in some specific places due to mine activity (e.g., Osor stream, tributary located in the mid Ter river). Nickel values are detected over the quality standard (EQS: 20 μg/L) in 8 water bodies located close to industrial zones, mainly in the Besòs river, the lower Llobregat river (close to Barcelona industrial area), and Onyar stream (tributary of the lower Ter river, located close to Girona industrial area), with an average value of 30 μg/L. Also values of cadmium (EQS: 0.08–0.2 μg/L) are slightly detected in two water bodies, in the Osor stream due a old mine activity, and in the mid-Tordera stream, with an average concentration of 0.15 μg/L, close to the EQS value and its threshold detection.

A total of 21 water bodies, 5.7% of whole water bodies in Catalan rivers, were diagnosed with a good biological quality, using diatoms, macroinvertebrate, and fish fauna quality indices (Table 1), but did not achieve good chemical status

Table 6 Number of water bodies (rivers) classified as good biological quality, but do not achieve good chemical status

	Good biological quality and bad chemical status	
	Number of water bodies	Percentage of total water bodies(%)
Catalan River Basins District	18	7.1
Ebro watershed in Catalonia	3	2.7
All Catalan rivers	21	5.7

according to the quality standards provided by the 105/2008/EC Directive (Table 6). Main water bodies with a bad chemical status but good biological quality were detected in small watersheds close to agricultural areas, basically located in the Catalan River Basin District. Water bodies were mainly impacted by triazines, organophosphates, and miscellaneous pesticides, as endosulfan and clorpyrifos with low concentrations.

4 Discussion and Conclusions

Results from the monitoring program carried out in Catalan rivers (2007–2009) show a higher percentage of nonfulfillment quality objectives due to ecological status rather than chemical status. A total of 144 river water bodies (39%) do not achieve the good ecological status according to the 2000/60/EC Directive, whereas 68 river water bodies (19%) do not achieve the EQS for priority and hazardous substances provided by the 105/2008/EC Directive (chemical status). This ratio (2:1) between the number of river water bodies which do not fulfill ecological status versus chemical status can be easily explained. Ecological status integrates the effects of a comprehensive range of pressures on aquatic ecosystems, which result in a complex impact assessment procedure using biological elements. Biological indices analyze changes in the function and the structure of aquatic ecosystems, which can be caused by hazardous chemical compounds, but also by several other pressures such as high organic matter or nutrient concentrations, hydrologic and morphology alterations, habitat lost, riparian vegetation damages, alien species invasions, etc. Furthermore, ecological status assessment covers cumulative impacts over time; that is why it shows an integrated measurement.

Several authors have found significant correlations between biological indices and chemical elements [21, 34], even with priority substances [48–50]. Therefore, a percentage of bad biological quality could be explained by chemical quality standard nonfulfillment. However, some sites with bad chemical status but good ecological status were also found in Catalan rivers. A total of 21 river water bodies (6% of total river water bodies) showed priority substance concentrations over the EQS thresholds provided by the 105/2008/EC Directive, whereas biological elements showed good quality. Biological indices based on community structure cannot detect specific chemical alterations at very low concentrations. EQS thresholds for some chemicals have been very restrictively set out, and low concentrations, even over the EQS thresholds, cannot affect the community structure and biological indices, at least to a short time [49, 50]. Some studies have found that general contaminants, as high salinity or nutrient concentrations, highly affect the biological structure and composition than priority substances [51]. Low priority substances concentrations, close to threshold detection and their EQS values, cannot be detected by using current biological indices. Traditionally, in Europe and other industrialized countries, biomonitoring of freshwaters has been based on community structure measures, richness, or sensitive taxa, mainly focused on diatom and macroinvertebrate communities [6, 7]. Thus, low persistent chemical concentrations or sudden inputs may be overlooked using standard biological indices provided according to the WFD (Annex V, 2000/60/EC) [51, 52]. Additional tools for risk assessment analysis (e.g., biomarkers) could be required in order to identify coming problems. In this case, a recent Marie Curie project (KeyBioeffects project) allows us to test a new risk assessment method in Catalan rivers in order to provide additional information and define possible hot spots for forthcoming quality problems [53].

Also, rivers with good chemical status should not be completely rejected for ecosystem risk assessment, and a complementary analysis could be suggested, even more when ecological measurements do not achieve a good status. A recent combined toxicity test analysis, using new tools through theoretical toxicity data on biota, was carried out in Catalan rivers [54]. This study showed that chemical elements found in Catalan rivers could potentially impact about 50% of biota in 90% of sites, and less than 10% of biota in 10% of sites. Potential effects on freshwater ecosystems of cumulative individual toxicants, or/and their mixtures, are higher than the effective nonachievement of quality standards for priority substances found. Moreover, some EQS thresholds for priority substances provided by the 105/2008/EC Directive are too low to be properly detected using standard procedures and laboratories submitted to a quality certification. For instance, brominated diphenyl ether and tributyltin quality standard thresholds are extremely low, and they are not easily achieved by standard methods without high costs and expensive specific equipment. Also, current European regulations have not clearly defined specific methods and right techniques to properly quantify some priority substances (e.g., Chloroalkanes $C_{10\text{-}13}$). The difficulty to properly reach some EQS thresholds for certain priority substances makes the chemical status classification uncertain, even more at low concentration values, close to threshold detection. Besides, not all possible hazardous substances that can potentially affect the ecological status are continuously analyzed. Therefore, additional tools can result in a very powerful risk assessment complement, which can be recommended to better evaluate the risk of chemicals in rivers with high human pressures (downstream industrial discharges or agricultural diffuse contamination sources).

As a conclusion, ecological status and chemical status must be properly combined for surveillance monitoring in river ecosystems. Ecological status offers a comprehensive quality assessment for the integrated aquatic ecosystem using the biological community structure analysis. Additionally, chemical status analyzing priority and hazardous compounds (pesticides, chlorinated and non-chlorinated solvents, PAHs, heavy metals, endocrine disruptors, etc.) must be considered and carefully combined with the ecological status output in order to completely diagnose ecosystem damages. Biological elements' response to a broad range of pressures, and specific chemical elements, could not be detected by biological indices (basically focused on structure at community level) at least in a short time. Therefore, both ecological and chemical status must be combined by selecting the worst quality class (using the "one out, all out" criteria) to properly classify the quality status in rivers. However, two things should be taken into account. (1) On the one hand, some chemical values must be carefully considered, even more when concentrations are found near the EQS and their threshold detection values. As we explained before, EQS values for some priority substances (mainly heavy metals and organic compounds) were extremely decreased by the new 105/2008/EC Directive, up to threshold detection levels. Those levels are set out under a precautionary principle according to several toxicity test on biota. However, the difficulty to reach such EQS levels by proper standard methods and suitably accredited laboratories makes those results uncertain. In these cases, results should be considered under a precautionary principle, and further analysis can be recommended in order to verify the bad chemical status. (2) On the other hand, good chemical and ecological status do not necessarily imply no risk for ecosystem damages. As we explained before, not all chemical and hazardous compounds are currently and continuously analyzed. Therefore, background concentrations (under the threshold detection), and/or emergent pollutants currently not taken into account, can potentially affect the water ecosystem and their biological communities. That is why, river sites under high human pressures, mainly close to industrial and heavily urbanized areas, or near agricultural activities, need to be additionally monitored in order to avoid ecosystem damages, evenly if good ecological status is found. Biological indices cannot detect some hazardous chemical compounds, even more at low concentrations. In this regard, complementary tools for risk assessment using biomarkers, species sensitivity distribution toxicity test, or other emerging analysis can provide additional information of possible coming problems, which should be considered for investigative monitoring.

References

1. Allan IJ, Vranaa B, Greenwooda R, Millsb GA, Knutssonc J, Holmbergd A, Guiguese N, Fouillace AM, Laschif S (2005) Strategic monitoring for the European water framework directive. Trends Anal Chem 25(7):704–715
2. Bloch H (1999) European water policy facing the new millennium: the water framework directive. Assessing the ecological integrity of running waters. Vienna, Austria, pp 9–11
3. Coquery M, Morin A, Bécue A, Lepot B (2005) Priority substances of the European Water Framework Directive: analytical challenges in monitoring water quality. Trends Anal Chem 24 (2):117–127

4. Irvine K (2004) Classifying ecological status under the European Water Framework Directive: the need for monitoring to account for natural variability. Aquat Conservat Mar Freshwat Ecosyst 14(2):107–112
5. Carsten von der Ohe P, Prüß A, Schäfer RB, Liess M, de Deckere E, Brack W (2007) Water quality indices across Europe, a comparison of the good ecological status of five river basins. J Environ Monit 9:970–978
6. Hellawell JM (1986) In: Mellanby K (ed) Biological indicators of freshwater pollution and environmental management. Elsevier, London, p 546
7. Hering D, Johnson R, Kramm S, Schmutz S, Szoszkiewicz K, Verdonschot PFM (2006) Assessment of European streams with diatoms, macrophytes, macroinvertebrates and fish: a comparative metric-based analysis of organism response to stress. Freshwat Biol 51(9):1757–1785
8. Johnson R, Hering D (2010) Spatial concurrency of benthic diatom, invertebrate, macrophyte, and fish assemblages in European streams. Ecol Appl 20(4):978–992
9. Bonada N, Prat N, Resh VH, Statzner B (2006) Developments in aquatic insect biomonitoring: a comparative analysis of recent approaches. Annu Rev Entomol 51:495–523
10. Cairns J, Pratt JR (1993) A history of biological monitoring using benthic macroinvertebrates. In: Rosenberg DM, Resh V (eds) Freshwater biomonitoring and benthic macroinvertebrates. Chapman and Hall, New York, pp 10–27
11. Metcalfe JL (1989) Biological water quality assessment of running waters based on macroinvertebrates communities: history and present status in Europe. Environ Pollut 60:101–139
12. Sandin L, Hering D (2004) Comparing macroinvertebrate indices to detect organic pollution across Europe: a contribution to the EC Water Framework Directive intercalibration. Hydrobiologia 516(1–3):55–68
13. Kelly MG, Cazaubon A, Coring E, Dell'Uomo A, Ector L, Goldsmith B, Guasch H, Hürlimann J, Jarlman A, Kawecka B (1998) Recommendations for the routine sampling of diatoms for water quality assessments in Europe. J Appl Phycol 10(2):215–224
14. Prygiel J, Coste M, Bukowska J (1999) Review of the major diatom-based techniques for the quality assessment of rivers – State of the art in Europe. Agence de l'eau Artois Picardie, pp 224–238
15. Szoszkiewicz K, Ferreira T, Korte T, Baattrup-Pedersen A, Davy-Bowker J, O'Hare M (2006) European river plant communities: the importance of organic pollution and the usefulness of existing macrophyte metrics. Hydrobiologia 566(1):211–234
16. Pont D, Hugueny B, Rogers C (2007) Development of a fish-based index for the assessment of river health in Europe: the European Fish Index. Fish Manage Ecol 14(6):427–439
17. Muñoz I, Prat N (1994) A comparación between different biological water quality indexes in the Llobregat Basin (NE Spain). Ver Int Verein Limnol 25:1945–1949
18. Prat N (1991) Present trends in river studies. Oecol Aquat 10:1–12
19. Prat N, Rieradevall M (2006) 25-years of biomonioring in two mediterranean streams (Llobregat and Besòs basins, NE Spain). Limnetica 25(1–2):541–550
20. Sabater S, Guasch H, Picon A, Romaní A, Muñoz I (1996) Using diatom communities to monitor water quality in a river after the implllementation of a sanitation plan (river Ter, Spain). In: Proceedings of an International Symposium held at the Volksbildugsheim Grillhof, Austria (17-19 September 1995). Whiton BA, Rott E (eds) Use of algae for monitoring rives II, pp 97–103
21. Munné A, Prat N (2009) Use of macroinvertebrate-based multimetric indices for water quality evaluation in Spanish Mediterranean rivers. An intercalibration approach with the IBMWP index. Hydrobiologia 628:203–225
22. Munné A, Prat N (2011) Effects of Mediterranean climate annual variability on stream biological quality assessment using macroinvertebrate communities. Ecol Indic 11:651–662
23. Sostoa A, Casals F, Caiola NM, Vinyoles D, Sánchez S, Franch C (2003) Desenvolupament d'un índex d'integritat biòtica (IBICAT) basat en l'ús dels peixos com a indicadors de la qualitat ambiental dels rius a Catalunya. Documents tècnics de l'Agència Catalana de l'Aigua, p 203

24. Buffagni A, Erba S (2004) A simple procedure to harmonize class boundaries of European assessment systems. Discussion paper for the Intercalibration process – WFD CIS WG 2.A ECOSTAT, 6 February 2004, p 21
25. Buffagni A, Erba S, Birk S, Cazzola M, Feld C, Ofenbock T, Murray-Bligh J, Furse MT, Clarke R, Hering D, Soszka H, Van de Bund W (2005) Towards european inter-calibration for the water framework directive: procedures and examples for different river types from the E.C. project STAR, 11th STAR deliverable. STAR contract No: EVK1-CT 2001–00089. Rome (Italy). Quad Ist Ric Acque 123, Rome (Italy), IRSA, p 460
26. Heiskanen AS, Van de Bund W, Cardoso AC, Noges P (2004) Towards good of ecological status of surface waters in Europe – interpretation and harmonisation of the concept. Water Sci Technol 49:169–177
27. Pollard P, van de Bund W (2005) Template for the development of a boundary setting protocol for the purposes of the intercalibration exercise. Agreed version of WG 2.A Ecological Status. Version 1.2. 6 June 2005, Ispra, p 24
28. CEC (1999) Revised proposal for a list of priority substances in the context of the water framework directive (COMMPS procedure). Fraunhofer-Institut Umweltchemie und Ökotoxicologie. 98/788/3040/DEB/E1
29. Teixidó E, Terrado M, Ginebreda A, Tauler R (2010) Quality assessment of river waters using risk indexes for substances and sites based on the COMMPS procedure. J Environ Monit 12:2120–2127
30. Daginus K, Gottardo S, Mostrag-Szlichtyng A, Wilkinson H, Whitehouse P, Paya-Perez A, Zaldívar JM (2010) A modelling approach for the prioritisation of chemicals under the water framework directive. European Commission – Joint Research Centre, Italy; Environment Agency, UK. EUR 24292 EM
31. James A, Bonnomet V, Morin A, Fribourg-Blanc B (2009) Implementation of requirements on Priority Substances within the context of the Water Framework Directive. Prioritisation process: Monitoring-based ranking. Contract Nº 07010401/2008/508122/ADA/D2
32. ACA – Catalan Water Agency (2009) Pla de Gestió del Districte de Conca Fluvial de Catalunya (Catalan River Basin District Management Plan). Memòria, p 393
33. ACA – Catalan Water Agency (2010) Estat de les Masses d'Aigua a Catalunya 2007–2009. Resultats del Programa de Seguiment i Control. Barcelona, p 65
34. Prat N, Munné A, Rieradevall M, Solà C, Bonada N (1999) La qualitat ecològica del Llobregat, el Besòs, el Foix i la Tordera. Informe 1997. Estudis de la qualitat ecològica dels rius. 7. p 154. Àrea de Medi Ambient de la Diputació de Barcelona.
35. ACA – Catalan Water Agency (2005) Caracterització de les masses d'aigua i anàlisi del risc d'incompliment dels objectius de la Directiva Marc de l'Aigua (2000/60/CE) a Catalunya. Agència Catalana de l'Aigua. Departament de Medi Ambient i Habitatge de la Generalitat de Catalunya. Octubre, 2005, p 860
36. European Commission (2003) Overall approach to the classification of the ecological status and ecological potential. Water Framework Directive. Common Implementation Strategy. Working Group 2A. Ecological Status (ECOSTAT). 27 Nov 2003, p 47
37. Williams DD (2006) The biology of temporary waters. Oxford University Press, New York, p 337
38. Sporka F, Vlek HE, Bulánková E, Krno I (2006) Influence of seasonal variation on bioassessment of streams using macroinvertebrates. Hydrobilologia 566:543–555
39. Alba-Tercedor J, Jáimez-Cuéllar P, Álvarez M, Avilés J, Bonada N, Casas J, Mellado A, Ortega M, Pardo I, Prat N, Rieradevall M, Robles S, Sáinz-Cantero CE, Sánchez-Ortega A, Suárez ML, Toro M, Vidal-Albarca MR, Vivas S, Zamora-Muñoz C (2004) Caracterización del estado ecológico de los ríos mediterráneos ibéricos mediante el índice IBMWP (antes BMWP'). Limnetica 21(3–4):175–185
40. UNE-EN ISO 10301 (1997) Water quality. Determination of highly volatile halogenated hydrocarbons. Gas-chromatographic methods

41. Lee H, Weng L, Chau AS (1984) Chemical derivatization analysis of pesticides residues. VIII. Analysis of 15 chlorophenols in natural water by in situ acetylation. J Assoc Off Anal Chem 67 (4):789–794
42. León VM, Llorca-Pórcel J, Álvarez B, Cobollo MA, Muñoz S, Valor I (2006) Analysis of 35 semivolatile compounds in water by stir bar sorptive extraction-thermal desorption-gas chromatography–mass spectrometry. Part II: Method validation. Anal Chim Acta 558:261–266
43. Grimalt JO, Vilanova R, Fernández P, Martínez C (2001) Organochlorine pollutants in remote mountain lake waters. J Environ Qual 30:1286–1295
44. Grimalt JO, Fernández P, Carrera G (2002) Atmospheric deposition of organochlorine compounds to remote high mountain lakes of Europe. Environ Sci Technol 36:2581–2588
45. Shen L, Wania F (2005) Compilation, evaluation, and selection of physical-chemical property data for organochlorine pesticides. J Chem Eng Data 50:742–768
46. Weber J, Halsall CJ (2006) Endosulfan and ç-HCH in the Arctic: an assessment of surface seawater concentrations and air-sea exchange. Environ Sci Technol 40:7570–7576
47. Barceló D, Petrovic M (2007) Under the analytical spotlight, contaminants emerge: Report on the 2nd EMCO Workshop. Emerging contaminants in wastewaters: monitoring tools and treatment technologies. Belgrade (Serbia), 26 and 27 April 2007. Trac-Trends Anal Chem 26:647–649
48. Ginebreda A, Muñoz I, López de Alda M, Brix R, López-Doval J, Barceló D (2010) Environmental risk assessment of pharmaceuticals in rivers: relationships between hazard indexes and aquatic macroinvertebrate diversity indexes in the Llobregat river (NE Spain). Environ Int 36:153
49. Ricart M, Guasch H, Barceló D, Brix R, Conceição MH, Geiszinger A, López de Alda MJ, López-Doval JC, Muñoz I, Postigo C, Romaní AM, Villagrasa M, Sabater S (2010) Primary and complex stressors in polluted Mediterranean rivers: pesticide effects on biological communities. J Hydrol 383(1–2):52–61
50. Ricart M, Guasch H, Barcelo D, Brix R, Conceicao MH, Geiszinger A, de Alda MJL, Lopez-Doval JC, Munoz I, Postigo C, Romani AM, Villagrasa M, Sabater S (2010) Primary and complex stressors in polluted mediterranean rivers: pesticide effects on biological communities. J Hydrol 383:52–61
51. Damásio J, Taulera R, Teixidó E, Rieradevall M, Prat N, Riva MC, Soares A, Barata C (2008) Combined use of Daphnia magna in situ bioassays, biomarkers and biological indices to diagnose and identify environmental pressures on invertebrate communities in two Mediterranean urbanized and industrialized rivers (NE Spain). Aquat Toxicol 87(4):310–320
52. Damásio JB, Barata C, Munné A, Ginebreda A, Guasch H, Sabater S, Caixach J, Porte C (2006) Comparing the response of biochemical indicators (biomarkers) and biological indices to diagnose the ecological impact of an oil spillage in a Mediterranean river (NE Catalunya, Spain). Chemosphere 66(7):1206–1216
53. Carafa R, Real M, Munné A, Ginebreda A, Guasch H, Tirapu L (2010) Recommendations for Water Monitoring Programs of priority and emerging pollutants. Contribution from newly developed methods. KEYBIOEFFECTS project: cause-effect relations of key pollutants on the biodiversity of European rivers, p 67
54. Carafa R, Fanggiano L, Real M, Munné A, Ginebreda A, Guasch H, Flo M, Tirapu L, Carsten von der Ohe P (2011) Water toxicity assessment in Catalan rivers (NE Spain) using Species Sensitivity Distribution and Artificial Neural Networks. Science of the Total Environmental 409:4269–4279

Index

H. Guasch et al. (eds.), *Emerging and Priority Pollutants in Rivers*,
Hdb Env Chem (2012) 19: 267–270, DOI 10.1007/978-3-642-25722-3,